# Digital Textile Printing

## The Textile Institute Book Series

Incorporated by Royal Charter in 1925, The Textile Institute was established as the professional body for the textile industry to provide support to businesses, practitioners and academics involved with textiles and to provide routes to professional qualifications through which Institute Members can demonstrate their professional competence. The Institute's aim is to encourage learning, recognise achievement, reward excellence and disseminate information about the textiles, clothing and footwear industries and the associated science, design and technology; it has a global reach with individual and corporate members in over 80 countries.

The Textile Institute Book Series supersedes the former 'Woodhead Publishing Series in Textiles' and represents a collaboration between The Textile Institute and Elsevier aimed at ensuring that Institute Members and the textile industry continue to have access to high calibre titles on textile science and technology.

Books published in The Textile Institute Book Series are offered on the Elsevier web site at: store.elsevier.com and are available to Textile Institute Members at a substantial discount. Textile Institute books still in print are also available directly from the Institute's web site at: www.textileinstitute.org

To place an order, or if you are interested in writing a book for this series, please contact Matthew Deans, Senior Publisher: m.deans@elsevier.com

## Recently Published Titles in The Textile Institute Book Series:

*Design of Clothing Manufacturing Processes*, Jelka Geršak, 978-0-08-102648-9
*Advances in Bio-Based Fiber*, Sanjay Mavinkere Rangappa, Madhu Puttegowda, Jyotishkumar Parameswaranpillai and Suchart Siengchin, Sergey Gorbatyuk, 978-0-12-824543-9
*Characterization of Polymers and Fibers*, Mukesh Singh and Annika Singh, 978-0-12-824239-1
*Applications of Biotechnology for Sustainable Textile Production*, O.L. Shanmugasundaram, 978-0-323-85651-5
*Advanced Knitting Technology*, Subhankar Maity, Sohel Rana, Pintu Pandit and Kunal Singha, 978-0-323-85534-1
*Green Chemistry for Sustainable Textiles*, Nabil Ibrahim and Chaudhery Hussain, 978-0-323-85204-3
*Engineered Polymeric Fibrous Materials*, Masoud Latifi, 978-0-12-824381-7
*Advances in Modeling and Simulation in Textile Engineering*, Nicholus Akankwasa and Dieter Veit, 978-0-12-822977-4
*Sustainable Technologies for Textile Wastewater Treatments*, Subramanian Muthu, 978-0-323-85829-8
*Fundamentals of Natural Fibres and Textiles*, Md. Ibrahim H. Mondal, 978-0-12-821483-1
*Antimicrobial Textiles from Natural Resources*, Md. Ibrahim H. Mondal, 978-0-12-821485-5
*Lean Tools in Apparel Manufacturing*, Prabir Jana and Manoj Tiwari, 978-0-12-819426-3
*Handbook of Footwear Design and Manufacture*, 2nd Edition, A. Luximon, 978-0-12-821606-4
*Waste Management in the Fashion and Textile Industries*, Rajkishore Nayak and Asis Patnaik, 978-0-12-818758-6
*Chemical Management in Textiles and Fashion*, Subramanian Muthu, 978-0-12-820494-8
*Advances in Functional and Protective Textiles*, Shahid ul-Islam and Bhupendra Singh Butola, 978-0-12-820257-9
*Latest Material and Technological Developments for Activewear*, Joanne Yip, 978-0-12-819492-8
*Cut Protective Textiles*, Daniel Li, 978-0-12-820039-1
*Thermal Analysis of Textiles and Fibers*, Michael Jaffe and Joe Menczel, 978-0-08-100572-9
Assessing the Environmental Impact of Textiles and the Clothing Supply Chain, Subramanian Senthilkannan Muthu, 978-0-12-819783-7
*Handbook of Natural Fibres Volume 1: Types, Properties and Factors Affecting Breeding and Cultivation*, 2nd Edition, Ryszard Kozlowski Maria Mackiewicz-Talarczyk, 978-0-12-818398-4
*Handbook of Natural Fibres: Volume 2: Processing and Applications*, 2nd Edition, Ryszard Kozlowski Maria Mackiewicz-Talarczyk, 978-0-12-818782-1
*Advances in Textile Biotechnology*, Artur Cavaco-Paulo, 978-0-08-102632-8
*Woven Textiles: Principles, Technologies and Applications*, 2nd Edition, Kim Gandhi, 978-0-08-102497-3
*Auxetic Textiles*, Hong Hu, 978-0-08-102211-5

The Textile Institute Book Series

# Digital Textile Printing

Science, Technology, and Markets

*Edited by*

## Hua Wang

## Hafeezullah Memon

The Textile Institute

Woodhead Publishing is an imprint of Elsevier
50 Hampshire Street, 5th Floor, Cambridge, MA 02139, United States
The Boulevard, Langford Lane, Kidlington, OX5 1GB, United Kingdom

Copyright © 2023 Elsevier Ltd. All rights reserved.

No part of this publication may be reproduced or transmitted in any form or by any means, electronic or mechanical, including photocopying, recording, or any information storage and retrieval system, without permission in writing from the publisher. Details on how to seek permission, further information about the Publisher's permissions policies and our arrangements with organizations such as the Copyright Clearance Center and the Copyright Licensing Agency, can be found at our website: www.elsevier.com/permissions.

This book and the individual contributions contained in it are protected under copyright by the Publisher (other than as may be noted herein).

MATLAB® is a trademark of The MathWorks, Inc. and is used with permission. The MathWorks does not warrant the accuracy of the text or exercises in this book. This book's use or discussion of MATLAB® software or related products does not constitute endorsement or sponsorship by The MathWorks of a particular pedagogical approach or particular use of the MATLAB® software.

**Notices**
Knowledge and best practice in this field are constantly changing. As new research and experience broaden our understanding, changes in research methods, professional practices, or medical treatment may become necessary.

Practitioners and researchers must always rely on their own experience and knowledge in evaluating and using any information, methods, compounds, or experiments described herein. In using such information or methods they should be mindful of their own safety and the safety of others, including parties for whom they have a professional responsibility.

To the fullest extent of the law, neither the Publisher nor the authors, contributors, or editors, assume any liability for any injury and/or damage to persons or property as a matter of products liability, negligence or otherwise, or from any use or operation of any methods, products, instructions, or ideas contained in the material herein.

ISBN: 978-0-443-15414-0 (print)

ISBN: 978-0-443-15415-7 (online)

For information on all Woodhead Publishing publications
visit our website at https://www.elsevier.com/books-and-journals

Publisher: Matthew Deans
*Acquisitions Editor:* Sophie Harrison
*Editorial Project Manager:* Tessa Kathryn
*Production Project Manager:* Surya Narayanan Jayachandran
*Cover Designer:* Matthew Limbert

Typeset by MPS Limited, Chennai, India

# Contents

| | |
|---|---|
| List of contributors | xiii |
| About the authors | xvii |
| Foreword | xxix |
| Preface | xxxi |

**1  Introduction to digital textile printing** — 1
*Hua Wang and Hafeezullah Memon*
   1.1  History of printing — 1
   1.2  Overview of the evolution of printing tools and equipment — 1
         1.2.1  Block printing and engraved printing — 2
         1.2.2  Drum and roller printing machine — 2
         1.2.3  Screen printing machine — 4
         1.2.4  Transfer printing machine — 7
   1.3  Key historical inventions of today's digital printing technology — 8
   1.4  Advantages of the digital printing technology — 11
         1.4.1  Promote the continuous development of clothing fabric printing — 11
         1.4.2  Promote the diversified development of clothing design — 11
         1.4.3  Color selection is freer and richer — 12
         1.4.4  Pattern design is more fashionable and cutting-edge — 12
   1.5  Digital printing breaks through the bottleneck of textile printing — 13
         1.5.1  Meet the needs of cleaner production — 13
         1.5.2  Breakthrough the bottleneck of traditional textile printing — 13
         1.5.3  Digital printing technology has a vast space for development in the international application — 14
         1.5.4  Products meet market demand — 14
         1.5.5  Production costs are relatively low — 15
         1.5.6  Attach importance to the concept of technological innovation — 16
   References — 18

**2  Digital printing mechanisms** — 21
*Hanur Meku Yesuf, Abdul Khalique Jhatial,*
*Pardeep Kumar Gianchandani, Amna Siddique and Altaf Ahmed Simair*
   2.1  Introduction — 21
   2.2  Digital printing methods — 22
         2.2.1  Fine art inkjet printing — 22

|   |     | 2.2.2 | Notable digital laser exposure | 23 |
|---|---|---|---|---|
|   |     | 2.2.3 | Digital cylinder printing | 24 |
|   | 2.3 | Generation of droplets | | 25 |
|   |     | 2.3.1 | Continuous inkjet | 25 |
|   |     | 2.3.2 | Drop-on-demand | 26 |
|   | 2.4 | Printing parameters and ink formulation | | 27 |
|   |     | 2.4.1 | Aqueous inks | 28 |
|   |     | 2.4.2 | Solvent inks | 28 |
|   |     | 2.4.3 | UV-curable inks | 29 |
|   |     | 2.4.4 | Dye sublimation inks | 29 |
|   |     | 2.4.5 | Solid ink | 30 |
|   |     | 2.4.6 | Metal nanoparticle ink | 30 |
|   | 2.5 | Printing heads | | 30 |
|   |     | 2.5.1 | Fixed head | 30 |
|   |     | 2.5.2 | Disposable head | 31 |
|   | 2.6 | Cleaning mechanisms | | 31 |
|   | 2.7 | Advantages and disadvantages of inkjet printing | | 33 |
|   |     | 2.7.1 | Advantages of digital printing | 33 |
|   |     | 2.7.2 | Disadvantages of digital printing | 34 |
|   | 2.8 | Solidification mechanisms | | 34 |
|   | 2.9 | Applications | | 35 |
|   | References | | | 35 |
| **3** | **Overview of different digital textile printing machines** | | | **41** |
|   | *Aijaz Ahmed Babar, Pardeep Kumar Gianchandani and Abdul Khalique Jhatial* | | | |
|   | 3.1 | Introduction | | 41 |
|   | 3.2 | Classification of digital printing machines | | 42 |
|   | 3.3 | Drop-on-demand inkjet technology | | 44 |
|   | 3.4 | Continuous inkjet technology | | 46 |
|   | 3.5 | Print head selection | | 48 |
|   | 3.6 | Companies active in print head technology | | 49 |
|   | 3.7 | Outlook | | 52 |
|   | 3.8 | Conclusion | | 53 |
|   | References | | | 53 |
| **4** | **Color management and design software for textiles** | | | **57** |
|   | *Pardeep Kumar Gianchandani, Abdul Khalique Jhatial, Aijaz Ahmed Babar and Hanur Meku Yesuf* | | | |
|   | 4.1 | Introduction | | 57 |
|   | 4.2 | Elements of color management | | 59 |
|   | 4.3 | Classification of colors | | 59 |
|   | 4.4 | Color measurement | | 60 |
|   |     | 4.4.1 | Instrumental color measurement | 62 |
|   |     | 4.4.2 | Measurement instruments | 62 |

|     | 4.5    | Color management                                         | 64 |
| --- | ------ | -------------------------------------------------------- | -- |
|     |        | 4.5.1 Agfa                                               | 65 |
|     |        | 4.5.2 Aleph                                              | 65 |
|     |        | 4.5.3 Chromix                                            | 66 |
|     |        | 4.5.4 Aquario design                                     | 67 |
|     |        | 4.5.5 Textile print by adobe portable document format print engine | 67 |
|     |        | 4.5.6 Colorburst systems—spectracore                     | 67 |
|     |        | 4.5.7 Ergosoft                                           | 68 |
|     |        | 4.5.8 X-rite                                             | 68 |
|     |        | 4.5.9 MatchPrint II by kodak polychrome                  | 69 |
|     |        | 4.5.10 Datacolor                                         | 69 |
|     |        | 4.5.11 Atexco digital                                    | 69 |
|     | 4.6    | Future trends                                            | 70 |
|     | References                                                        | 71 |

## 5 Digital image design and creation of printed images on textile fabrics 73

*Bewuket Teshome Wagaye, Degu Melaku Kumelachew and Biruk Fentahun Adamu*

|     | 5.1 | Introduction                                             | 73 |
| --- | --- | -------------------------------------------------------- | -- |
|     | 5.2 | Image capture and display                                | 76 |
|     |     | 5.2.1 Image capture                                      | 77 |
|     |     | 5.2.2 Pattern data encoding compression and storage      | 79 |
|     |     | 5.2.3 Digital color management systems and color communication | 81 |
|     |     | 5.2.4 Input and output devices color calibration         | 83 |
|     | 5.3 | Pixel and image creation using inkjet printers           | 84 |
|     | 5.4 | UV and latex curing methods                              | 86 |
|     | 5.5 | Printing machine control                                 | 87 |
|     | 5.6 | Printing head performance monitoring                     | 87 |
|     | 5.7 | Future prospects                                         | 88 |
|     | 5.8 | Conclusion                                               | 89 |
|     | References                                                     | 89 |

## 6 Recent developments in the preparatory processes for the digital printing of textiles 93

*Sharjeel Abid, Jawad Naeem, Amna Siddique, Sonia Javed, Sheraz Ahmad and Hanur Meku Yesuf*

|     | 6.1 | Introduction                                             | 93 |
| --- | --- | -------------------------------------------------------- | -- |
|     | 6.2 | The difference between conventional and digital printing pretreatment processes | 94 |
|     | 6.3 | Recent developments in pretreatment processes            | 95 |
|     |     | 6.3.1 Mercerization                                      | 95 |
|     |     | 6.3.2 Cationization for digital prints                   | 96 |

|  |  |  | |
|---|---|---|---|
| | 6.3.3 | Plasma pretreatment for digital prints | 98 |
| | 6.3.4 | Enzymatic treatment for digital prints | 99 |
| | 6.3.5 | Other pretreatment processes for digital prints | 100 |
| 6.4 | | Pretreatment processes for pigment digital printing | 102 |
| 6.5 | | Pretreatment of cotton fabrics for digital printing | 103 |
| | 6.5.1 | Impact of one-bath pretreatment for increased color yield of inkjet prints using reactive inks | 104 |
| | 6.5.2 | Impact of concentration of glycidyl trimethylammonium chloride on color yield of magenta ink | 105 |
| | 6.5.3 | Impact of concentration of NaOH | 105 |
| | 6.5.4 | Impact of chitosan and acetic acid on color yield | 105 |
| | 6.5.5 | Impact of sodium bicarbonate on color yield | 106 |
| | 6.5.6 | Impact of urea | 106 |
| | 6.5.7 | Effect of cationization of cotton in digital printing | 107 |
| | 6.5.8 | Effect of hydroxypropyl methylcellulose pretreatment on cotton and polyamide fabric in inkjet printing | 107 |
| | 6.5.9 | The color performance of inkjet printing | 108 |
| | 6.5.10 | Use of Plasma pretreatment for depositing printing paste on the cotton substrate for digital inkjet printing | 108 |
| | 6.5.11 | Determination of color yield | 109 |
| 6.6 | | Pretreatments of wool fabrics for digital printing | 109 |
| | 6.6.1 | Impact of sodium alginate on color yield of wool fabric | 109 |
| | 6.6.2 | Impact of ammonium tartrate on color yield of wool fabric | 110 |
| | 6.6.3 | Impact of urea on the color yield of wool fabric | 110 |
| 6.7 | | Pretreatment of polyester fabrics for digital printing | 110 |
| | 6.7.1 | Impact of pretreatment on color strength | 110 |
| | 6.7.2 | Impact of pretreatment on color saturation | 111 |
| 6.8 | | Pretreatment of silk fabric for digital printing | 111 |
| | 6.8.1 | Effect of steaming time | 112 |
| | 6.8.2 | Amount of alkali | 112 |
| | 6.8.3 | Amount of urea | 112 |
| | 6.8.4 | The pH of the pretreatment paste | 112 |
| | 6.8.5 | Concentration of thickener | 112 |
| 6.9 | | Pretreatment of blended fabrics for digital printing | 113 |
| 6.10 | | Pretreatment free digital printing | 113 |
| 6.11 | | Conclusions | 113 |
| | | References | 114 |

**7 Colorants for digital textile printing and their chemistry**     **119**
*Abdul Khalique Jhatial, Pardeep Kumar Gianchandani, Biruk Fentahun Adamu, Aijaz Ahmed Babar and Hanur Meku Yesuf*

| | | |
|---|---|---|
| 7.1 | Introduction | 119 |
| 7.2 | Why digital printing? | 120 |
| 7.3 | Colorants for digital printing | 120 |
| 7.4 | Preparation of substrate for digital printing | 122 |

|       |        |                                                                    |      |
|-------|--------|--------------------------------------------------------------------|------|
| 7.5   |        | Post-treatment for digital printing                                | 123  |
| 7.6   |        | Types of inks for digital printing                                 | 123  |
|       | 7.6.1  | Reactive dye inks                                                  | 123  |
|       | 7.6.2  | Acid dye inks                                                      | 126  |
|       | 7.6.3  | Disperse dye inks                                                  | 127  |
|       | 7.6.4  | Pigment inks                                                       | 128  |
| 7.7   |        | Important properties of inks                                       | 130  |
|       | 7.7.1  | Viscosity                                                          | 130  |
|       | 7.7.2  | Surface tension                                                    | 130  |
|       | 7.7.3  | Particle size                                                      | 131  |
|       | 7.7.4  | pH and electrolytes                                                | 131  |
|       | 7.7.5  | Dielectric properties and conductivity                             | 132  |
|       | 7.7.6  | Ink storage and stability                                          | 132  |
| 7.8   |        | Challenges in digital printing of textiles                         | 132  |
|       |        | References                                                         | 133  |

## 8  Inkjet printing of textiles enhanced by sustainable plasma technology — 137
*Alka Madhukar Thakker, Danmei Sun and Muhammad Owais Raza Siddiqui*

| 8.1 | Introduction | 137 |
| 8.2 | Plasma treatment of varied fabrics to facilitate inkjet printing | 139 |
| 8.3 | Conclusions | 153 |
|     | References | 153 |

## 9  Technological barriers to digital printing in textiles: a study — 157
*Md Aktarul Hasan*

|      |        |                                                        |     |
|------|--------|--------------------------------------------------------|-----|
| 9.1  |        | Introduction                                           | 157 |
| 9.2  |        | Pretreatment and posttreatment of the fabric           | 157 |
| 9.3  |        | Printheads                                             | 161 |
|      | 9.3.1  | Single-nozzle and multinozzle printheads               | 162 |
|      | 9.3.2  | Drop-on-demand and continuous inkjet printheads        | 163 |
| 9.4  |        | Inks                                                   | 165 |
| 9.5  |        | Ink supply system                                      | 165 |
| 9.6  |        | Pigments and dyes                                      | 166 |
| 9.7  |        | Textile printing and ink depository                    | 168 |
| 9.8  |        | Clogging and monitoring of nozzles                     | 168 |
| 9.9  |        | Automated maintenances                                 | 169 |
| 9.10 |        | Handling motion systems and substrates                 | 170 |
| 9.11 |        | Single-pass printing                                   | 171 |
| 9.12 |        | Software and electronics                               | 171 |
| 9.13 |        | Concentrating on printing technology                   | 174 |
|      | 9.13.1 | Learning curve                                         | 177 |
|      | 9.13.2 | What can go wrong? a real-world example                | 178 |
|      | 9.13.3 | Digital textile printing-a threat, an opportunity, or a risk? | 178 |

|  |  | 9.14 | Awaiting chemistry | 179 |
|---|---|---|---|---|
|  |  |  | 9.14.1 Digital finishing | 179 |
|  |  |  | 9.14.2 Digital dyeing | 179 |
|  |  |  | 9.14.3 Technology advancements in deposition | 180 |
|  |  | 9.15 | Conclusion | 180 |
|  |  | References |  | 181 |

**10 Quality of digital textile printing** — 185

*Biruk Fentahun Adamu, Esubalew Kasaw Gebeyehu, Bewuket Teshome Wagaye, Degu Melaku Kumelachew, Melkie Getnet Tadesse and Abdul Khalique Jhatial*

- 10.1 Introduction — 185
- 10.2 Digital textiles print quality attributes — 186
- 10.3 Factors that affect digital textile print quality — 186
  - 10.3.1 Effect of fiber type and its properties on digital textile print quality — 186
  - 10.3.2 Effect of yarn type and its properties on digital textile print quality — 189
  - 10.3.3 Effect of fabric type and its properties on digital textile print quality — 191
  - 10.3.4 Effect of fabric pretreatment on digital textile print quality — 192
  - 10.3.5 Effect of colorants on digital textile print quality — 194
  - 10.3.6 Effect of printing head on digital textile print quality — 196
- 10.4 Methods of measuring digital textile print quality — 196
  - 10.4.1 Subjective evaluation method of digital textile print quality — 197
  - 10.4.2 Objective evaluation method of digital textile print quality — 198
- 10.5 Digital print quality evaluation standards — 201
- 10.6 Conclusions — 202
- References — 202

**11 Western markets for digitally printed textiles** — 207

*Muhammad Ayyoob and Muhammad Khan*

- 11.1 Background — 207
- 11.2 Digital textiles printing and sustainability — 209
- 11.3 Market overview — 212
  - 11.3.1 Home furnishing and interior decoration — 212
  - 11.3.2 Apparel and fashion — 213
  - 11.3.3 Direct-to-garment printing — 213
  - 11.3.4 Soft signage — 213
- 11.4 COVID-19 impact on digital textiles printing market — 214
- 11.5 Market driving forces — 215

|  |  |  |  |
|---|---|---|---|
| | 11.6 | Share of digital printed textiles in printed textiles | 216 |
| | 11.7 | Western markets | 217 |
| | | 11.7.1 Europe | 218 |
| | | 11.7.2 North America | 219 |
| | | 11.7.3 South America and the caribbean | 220 |
| | 11.8 | Summary | 220 |
| | References | | 221 |
| **12** | **Emerging market trends: the cultural designs printed with digital printing technology: an overview of Ajrak design** | | **225** |
| | *Sippi Pirah Simair, Nuzhat Baladi, Hanur Meku Yesuf and Altaf Ahmed Simair* | | |
| | 12.1 | Introduction | 225 |
| | 12.2 | History of textile in Pakistan | 226 |
| | 12.3 | The history of traditional clothing in Sindh | 228 |
| | 12.4 | Digital printing: a fast-growing vibrant sector in Pakistan | 233 |
| | | 12.4.1 Technology takes digital printing to a new level | 235 |
| | | 12.4.2 Starting to make wise decisions | 236 |
| | 12.5 | Culture is not a trend, but it has become a significant source of profit worldwide | 236 |
| | | 12.5.1 Ezri collection, 2016 | 237 |
| | | 12.5.2 Urban outfitters, 2016 | 237 |
| | | 12.5.3 Forever 21, 2016 | 237 |
| | 12.6 | Conclusion | 237 |
| | References | | 238 |
| **13** | **Digital textile printing innovations and the future** | | **241** |
| | *Degu Melaku Kumelachew, Bewuket Teshome Wagaye and Biruk Fentahun Adamu* | | |
| | 13.1 | Digital textile printing advancement history | 241 |
| | | 13.1.1 Machine innovative advancements | 243 |
| | | 13.1.2 Ink formulation technology | 243 |
| | | 13.1.3 Inkjet heads/print heads | 246 |
| | | 13.1.4 Limitless design (media and color management software) | 247 |
| | | 13.1.5 Surface imaging | 248 |
| | 13.2 | Printing process innovative advancements | 250 |
| | | 13.2.1 Fabric preparation | 250 |
| | | 13.2.2 Product development creativity and inspiration | 251 |
| | | 13.2.3 Volume and flexibility | 251 |
| | | 13.2.4 Speed of printing | 252 |
| | 13.3 | Benefits of innovation | 252 |
| | | 13.3.1 The economics | 252 |
| | | 13.3.2 The environment | 252 |
| | | 13.3.3 Optimizing the supply chain | 254 |

| | | | |
|---|---|---|---|
| 13.4 | The future of digital textile printing | | 255 |
| | 13.4.1 | Technology adaption | 256 |
| | 13.4.2 | Specialization | 256 |
| | 13.4.3 | Colors | 256 |
| 13.5 | Global market projections on digital textile printing | | 257 |
| 13.6 | Conclusion | | 258 |
| References | | | 258 |

**Index**     **261**

# List of contributors

**Sharjeel Abid** School of Engineering and Technology, National Textile University, Faisalabad, Pakistan

**Biruk Fentahun Adamu** Textile Engineering Department, Ethiopian Institute of Textile and Fashion Technology, Bahir Dar University, Bahir Dar, Ethiopia; Key Laboratory of Textile Science and Technology, Ministry of Education, College of Textiles, Donghua University, Shanghai, P.R. China

**Sheraz Ahmad** School of Engineering and Technology, National Textile University, Faisalabad, Pakistan

**Muhammad Ayyoob** Department of Polymer Engineering, National Textile University, Karachi Campus, Karachi, Pakistan

**Aijaz Ahmed Babar** Department of Textile Engineering, Mehran University of Engineering and Technology, Jamshoro, Sindh, Pakistan

**Nuzhat Baladi** Department of English, Government College University, Hyderabad, Sindh, Pakistan

**Esubalew Kasaw Gebeyehu** Textile Chemical Process Engineering, Ethiopian Institute of Textile and Fashion Technology, Bahir Dar University, Bahir Dar, Ethiopia; Key Lab of Science and Technology of Eco-textile, Ministry of Education, College of Chemistry, Chemical Engineering, and Biotechnology, Donghua University, Shanghai, P.R. China

**Pardeep Kumar Gianchandani** Department of Textile Engineering, Mehran University of Engineering and Technology, Jamshoro, Sindh, Pakistan

**Md Aktarul Hasan** School of Informatics, Zhejiang Sci-Tech University, Hangzhou, P.R. China

**Sonia Javed** Government College Women University, Faisalabad, Pakistan

**Abdul Khalique Jhatial** Department of Textile Engineering, Mehran University of Engineering and Technology, Jamshoro, Sindh, Pakistan

**Muhammad Khan** Nanotechnology Research Lab, Department of Textile & Clothing, National Textile University, Karachi Campus, Karachi, Pakistan

**Degu Melaku Kumelachew** Textile Engineering Department, Ethiopian Institute of Textile and Fashion Technology, Bahir Dar University, Bahir Dar, Ethiopia; Key Laboratory of Textile Science and Technology, Ministry of Education, College of Textiles, Donghua University, Shanghai, P.R. China

**Hafeezullah Memon** College of Textile Science and Engineering, International Institute of Silk, Zhejiang Sci-Tech University, Hangzhou, P.R. China

**Jawad Naeem** School of Engineering and Technology, National Textile University, Faisalabad, Pakistan

**Amna Siddique** School of Engineering and Technology, National Textile University, Faisalabad, Pakistan

**Muhammad Owais Raza Siddiqui** School of Textiles and Design, Heriot-Watt University, Galashiels, United Kingdom; Department of Textile Engineering, NED University of Engineering & Technology, Karachi, Pakistan

**Altaf Ahmed Simair** Department of Botany, Government College University, Hyderabad, Sindh, Pakistan

**Sippi Pirah Simair** Key Lab of Science and Technology of Eco-Textile, Ministry of Education, College of Chemistry, Chemical Engineering and Biotechnology, Donghua University, Shanghai, P.R. China

**Danmei Sun** School of Textiles and Design, Heriot-Watt University, Galashiels, United Kingdom

**Melkie Getnet Tadesse** Textile Chemical Process Engineering, Ethiopian Institute of Textile and Fashion Technology, Bahir Dar University, Bahir Dar, Ethiopia

**Alka Madhukar Thakker** School of Textiles and Design, Heriot-Watt University, Galashiels, United Kingdom

**Bewuket Teshome Wagaye** Textile Engineering Department, Ethiopian Institute of Textile and Fashion Technology, Bahir Dar University, Bahir Dar, Ethiopia; Key Laboratory of Textile Science and Technology, Ministry of Education, College of Textiles, Donghua University, Shanghai, P.R. China

**Hua Wang** Western Research Institute, Chinese Academy of Agricultural Science, Changji, P.R. China

**Hanur Meku Yesuf** Ethiopian Institute of Textile and Fashion Technology, Bahir Dar University, Bahir Dar, Ethiopia; Key Laboratory of Textile Science and Technology, Ministry of Education, College of Textiles, Donghua University, Shanghai, P.R. China

# About the authors

**Prof. Dr. Hua Wang** received his bachelor's degree in dyeing and finishing engineering from Tianjin Textile Institute of Technology, China, in 1984. In 1994 he completed his postgraduation in management engineering at China Textile University (now Donghua University, China). In 2006 he completed his doctoral degree in textile science and engineering from Donghua University, China. He has long-term working experience in cotton and wool textile production, printing and dyeing industry, and international trade. In 2012 he was appointed as a senior visiting scholar at Deakin University in Australia and studied cotton and wool  fibers. In 2018 he was appointed as an honorary professor by the Tashkent Institute of Textile and Light Industry, Uzbekistan, and also by the Ministry of Education and Science and the Ministry of Industrial Innovation and Development of Tajikistan. In 2019 he was a visiting professor at the Novi Sad University of Serbia as an expert committee of the International Silk Union.

At present, he has remained engaged in teaching and researching textile intelligent manufacturing technology, digital printing technology, and textile intangible cultural heritage at Donghua University. His main research directions include but are not limited to the manufacturing and application technology of raw materials for wool textiles, digital printing of textiles, and research on world textile history. He has completed five provincial and ministerial level projects, two individual research projects, and three joint research projects. He has authored four invention patents and published more than 50 papers. Also, he has published three textbooks in the field of textiles as an editor, including "Textile Digital Printing Technology." He has taught five courses for undergraduate, master, and doctoral students and one full English course for international students at Donghua University. He has also been a chief member in establishing joint laboratories and research bases for natural textile fiber and processing in Xinjiang Autonomous Region and Central Asian countries. In 2018 he won the only "Golden Sail Golden Camel" award from Donghua University. In 2019 he won the second prize in the science and technology progress of the China Textile Federation. He has been awarded the title of "Best Teacher and Best Tutor" by overseas students of Donghua University for the last three consecutive years. Currently, he has been deputed to the Western Research Institute of the Chinese Academy of Agricultural Science, Xinjiang, China.

Dr. Hafeezullah Memon received his BE in Textile Engineering from Mehran University of Engineering and Technology, Jamshoro, Pakistan, in 2012. He served at Sapphire Textile Mills as an assistant spinning manager for more than 1 year while earning his master's in business administration from the University of Sindh, Pakistan. He completed his master's in Textile Science and Engineering from Zhejiang Sci-Tech University, China, and a PhD in Textile Engineering from Donghua University in 2016 and 2020.

He focuses on fiber-reinforced composites, textiles and management, and bio-based materials. Since 2014, he has published more than 60 peer-reviewed technical papers in international journals and conferences, filed more than 10 patents, edited more than 10 books and special issues, and worked on more than 10 industrial projects.

He is a fellow of the textile institute, a full professional, and an active member of the society for the Advancement of Material and Process Engineering (SAMPE), the Society of Wood Science and Technology (SWST), and the International Textile and Apparel Association (ITAA). Moreover, he is a registered engineer of the Pakistan Engineering Council. He has served as a reviewer and editorial board member of several international journals and has reviewed more than 600 papers. He is a recipient of several national and international awards. Currently, he is serving as a foreign talent expert after completing his postdoc at Zhejiang Sci-Tech University. Moreover, he is elected as a chairman of the Alumni Association of International Students at Zhejiang Sci-Tech University (https://orcid.org/0000-0001-5985-5394).

**Engr. Hanur Meku Yesuf** is currently a lecturer at Bahir Dar University, Ethiopia, and pursuing his PhD in Nonwoven Materials and Engineering at Donghua University, China. He received his BSc in textile engineering and MSc in textile manufacturing from Bahir Dar University. His investigative work mainly involves nanofiber electrospinning theory and methods. He has worked and learned with various types of teams with people from diverse cultural backgrounds and has significant exposure working with experts. He is a hard-working, quick learner, dedicated to bringing change, and able to handle complex

and hard-working conditions. He can do mechanical work and is an expert in the textile sector. He worked at the Kombolcha Textile Share Company and the Ethiopian Textile Industry Development Institute in Ethiopia. He has experience in controlling processes in rewinding, warping, sizing, drawing-in, tying-in, and loom shed, identifying new fabric structures and making suitable for the loom, controlling machine efficiency, managing and supervising, consulting, delivering training, doing different research, and implementing benchmarking for weaving and knitting factories, participating in marketing issues, machine installation and commissioning, teaching-learning activities, research and development, community service, and technology transfer.

**Engr. Abdul Khalique Jhatial** is a PhD scholar at the College of Textiles Donghua University, China. He received his Bachelor of Textile Engineering in 2009 and Masters of Textile Engineering in 2015 from Mehran University of Engineering & Technology (MUET). He has 10 years of teaching and research experience and is a lecturer at the Department of Textile Engineering, MUET Jamshoro, Sindh, Pakistan. He has served in the Department of Textile Engineering, MUET Jamshoro, since 2010. He has served as a laboratory supervisor in the textile chemistry and wet processing laboratory, color measuring laboratory, and yarn manufacturing laboratory of the Textile Engineering Department at MUET. His research interests are conductive biopolymers, smart textiles, functional nanofibers, multifunctional textiles, textile coloration, and yarn manufacturing. He has published five SCI journal articles as the first coauthor. He has attended several workshops and training sessions in Pakistan. He has participated in national and international conferences. His current research focus is conductive biopolymers for biomedical textile applications.

**Dr. Pardeep Kumar Gianchandani** works in the Department of Textile Engineering at Mehran University of Engineering & Technology Jamshoro. He received his Bachelor of Engineering in Textile Engineering in 2009. He served Lucky Cotton Mills Ltd-Nooriabad as an assistant spinning manager from February 1, 2009, to September 28, 2009, and Renfro Crescent Pvt Ltd, Karachi, as a quality controller from September 29, 2009, to April 6, 2010. He has been involved in research and teaching various subjects of textile engineering since February 2010. In 2012 he was awarded a scholarship by the Higher Education Commission of Pakistan (HEC) for higher studies (master's leading to PhD). In October 2012, he joined Politecnico Di Torino (Polito), Italy, for a master's leading to PhD studies. During his study at Polito, the author has worked on various national and international (Italian and European Union) projects on aerospace and energy applications. After completing his PhD, in August 2018, he joined the Department of Textiles Engineering MUET, Jamshoro. The author teaches various textile engineering subjects and is actively involved in research activities, particularly in technical textiles.

**Dr. Amna Siddique** is currently an assistant professor at National Textile University, Faisalabad, Pakistan. She has done her PhD at Donghua University, China. She has more than 30 scientific research articles and three book chapters. Her research interests include textile composites, textile processing, polymeric materials, and finite element modeling.

**Prof. Dr. Altaf Ahmed Simair** has been a constant source of inspiration for everyone out there. He was born in a small village of district Sukkur; after getting his early education in his hometown, he spent 4 years at Shah Abdul Latif University, Khairpur, from where he did his master's in plant sciences with distinction. After graduation, he was appointed as a lecturer and has been teaching undergraduates and graduates. He also got a PhD degree from the University of Sindh, Jamshoro, Biotechnology; later, he went for postdoctoral studies, worked under the advisement of Prof. Dr. Changrui Lu, Donghua University

Shanghai, China, and published 16 research articles. He has already rendered his valuable services to the Government of Pakistan as a professor for more than 25 years. He worked as an editorial member, a member scientific committee, and a member of the review committee of various national and international journals and participated in national and international workshops and short training courses and international symposiums. Recently, He is serving as a chairman of the Department of Plant Sciences at Government Collage University, (GC University), Hyderabad. More than 40 (national and international) research articles are on his credit, which shows his proficiency and scholarly erudition in the realm of biotechnology and biochemistry. Already, he was honored to write excellent biology books for Sindh Text Book Board from matriculation to intermediate level. His contribution in the form of these books is marvelous and exceptional. His work has always been as helpful and all-embracing as he is and his published work on "Advances in Genomic Tools to Study Phylogeny, Diversity, and Classification" is undoubtedly a masterly distillation of his well-equipped pen and his ceaseless efforts for the compilation of this book (https://link.springer.com/chapter/10.1007/978-981-15-9169-3_2). This one is second to his current work "Digitally printed textiles and modern trends in local markets" (https://orcid.org/0000-0001-9455-5207).

# About the authors

**Dr. Aijaz Ahmed Babar** graduated with BE and ME in Textile Engineering from the Mehran University of Engineering & Technology, Jamshoro, Pakistan (2007−14). He joined the Textile Engineering Department of Mehran University as a laboratory supervisor in 2013. He was then awarded a Chinese Government Scholarship and went on to pursue his Doctor of Engineering degree in Nanofibers and Hybrid Materials from Donghua University China. There, he was awarded the "Outstanding International Student Scholarship/ Award" thrice in 2017−19, the "Academic Competition and Scientific Research Award" in 2019, and "Excellent International Graduate of Donghua University" in 2020. After acquiring a doctoral degree in 2020, he resumed his duties as a laboratory supervisor at Mehran University and is currently working there in the said post and teaching various subjects of textile engineering there. With his educational background in textile engineering, materials science, and nanotechnology, he has been working in textile coloration, moisture management textiles, nonwoven materials, and carbon capture technologies, and developed functional nanofibers for the application of directional moisture transport, fog collection, air filtration, $CO_2$ adsorption, and flexible electrodes for supercapacitors. Since 2016, he has authored or coauthored more than 30 SCI-indexed peer-reviewed journal articles in esteems journals including ACS Nano, Small Methods, Nano Letters, and ACS Appl. Mater. Interfaces., etc., and two book chapters (Andrew Williams Publishing). His Google Scholar citation is more than 800 times with a current h-index of 16.

**Bewuket Teshome Wagaye** is currently a Ph.D. student at the College of Textiles, Donghua University, China. His research area is on natural fiber-reinforced composite materials. He received his degree of Bachelors of Textile Engineering in 2009 and a postgraduation in Textile Manufacturing in 2015 from the Ethiopian Institute of Textile and Fashion Technology, Bahir Dar University, Bahir Dar. He has 12 years of industry, research, and teaching experience. He is a lecturer at Bahir Dar University. His career started as a junior textile engineer at Bahir Dar Textile Share Company. He also participated in data collection, consultancy, and technical support at the Ethiopian Textile Industry Development Institute.

**Degu Melaku Kumelachew** is currently doing his PhD in biomedical Textiles from the College of Textiles, Donghua University, China. He received his BSc degree in Textile Engineering in 2010 and an MSc degree in Textile Manufacturing in 2015 from the Ethiopian Institute of Textile and Fashion Technology-Bahir Dar University, Bahir Dar, Ethiopia. He is a lecturer with two and half years of experience in industry and eight-plus years of teaching in textile testing, fabric structures, and research experience. He joined work as a production supervisor in the textile industry and shifted his career to Bahir Dar University, Bahir Dar, upon interest. He took a number of skill training, including higher diploma programs and testing (both yarn and fabric testing skills). He worked on different projects.

**Biruk Fentahun Adamu** is currently doing his PhD in Textile Engineering at the College of Textiles, Donghua University, China. He received his BSc degree in Textile Engineering in 2008 and an MSc degree in Textile Manufacturing in 2016 from Bahir Dar University, Bahir Dar, Ethiopia. He is a lecturer at Bahir Dar University, has more than 12 years of experience teaching in textile testing and fabric manufacturing, and also has research experience. His career started as an instructor at Technical and Vocational Training College in Ethiopia and then joined Bahir Dar University, Bahir Dar. His research interest is biomedical textiles, biomaterials, fibers, and polymers. He has more than 12 publications. He worked on different projects.

**Dr. Sharjeel Abid** is currently an assistant professor at the Department of Textile Engineering, National Textile University, Faisalabad, Pakistan. He has more than forty international publications and three book chapters with an h-index of 15. His research areas include textile processing, sustainable textiles, and biomedical textiles.

# About the authors

**Dr. Jawad Naeem** is a textile engineer from National Textile University, Faisalabad. He has done his PhD in textile from the Technical University of Liberec, Czech Republic. He has expertise in thermal protective performance and thermal comfort properties of textile substrates. Currently, he is serving as an assistant professor at National Textile University, Faisalabad.

**Sonia Javed** graduated in 2021 from Government College Women University, Pakistan, in chemistry. From 2019 to 2021, she was a research assistant for the Higher Education Commission Pakistan for a funded project titled "Development of Green Electrospun Nano Composites for Wound Dressing" at GCWUF. She is the author of two book chapters titled "Natural Plant Extract Treated Bio-Active Textiles for Wound Healing," in the Textile Institute Book Series, 2021, and "Protective Facemask Made of Electrospun Fibers," in the Electrospun Nanofibers, Springer, Cham, 2022. Her research interests include the development of eco-friendly, sustainable extraction media, and the production of nanostructured materials incorporated with natural extracts and applications in drug delivery systems.

**Sheraz Ahmad** graduated master's in Fiber Technology from the University of Agriculture and PhD in Textile Engineering with a focus on textile materials from Université de Haute Alsace France. He is working as an associate professor at National Textile University since October 2012, a chairperson of the textile technology department, teaching undergraduate, Ms, and PhD level classes, doing research on textile fibers, hydrogels, and recycled materials, and has authored over 50 peer-reviewed journal articles, three books, and 20 conference communications. He is fluent in English, French, Urdu, and Punjabi and experienced in developing course curricula as well as executing field trips, laboratory exercises, and other activities beyond traditional lectures.

**Dr. Danmei Sun** has nearly 30 years of work experience in textile materials and technology. She has been supervising PhD students and postdoctoral research assistants with various research projects, including smart polymeric materials and functional fibers, natural dyes for sustainable textiles, etc. Her research has been funded by various funding bodies and numerous company partners. The collaborative research with partner companies has expanded her research extensively over the years, and in turn, has attracted more funding applications. She is an editor-in-chief of Journal of Textile Science & Fashion Technology.

**Dr. Alka Madhukar Thakker** has recently completed her PhD in "Developing Sustainable Fabrics with Plant-Based Formulations" in the United Kingdom. Her subject areas of interest are experimentations with circular materials and life cycle assessment, as well as, surface engineering, textile testing, and analysis. She has both the textile industry and academic research work experience of 15 years. She had received sponsorship for her master's and PhD degrees from India and United Kingdom.

**Dr. Muhammad Owais Raza Siddiqui** is an associate professor in the Department of Textile Engineering, NED University of Engineering & Technology, Karachi, Pakistan. He received a PhD degree from Herriot Watt University, Galashiels, Scotland. He has been extensively involved in research in the development of digital and computational analysis of smart/functional textile materials.

**Md. Aktarul Hasan** received a bachelor's degree in Computer Science and Engineering from Daffodil International University, Dhaka, Bangladesh, and is currently enrolled as a master's degree student in computer technology at Zhejiang Sci-Tech University (ZSTU), Hangzhou, Zhejiang, China. He is a member of the Institution of Engineers, Bangladesh (IEB). His interest is embedded systems, wireless sensor networks, and computer vision.

**Esubalew Kasaw Gebeyehu** is currently working as a lecturer and researcher at the Ethiopian Institute of Textile and Fashion Technology, Bahir Dar University, Bahir Dar, Ethiopia. He has rich experience in teaching and research in the areas of smart textiles, smart hydrogel, biomaterials, scaffolds, pretreatment processes of textiles, textile dyeing, printing, textile finishing, textile polymer science and engineering, textile fiber science and engineering, color and color measurement, and technical textiles. He has published a number of research papers in various national and international journals. He has administered several research projects on smart textiles, dyeing, and printing. Perhaps, he also has experience in leading higher education institutions in the capacity of a department manager, research unit coordinator, and head for international relations.

**Dr. Melkie Getnet Tadesse** is an award winner in the Alexander Von Humboldt-Stiftung scholarship in the Georg Forster scheme and a postdoctoral researcher at Albstadt-Sigmaringen University, Germany, who obtained his bachelor's degree in Textile Engineering and Master of Science in Textile Technology at Bahir Dar University since 2007 and 2013, respectively. He received an award in the SMDTex scholarship sponsored by the European commission. He obtained his PhD in Textile Materials Technology at the University of Boras (Sweden), Gheorghe Asachi Technical University of Iasi (Romania), and Soochow University (China) in 2019 with excellent publication records and excellent thesis results with the title "Quality Inspection and Evaluation of Smart or Functional Textile Fabric Surface by Skin Contact Mechanics." He worked as an assistant professor in EiTEX, Bahir Dar University, from 2019 to 2022 and taught the functional textile development and quality control course. He cosupervised two PhD students and more than 10 MSc students in textile areas.

He is a reviewer board member of MDPI and has reviewed more than 90 times since 2018; he is a topical advisory panel member of polymers and micromachines (MDPI) journals. He is a verified journal reviewer in Publons and has reviewed more than 150 times since 2018 for more than 10 journals. He presented his scientific results at international conferences such as AUTEX, Tunisian−German Technical Textile Events, ITMC2017, Aachen−Dresden−Denkendorf International Textile Conference, and many more (more than eight oral and six poster presentations since 2016).

He has published over 37 peer-reviewed articles, book chapters, and conference papers in internationally prestigious journals since 2016. His works have an excellent reputation in Google scholar, ResearchGate, Orchid, and web of sciences indexes.

**Dr. Engr. Muhammad Qamar Khan** is currently serving as a chairman of the Department of Textile and Clothing. He has done graduation in Textile Engineering with a specialization in garments manufacturing in 2013 from National Textile University Faislabad; he pursued his postgraduation and PhD from Shinshu University Nagano, Japan. In 2019 he rejoined National Textile University Karachi Campus as a chairman of the Textile and Clothing Department. He has more than 60 impact factor research publications, two patents, and 12 book chapter publications. He has won 12 million rupees (MRS) research funding from the Higher Education Commission of Pakistan through Technology  Transfer Support Fund. He made eight memorandum of understanding (MOUs) between National Textile University (NTU) and industries and two international universities; one with Shinshu University, Japan, another one with Hanoi University of Engineering & Technology, Vietnam, and one MOU with an international company which is Hangho Chemical Ltd, China. During his tenure as a chairman, he has established four new labs for undergraduate programs and launched two new undergraduate programs and one graduate program named Ms Textile & Apparel.

**Dr. Muhammad Ayyoob** is currently serving as a chairman of the Department of Polymer Engineering. He completed his PhD study in Chemical Engineering from SKKU Korea in 2019. In 2020 he joined National Textile University Karachi Campus as an assistant professor of Polymer Engineering in the Department of Polymer Engineering. He has more than 10 impact factor research publications. He has won nine MRS research funding from the Higher Education Commission of Pakistan through National Research Program for Universities. He made two MOUs between NTU and industries.

**Sippy Pirah Simair** is pursuing her PhD in biological material science at the College of Biological Science and Medical Engineering, Donghua University, Shanghai, China. She has many feathers in her cap; her outstanding achievements bear out her credentials, and she completed her early education with excellent grades. She spent four highly productive years at the University of Sindh, Jamshoro, culminating with the degree of BS (Hons) in biochemistry, where she mastered the skills of scientific research and experiments. She wrote three excellent articles on xylanase production and the antifungal potential of shell extract. After that, she got a Chinese government scholarship and completed her master's in Chemical Engineering and Technology at Donghua University, Shanghai, China. Her contribution to scientific research is marvelous and exceptional. She is a glaring example of hard work, diligence, and assiduousness. She published two more articles in high-impact journals and one book chapter on "Advances in Genomic Tools to Study Phylogeny, Diversity, and Classification, and she put ceaseless effort to compile this book (https://link.springer.com/chapter/10.1007/978-981-15-9169-3_2). Her current work is "Digitally Printed Textiles and Modern Trends in Local Markets."

**Nuzhat Baladi** is currently rendering her services as a lecturer at the Department of English, Government College University, Hyderabad. From a young age, she has learned the value and love of education, which has led her to the profession of teaching. Her development in the educational realm has shaped her ideas, becoming what she has achieved in her educational career. She completed her BS in English Literature in 2016 from IELL. During her undergraduate studies, she was awarded a fully funded scholarship by USEFP (United States Education Foundation Program). Where she was offered to study one semester at the University of Wisconsin, United States, she has diverse experience serving various organizations, that is, academic and development sectors. She served as a visiting faculty member at IELL, University of Sindh, Jamshoro, for 3 years and later started her professional journey at Government College University, Hyderabad as a lecturer. She is an aspiring teacher and has mastered skills of communication, listening, collaboration, adaptability, empathy, and patience, which are required of a good teacher.

# Foreword

As an associate professor at the Amirkabir University of Technology (Tehran Polytechnic) working in textile engineering, I would like to provide this foreword for this high-value book investigating different aspects of digital printing as one of the most important and rarely investigated subjects in the textile industry.

Digital textile printing is a process of printing on textiles substrates and garments using inkjet technology to print colorants onto fabric. This process allows for single pieces, mid- to small-run cycle production, and even long runs as an alternative option to traditional printed fabric. This technique is considered one of the most promising developments in the textile industry. Providing the possibility of printing a digital-based image directly onto various textile-based media and creating a high-quality design has opened the door to many possibilities to meet the growing demand for textile printing.

Dr. Hua Wang, the editor of this book, is one of the leading scientists in the international correlation between the textile industry and state-of-art science and technology. Dr. Wang has spent more than 40 years as a university professor, as general manager of different textile companies, and as a visiting scholar exploring different aspects of textile materials and processes in industry and academia. One of the main characteristics of this book is the full coverage of digital printing topics based on the experiences of Dr. Wang and his coeditor, that is, Dr. Hafeezullah Memon, who have been involved in cooperation with diverse external companies and research institutes. This attribute of the book, along with the diversity of the topics related to the science and technology of digital printing covered by other contributors, makes this publication a hands-on experience manual for those working in this area. It is expected that, as a reference manual, the book would attract many audiences from industry and academia.

**Roohollah Bagherzadeh**
*Institute for Advanced Textile Materials and Technologies,*
*Amirkabir University of Technology (Tehran Polytechnic), Tehran, Iran*

# Preface

Digital textile printing technology combines mechanical, computer, electronics, textiles, and information technology. It is one of the high-tech products gradually formed with the continuous development of computer technology. This technology's emergence and constant improvement have brought a new concept to the textile printing and dyeing industry. Its advanced production principles and methods have brought unprecedented development opportunities to printing, dyeing, and clothing manufacturing. Digital printing technology is the product of the combination of digital technology and traditional printing and dyeing technology, and it is a revolutionary breakthrough in the traditional printing and dyeing industry. In the context of digitalization, intelligence, and networking, digital printing represents one of the directions of technological development in the printing and dyeing industry.

Digital printing technology has the following undeniable advantages:

First, in textile printing production, digital printing significantly shortens the original process route, with faster order speed and dramatically reduced proofing cost. Because the production process of digital printing gets rid of the traditional printing in the production process of color separation sketching, filming, and mesh making, the production time is considerably shortened. The way to accept the pattern is through the CD-ROM, Email, and other advanced means; the general proofing time is not more than one working day, while the traditional proofing cycle is generally about 1 week.

Second, digital printing technology breaks the restrictions of limited colors and pattern repeat length, which can make textile fabrics achieve the printing effect of high-grade printing. Because digital printing technology can use digital patterns through the computer for color measurement, color matching, and inkjet printing itself, the color of digital printing products can theoretically reach 16.7 million, with high-precision patterns such as color gradients and moire.

Third, digital printing production truly understands the production process of small batches and fast production, and the production batch is not subject to any restrictions. It can bring rapid response to fashion trends and zero inventory production. Digital printing production has a consistent printing effect with the improved production environment. The use of digital printing technology makes it easy to design and check patterns and thus ensures accurate samples.

Fourth, the high-precision printing process makes the printing process without water or paste. The use of dyes in the printing process is "allocated on demand" by the computer so that in the printing process, there are no waste of dyestuffs and no wastewater generation. The computer-controlled printing process does not produce noise, so there is no noise pollution. The green production process is realized so that the production of textile printing gets rid of the postproduction process of high energy consumption, high pollution, and high noise and realizes the production process of low energy consumption and pollution, which brings a technological revolution to the production of textile printing and dyeing.

Digital printing is a combination of technology and art. This technology will lead the trend of future product development. Many well-known international fashion brands currently use the procurement and transportation system and "fast fashion, fast response" digital printing technology in their designs. The production process has made a breakthrough in concentration in the production chain according to the order of small and multiple batches with different needs, designs, and required characteristics. The world has a great demand for digital printing products; digital printing has penetrated design, workwear, home textiles, automotive decoration, advertising, personalized customization, online stores, and other fabric products in various fields. In recent years, it has shifted from the past model of design proofing to small batch production, and many brands have accepted the use of digital inkjet printing fabrics.

The promotion and application of digital printing technology will have a significant impact on the development of the world's textile printing and dyeing industry, as follows:

First, it will promote the reform of enterprise production methods and business models.

For a long time, due to the limitations of traditional technology, the production of printed products could not meet the diversified and personalized needs and requirements of the market at an economical cost. The emergence of digital printing technology has already somehow solved this problem and changed enterprise production methods and business models. In addition, digital textile printing technology combined with information technology would achieve a completely personalized, one-to-one tailor-made business model in a very economical and eco-friendlier way in the future.

Second, it will meet people's needs to the greatest extent possible and improve their quality of life.

For textile designers worldwide, it is an excellent tool for unleashing their creativity, enabling them to fully display their creative talents and turn their outstanding work into a realistically finished product. Digital textile printing is rich in colors, thus home decoration, clothing, and travel supplies can be personalized according to personal preferences for the majority of consumers using a wide range of possible colors. Hence, this new technology drives a new market and stimulates a new round of consumer demand.

Third, it will directly promote the development of "green textiles" and "green manufacturing."

As colorants in digital textile printing are sprayed directly on the fabric, neither the colors are wasted nor generate wastewater pollution, nor large rounds of sampling are required. Besides, the consumption of materials, such as film, wire mesh, silver cylinders, and many inventories, is also omitted. It reduces the burden on enterprises and impacts positively on the environment.

Soon, digital textile printing technology would develop into an intelligent, meta-universe consumption mode. The development of digital printing technology will continue to iterate and improve with a bright future.

In this book, all chapters were welcomed by the authors. The editor declares that there is no relation and responsibility of any of the editors in terms of any political, religious, cultural, or technological claim, thoughts, or ideas made by authors in their chapters. The corresponding author of each chapter is assumed to be responsible for this matter. The editors would like to acknowledge the funding, that is, Research Fund for International Scientists (RFIS-52150410416), the National Natural Science Foundation of China.

**Hua Wang**
**Hafeezullah Memon**

# Introduction to digital textile printing

*Hua Wang[1] and Hafeezullah Memon[2]*
[1]Western Research Institute, Chinese Academy of Agricultural Science, Changji, P.R. China,
[2]College of Textile Science and Engineering, International Institute of Silk, Zhejiang Sci-Tech University, Hangzhou, P.R. China

## 1.1 History of printing

Textile printing production and human life have been closely related since ancient times, and thousands of years of accumulation have resulted in a rich historical tradition. Printing of traditional textiles is a traditional folk art with profound cultural elements. In principle, the printing process refers to applying patterns on textiles using dyes or pigments, while traditional textile printing mainly refers to ancient handicraft printing closely related to ancient textile science and technology. Textile printing encompasses all traces left by dyes or pigments on textiles, including hand-painted and warp dyeing. In the narrow sense, textile printing refers to printing colored patterns or patterns of antidyeing (dye-resisting) material to make textiles display a regular pattern after the process.[1] A significant turning point in the history of textile printing occurred after the first industrial revolution, and textile printing entered the mechanical age. The era of textile printing automation began in the last decade with the development of electrical and computer chips. Textile printing has entered the digital and intelligent era due to the rapid development of information technology and the demand for fast fashion. To summarize the history of textile printing, it has experienced the development from hand painting, hand tool printing, mechanical printing, and digital printing to intelligent printing.

## 1.2 Overview of the evolution of printing tools and equipment

Printing has evolved, as have the printing tools and mechanical equipment used. In the past, printing was like hand painting, resulting in low production efficiency and inconsistent product quality. Later, a block printing method and an engraved printing method were developed. The roller printing machine was created due to the continuous development of science and technology and the advent of the industrial revolution, which significantly improved printing production efficiency. Later, flat-screen printing machines, circular-screen printing machines, transfer printing machines, and digital printing machines significantly contributed to the printing industry's development.

### 1.2.1 Block printing and engraved printing

Block printing refers to the carving of the pattern on a wooden block. The pattern embossed on the block is dipped into a printing paste and later stamped on the fabric. The printing effect of this printing technology is not ideal, the wooden block is difficult to clean, and the production efficiency is low. Block printing was used in the Indus Valley civilization in 2300–1750 BC.[2] Block printing has evolved into an embossed roller printing technology after a long development after the modern industrial revolution.

Meanwhile, engraved printing entails carving out the pattern to be printed on cardboard, metal plate, chemical board, or wooden block, covering it with fabric, adding printing paste, and scraping the paste to obtain the printed fabric. The engraved printing technology first appeared during the Warring States period, matured during the Tang Dynasty, and was replaced by modern mechanized large-scale production after the Industrial Revolution. According to literature, three pieces of colored silk were found in 1972 on the lid of tomb No. 48 of Wuwei Mozuizi in Gansu Province, dating from the end of the Western Han Dynasty to the middle of the Eastern Han Dynasty.[3] After careful examination, it was found that the printing marks on the colored silk had noticeable effects, and a local staining phenomenon had occurred, likely caused by the hollow version of the printing. The engraved printing technology has evolved into screen printing technology in modern times.

### 1.2.2 Drum and roller printing machine

Metal prints became popular in Europe from the 17th century onwards. Fabrics were produced using copperplate printing. France used a machine similar to that used for printing prints. This method was first used to print handkerchiefs, small floral cloth, and later monochrome or bi-color floral cloth. These floral patterns are varied, ranging from geometric patterns to floral motifs and typical European coat of arms motifs and themed paintings. Exotic "trees of life" and Turkish, Iranian, Indian, and Chinese motifs were also observed. Most of these patterns are, however, fake oriental patterns because similar patterns in oriental patterns are challenging to find. These floral fabrics are used to sew clothes and for interior decoration. For example, most of the cover furniture's flower cloth is a strong one printed with small flowers on a white background, the pattern known then as the "Thousand Flowers Diagram." These patterns of different forms and colors are quite crowded with the architectural style of the time. In the 18th and 19th centuries, the style of such patterns in Western European floral cloth was outstanding.

The printed fabrics produced in Germany, Switzerland, and France in the middle of the 18th century included cotton and silk brocade, and various "rose patterns" were popular. This pattern was extensively used on printed fabrics by the silk designer and entrepreneur, that is, Philippe de Lasalle from the late 18th century to the beginning of the 19th century.[4] At that time, the patterns created by LaSalle became a model for many artists to imitate and were popular among the locals. In the late 18th century, European printing began to use machines, and the patterns printed in Germany and Britain were mainly in the LaSalle popular style, with banded rose bouquets separated by the famous Valenciennes lace.[5] These patterns were printed on a white or dark background, and most of them were printed with a color followed by a band with blue, red, and snow cyan. At that time, multichromatic printing was rarely used in western European printing workshops.

Roller printing, also known as copper roller printing, is printing on the fabric with a copper roller engraved with a concave pattern. The flower tube is the roller that carved the flower. During printing, the surface of the flower tube is first coated with a color paste. Then, the surface paste of the uncarved part of the flower barrel is scraped off with a sharp and flat scraper to ensure that the concave pattern leaves a color paste, the color paste is transferred to the fabric, and the pattern is printed. The color pattern can be continuously printed if multiple cylinders are simultaneously installed on the printing equipment.

The roller printing machine was invented in 1780 by Jams Bell of Scotland, who patented his design in 1783, and his first six-set color printing machine was installed in a Scottish factory in 1785.[6] This machine could replace the operation of 40 letterpress printing workers, eliminating manual letterpress printing. The century-long Industrial Revolution in Britain was essentially complete from the mid-18th century to the mid-19th century. Machine printing used the carved roller continuous coloring method, which could simultaneously print various chromatic patterns. The pattern was delicate and dense, with large output and low cost. This method was used to produce many cheap and beautiful printed cotton fabrics. The chromatic color of machine printing could reach up to 20 types, while the ordinary color ranges from 1 to 10 distinct colors.

Today, roller printing is fast and can produce more than 6000 yards of printed fabric per hour because the copper roller can be carved with very delicate fine lines, resulting in a delicate and soft printed pattern. The roller engraving cycle is, however, long, with a high cost, and the operation of the engraving is complicated when printing, with high labor intensity; thus, it is not suitable for today's market demand for small batches and multiple colors; therefore, the roller printing has been replaced by screen printing.

### 1.2.2.1 Embossed roller printing machine

The embossed roller printing machine is primarily used for wool printing. The main components of embossed drum printing are the cloth feeding device, the printing locomotive head, the dryer, the cloth outlet device, and the transmission equipment. The general process is to pave, mark, and position the textile before the printing process; then, the dye paste is coated on the embossed part of the printing plate, and a certain pressure is applied to the pattern to transfer and fix it on the textile, presenting the pattern on the flower plate, which is somewhat similar to the stamp. The production efficiency of the embossed printing machine is, however, low, and the production of copper rollers is time-consuming and laborious. Thus, the market has eliminated this type of machine.

### 1.2.2.2 Groove roller printing machine

The groove roller printing machine is used in the factory more than the drum printing machine. This machine comprises a cloth feeding device, a printing machine head, an interlining and printing cloth drying device, and cloth out of the device. Some advanced groove roller printing machines also have interlining washing devices. The flute roller printing machine can be divided into four types based on the different heads of the printing machine: radial, vertical, inclined, and horizontal, with the radial type being the most common.

The color paste required for printing is stored in the pulp plate before printing. The paste roller is used during printing to transfer the color paste to the surface of the flower tube. Then, the scraper knife scrapes the paste and the color paste into the groove. After that, the interlining cloth and fabric to be printed simultaneously enter the rolling point between the pressure roller and the flower barrel due to the squeezing effect of the pressure-bearing roller. The color paste at the groove of the flower barrel can be printed on the fabric, and the excess printing paste is printed on the interlining.

Plate making is the most critical part of concave cylinder printing. The five main plate-making methods currently used are the shrinking engraving method, the photographic engraving method, the rolling engraving method, the hand engraving method, and the electronic engraving method. No matter which plate-making method is used, it is time-consuming and laborious, and it requires a significant investment, which limits the application of the groove roller printing machine in actual production. Thus, groove roller printing has been gradually eliminated from the market.

### 1.2.3 Screen printing machine

Screen printing was the predecessor of modern flatbed screen printing. This mechanism evolved from engraving. A simple flower plate is carved into cardboard, sprayed with paint or ink, passed through a small hole, and printed on the paper below. Commercial engraving is printed on waxed cardboard or sheet metal. The French proposed engraved paper plates with silk fabric as continuous brackets in the mid-19th century. The supporting fabric was tightened on a bracket, later called a sieve, for best results. This method is significant because it automatically provides a link and allows for controlling the number of colorants used. The engraved paper version was replaced by a durable lacquer applied to the screen fabric. Since then, the advantages of screen printing have been well demonstrated, especially in fashion stores that are becoming increasingly popular among fashion enthusiasts. The screen printing machine is classified into two types based on the screen's shape: flat and circular screen printing machines.

## 1.2.3.1 Flatbed screen printing machine

After a long development period, many types of flatbed screen printing machines have been produced, and their performance and printing quality vary. Flatbed screen printing is classified into three types based on the different processing methods and movements, that is, screen moving, cloth moving, and turntable flatbed screen printing machines. At present, these three types are commonly used in the market. The main models are CH-8000, HZ3612, FSM-A, ZX-250A, LHM5V, M1336, HLM2611, and BC-8000T. The development of flatbed screen printing toward modern high output concurrently occurs with the advancement of the screen itself. Also, it can be classified based on the automation involved, that is, manual flatbed screen printing, semi-automatic flatbed screen printing, and fully automatic flatbed screen printing. Although the production output is lower than rotary screen printing and more suitable for small batches of multivariety high-grade fabric printing, it has some advantages, as shown in Fig. 1.1.

## 1.2.3.2 Rotary screen printing machine

The rotary screen printing machine was first created in 1963 by the Dutch Stork company, and it has seen rapid development despite its short history. Circular screen printing has the characteristics of roller printing and flatbed screen printing, which can be continuously printed with rich color, strong adaptability, and high production efficiency.

The performance of the scraper is an important factor affecting the printing effect in rotary screen printing. Reasonable selection of printing spatula is one of the important contents of the production technology management of rotary screen printing, which directly affects the amount of color given to the rotary screen,

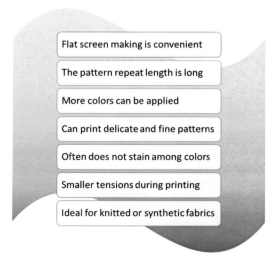

**Figure 1.1** Advantages of flatbed screen printing.

which will be permeable to the fabric, and the uniformity of the fabric printing scraper and other printing indicators. The squeegee used should have high reliability. The rotary screen printing scraper is divided into the coating scraper and the magnetic scraper.

Rotary screen printing is a printing method that uses a spatula to print the color paste in the rotary screen to the fabric under pressure. This process has several important aspects that differ from other screen printing methods. Rotary screen printing, like roller printing, is a continuous process, and the printed fabric is transported through a wide rubber belt to the underside of the moving rotary screen flower. The rotary screen printing machine comprises cloth feeding, printing, drying, fabric, and other devices. The mechanical components of the printing part can be divided into printing adhesive guide tape, rotary screen drive device, rotary screen, printing scraper holder, flower pairing device, rubber guide belt washing and wiping device, and fabric printing paste and slurry pump system. When the printed fabric comes into contact with the rubber guide belt, it adheres to it without moving because it is precoated with a layer of adhesive. When the fabric enters the drying device after printing, the rubber guide belt is transferred to the machine for water washing and scraping water droplets according to the reciprocating loop. The drying and fabric units are dried by open hot air. The printed fabric is separated from the rubber belt, relying on the polyester guide belt rotated by the active roller, and the pressure of the hot air nozzle makes the fabric smoothly attach to the polyester guide belt into the oven for drying. The fabric is dried by hot air and is sent out of the drying section with the appropriate tension at the exact speed. In screen printing, the rotary screen printing production speed is fast, more significant than 3500 yards per hour; a seamless perforated metal mesh or plastic mesh is used. The largest mesh circumference is greater than 40 in.; thus, the largest flower back size is also greater than 40 in.; more than 20 round printing machines with chromatic colors have also been produced, and this printing method is slowly replacing roller printing.

The screen of the rotary screen printing machine is cylindrical, usually made of metal nickel, and the mesh is hexagonal compared with the flatbed screen printing machine. The rotary screen printing machine can be divided according to the arrangement of the rotary screen: horizontal, vertical, and radial models. The horizontal rotary screen printing machine is the most used on the market. The following focuses on the horizontal rotary screen printing machine.

The printing unit is the core part of the rotary screen printing machine, mainly composed of a patch device, a scratching device, a flower pairing device, a guide belt full position device, a rotary screen, a guide belt transmission device, and a guide belt cleaning mechanism. The core part of the printing unit is the scratching device, and the main component of the scraping device is the scraper, which the expansion of the trachea can fix. In the early days of the development of rotary screen printing machines, people used stainless steel scraper blades, but their application range was small, and the printing effect was not ideal. Later, more superior polymer material airflow squeezing blades were developed. The mechanical pressure of the airflow scraper blade can be adjusted, the application range is wider, it is made of a polymer material, and the friction between the mesh is slight compared with the traditional stainless steel scraper blade, thereby significantly reducing the wear of the mesh and prolonging the service life.

The primary purpose of the drying device is to remove excess moisture at high temperatures and ensure the printing effect. The primary purpose of the falling device is to wind up the printed fabric and wait for the following process.

The outstanding characteristics of the rotary screenprinting machine are low labor intensity, high production efficiency, and strong adaptability to different types of fabrics. The printing effect is also good. Certain shortcomings, however, cannot print the flower type of the cloud pattern, snowflake, and other structures well, and the circumference of the circle net will also limit the size of the flower type.

### 1.2.4 Transfer printing machine

Transfer printing involves preprinting a pattern on paper and other materials and transferring the pattern to the fabric using hot pressing. Transfer printing began in the 1960s; it can realistically transfer colored patterns from printing paper to textile or polymer film. Four mature transfer printing modes are available, as shown in Fig. 1.2.

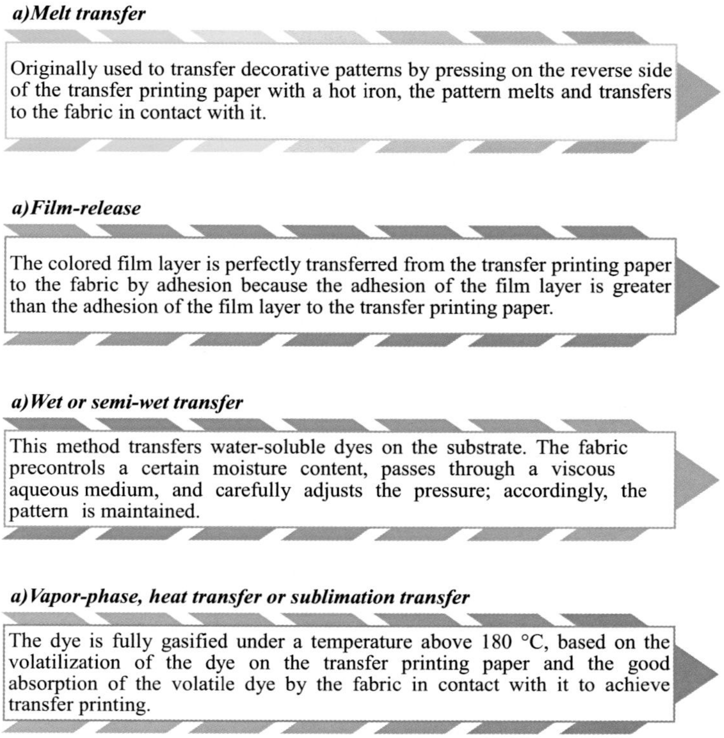

*a) Melt transfer*

Originally used to transfer decorative patterns by pressing on the reverse side of the transfer printing paper with a hot iron, the pattern melts and transfers to the fabric in contact with it.

*a) Film-release*

The colored film layer is perfectly transferred from the transfer printing paper to the fabric by adhesion because the adhesion of the film layer is greater than the adhesion of the film layer to the transfer printing paper.

*a) Wet or semi-wet transfer*

This method transfers water-soluble dyes on the substrate. The fabric precontrols a certain moisture content, passes through a viscous aqueous medium, and carefully adjusts the pressure; accordingly, the pattern is maintained.

*a) Vapor-phase, heat transfer or sublimation transfer*

The dye is fully gasified under a temperature above 180 °C, based on the volatilization of the dye on the transfer printing paper and the good absorption of the volatile dye by the fabric in contact with it to achieve transfer printing.

**Figure 1.2** Four different well-known modes of transfer printing.

The production cost of transfer printing is low, and the environmental pollution is negligible. Between 1967 and 1977, this printing method was commercially successful and widely adopted.

The research and development of heat transfer printing paper must be accelerated, and low-quantity, multiple transfers, cold transfer, and water transfer printing paper must be developed to provide a greater development space for modern textile, construction, electronics, and other industries.

The process of transferring printing is mainly divided into four steps: the first step is to make the dye into ink, then the printing transfer paper, followed by the transfer of the printing machine, and finally, the printing product is obtained. The pattern obtained by transfer printing is realistic, delicate, and compared with traditional roller and screen printing. The transfer printing process is simple and environmentally friendly. The transfer printing machine is mainly classified according to the transfer principle, that is, cold transfer printing machine and heat transfer printing machine.

### 1.2.4.1 Cold transfer printing machine

The basic principle of cold transfer printing is to make the dye into a color paste and then adjust the surface tension between the color paste and the paper. Then, the pattern is printed onto the paper, dried, and rolled. Subsequently, the pretreated fabric is aligned with the transfer printing paper in the transfer printing machine. Under certain pressure, the affinity of the dye to the fabric is greater than that of the transfer paper, which can be transferred to the fabric. Then, the transfer paper is separated from the fabric. The oven dries the fabric, and the cold transfer printing is completed. A cold transfer printing machine can be used for transfer printing of natural cellulose fibers, such as cotton.

In 1984, Dansk Tranfertryk in Denmark launched a new type of cold transfer printing process, that is, the Cotton Art transfer printing method, which was later developed and transformed by the German Kuhster Co. Ltd.[7]

### 1.2.4.2 Heat transfer printing machine

The principle of heat transfer printing is different from cold transfer printing because the dye on the transfer paper is sublimated and transferred to the printed fabric after the patterned side of the transfer paper is combined with the printed fabric through heat and pressure. The more typical heat transfer printing machine is the ZY-series heat transfer printing machine.

## 1.3 Key historical inventions of today's digital printing technology

Digital printing is a new type of printing technology that directly transmits graphic information to the printer. The digital printing system mainly comprises a computer and a digital printing machine. To understand the origins of inkjet technology, we must first go back to the 18th century in the fields of physics and electronics, which laid the foundation for the emergence and progress of inkjet technology, such as Jean-Antoine Nollet's publication of his research on the electrostatic effect of liquid droplets in 1749 and Michael Faraday's electromagnetic induction experiment in 1831. Félix Savart, a French scientist, conducted a droplet fracture experiment in 1833 to find the existence of liquid fracture law and used sound energy to form a uniform droplet. Thomas Young in 1804 and Pierre-Simon Laplace in 1805 described the thrust of a droplet after it breaks.

In 1884, Dr. Richard G. Sweet of Stanford University in the United States filed patents[8,9] related to inkjet technology using the principle of electric field control. Specifically, when pressure is applied to a syringe filled with ink, the droplets sprayed from the syringe's needle are charged, and an electric field controls the drop point. Accordingly, a rich image can be formed on the medium. Later, in 1964 Schneider and Hendricks conducted in-depth research on this topic and developed a practical continuous inkjet system.[10]

In 1858, Sir William Thomson, also known as Lord Kelvin, invented the world's first siphon recording device that worked like inkjet printing, patented in 1867, and could control the release of ink droplets by electrostatic electricity. In 1878, Lord Rayleigh discovered the technique of liquid droplet formation,[11] in which the droplet stream ejected from the nozzle could split into microdroplets of uniform size and distance by applying periodic energy or vibrating at the nozzle opening to form droplets. It was not until 1948 that Siemens-Elema in Sweden, however, applied for the world's first patent based on this principle and developed an inkjet chart recorder called the Mingograph.[12]

In 1880, the Curie brothers discovered that the piezoelectric effect (including the inverse piezoelectric effect) laid the foundation for the research and application of piezoelectric materials.[13] The piezoelectric on-demand inkjet technology is based on the principle of the piezoelectric effect. Clarence Hansell of the American radio company RCA invented the world's first piezoelectric inkjet device in 1946[14]; however, his radio company did not translate the results of this invention into valuable commercial equipment. In 1951, R Elmqvist of Siemens applied for a patent for the first inkjet equipment,[15] entitled "Measuring instrument of the recording type," based on the Rayleigh principle, and Siemens manufactured the first inkjet recorder. In 1972, Steven I. Zoltan working at Clevite Corp, patented[16] the idea of a reverse piezoelectric effect extrusion model, which remained their monopoly for a longer time.[17] Prof. N. Stemme of Chalmers University in Sweden patented[18] the bending mode technology of the inverse piezoelectric effect in 1973. In 1976, Edmond L. Kyser and Stephan B. Sears of the American Silonics company patented[19] a similar bending mode technology with some further advancements. In 1978, Silonics introduced an on-demand inkjet printer with bend mode technology. Based on continuous inkjet technology, IBM created the IBM4640 inkjet printer.[20] In 1977, Siemens' first drop-on-demand (DOD) printing, the Seimens PT-80, was introduced to the market.[21]

In 1962, Mark Naiman from Sperry Rand Corp, New York, invented the Sudden Steam Printing technology,[22] but it was not valued and adopted by his company.

Later, HP began to implement the development of the inkjet printer project in 1979. After unremitting efforts, HP's 2225 Thinkjet thermal inkjet printer was market-oriented in 1984.[23] In 1991, HP introduced the first color inkjet printer HPDeskJet 500C, and the first large-format monochrome printer HPDesignJet was launched.[24] Furthermore, Japan's Canon's Endo and Hara invented inkjet technology similar to Hewlett–Packard in 1979 and named it Bubblejet, which was developed faster than Hewlett–Packard; Endo et al.[25] also patented this technology. In August 1980, Canon applied bubblejet technology (Bubblejet) to the inkjet printer Y-80. Based on the results of Cambridge Consultancy in 1990, Xaar acquired the development and commercial use of digital inkjet printing technology. Xaar's business is focused on licensing these patents and technical expertise to office printer manufacturers. Xaar's major licensors are large multinational corporations that manufacture large quantities of inkjet heads used in their respective products or sell them to OEM customers. In 1993, the Japanese Epson Company introduced the Stylus 800 printer based on piezoelectric inkjet technology.[26] In addition, the phase change inkjet technology is to melt the solid ink into a liquid after spraying. In the 1960s, Charles R. Winston from Teletype Corporation, Chicago, developed a continuous inkjet device and later used it in the field of on-demand inkjet.[27] On this basis, in 1982, the United States Howtek, Inc. developed the Pixelmaster phase change color inkjet printer.[28] Recently, Atexco patented[29] the target pattern template obtained by the secondary design of the fabric with the original pattern, and accurate printing is achieved by digital printing to enrich the fabric's pattern and color. In addition, users can redesign existing fabrics according to customer requirements.

Inkjet heads and inkjet printer suppliers based on DOD technology are also increasing due to the technical advantages of DOD itself and the increasing market demand. For example, Xaar, Trident, Dimatix, Toshiba TEC, Silverbrook, Sharp, SII Printek, Agfa, Aprion, Brother, Domino, Epson, HP, Canon, Hitachi Koki, Imaje, Konica Minolta, Mimaki, Olympus, Panasonic, Atexco, Ricoh, and Kyocera, are also increasing to meet the application needs of more field's inkjet printer models. The development and application of inkjet technology are becoming gradually extensive and in-depth.

Computer technology gradually became popular, and continuous digital jet printing machines appeared in the 1990s. Carpets, wall hangings, and other rough textiles had mature applications by 1995. The Austrian company ZIMMER took the lead in exhibiting a digital jet printing prototype for carpets with a resolution of 9–18 dpi by using a valve spray technology, without using nozzles in the usual sense, and its production speed is fast.[30] This tool, however, cannot be applied to clothing fabrics due to a large number of spray droplets and low resolution. In 1995, the inkjet digital jet printing machine was in demand. This machine was mainly used as a hot foam nozzle with a resolution of 300 dpi, then gradually developed to 600dpi, using CMYK printing four-color mode, which could be applied to the fabric. From 1999 to 2000, the digital jet printing machine using piezoelectric nozzles was exhibited in more countries, including the Netherlands, Japan, Switzerland, the United States, Italy, and China, with a resolution of up to 360–720 dpi. The minimum inkjet amount of the piezoelectric nozzles was very small, the product pattern produced was finer, and the color was more uniform and natural. Thus, these nozzles have remained widely popular, and their application is still gradually expanding.

Digital jet printing has now become the industry's recognized focus of development. In the North American market, which accounts for half of the world's total printing scale, the inkjet printing revenue accounts for more than 70% of the annual revenue, and the inkjet printing market has crossed a total size of $100 billion. In the next 10 years, inkjet printing technology will be comparable to traditional printing technology in terms of print quality, printing speed, and production cost, leading to the future development trend of the printing industry. The sales volume of printing inks has also maintained continuous growth with the vigorous development of the printing market. According to statistics, the global sales volume of printed ink in 2018 was about 182,900 tons (approximately 4.889 billion US dollars) and will continue to expand in the next 5 years with a compound annual growth rate of 10.7%.

In the current stage of clothing design, digital printing textile fabrics have a significant application value. This type of fabric cannot only significantly highlight the appearance and beauty of a clothing design but also improve the overall structure and quality of clothing. Accordingly, the appearance and creation methods of clothing design show a diversified development trend. Digital printing technology is essential in the overall fabric and local clothing design. Thus, the application of digital printing textile fabrics must be continuously explored to promote the development of the clothing design industry.

## 1.4 Advantages of the digital printing technology

### 1.4.1 Promote the continuous development of clothing fabric printing

The application of digital printing technology in clothing design is an inevitable trend because it has the advantages of fast response speed and high design quality, which can meet not only people's design needs for daily clothing but also high-end customization and stage clothing, such as clothing design with high professional requirements. Digital printing can also reduce some unnecessary links in the original design, thereby directly improving the efficiency of clothing design. Based on the information digital printing technology, the designer in the clothing design can no longer use other equipment. Only the digital printing technology software can complete the design and preservation of the pattern. The efficiency of clothing production has been improved according to the computer's color separation technology operation.

### 1.4.2 Promote the diversified development of clothing design

Given the changes in fashion trends, clothing design must also develop according to the development of the trend to meet the esthetic needs of today's young consumers. This task requires clothing designers to explore and study the development trend and carry out uninterrupted innovation in clothing design to ensure its vitality. Many new fabric processing methods have been developed with the progress of science and technology. Some design patterns, however, could not be presented in past clothing designs due to the imperfection of printing technology. Digital printing technology can compensate for this shortcoming by realizing the simulation texture of the clothing pattern design and making the image present clearly with a better effect. Digital printing technology is based on the good treatment of the pattern, which makes the appearance of clothing show a diversified development.

Clothing is necessary for people's lives; hence, the beauty of the appearance design of the clothing and its comfort must be ensured. Designers apply a variety of fabrics to meet people's needs for clothing comfort. Now, we can use digital printing technology to combine clothing fabrics and patterns to form a new design style and promote the diversified development of clothing creation and design methods.

### 1.4.3 Color selection is freer and richer

Designers need to use the elements of color reasonably because the color is one of the main essential components of clothing design. Color is used not only to enhance the beauty of the appearance of clothing but also to convey the profound connotation and specific individual style in the clothing design to people. In clothing design, the rational use of color must be continuously strengthened, and the technical requirements of color in the use and selection must be improved. The changes brought about by the use of color by digital printing technology and high-tech equipment based on information technology have provided more possibilities for the use of color in clothing design that was not available in previous color designs. In clothing design, one can use advanced digital printing technology to more realistically present the gradient design sense that needs to be given in the printing and dyeing of fabric. Fig. 1.3 shows some unique printing possibilities only possible through inkjet printing in combination with the Atexco vision machine used by many top international brands, including Dior and Chanel. This design's printing and dyeing pattern bring new changes to people's vision, enhancing the color's vividness and providing. Furthermore, the pattern of a very realistic printed textile fabric presented with rich colors will bring a completely different visual effect to the clothing design.

### 1.4.4 Pattern design is more fashionable and cutting-edge

With the continuous improvement of living standards, people have a higher pursuit of quality of life, and the demand for clothing design has also changed with the change of esthetic concepts and current fashion trends, so there are more ideas for the choice of styles and patterns in clothing design. Based on this situation, the clothing design industry has strengthened the innovation of pattern design to adapt to people's esthetics and has also more actively used digital printing technology.

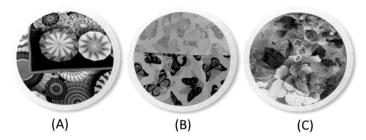

(A) (B) (C)

**Figure 1.3** Some unique print possibilities of digital inkjet printers are (A) double sided silk scarves, (B) jacquard embroidery counter printing, and (C) lace positioning printing.
*Source*: Atexco, Hangzhou, China, provided the photos.

The concept in clothing design can be more cutting-edge and more in line with people's requirements for clothing esthetics at this stage because of the superiority and high informatization of digital printing technology itself. Clothing designers need to use digital printing fabrics to make the clothing design more vivid and can also effectively make the corresponding pattern designs for various popular styles at this stage to further meet people's needs for different clothing designs.

## 1.5 Digital printing breaks through the bottleneck of textile printing

### 1.5.1 Meet the needs of cleaner production

Traditional textile screen printing uses screens to print the colors one by one—the more colors to be displayed, the more complicated the work; however, the screen-printed patterns are still quite monotonous, and the color gradient and natural tones of high-resolution images are impossible to be seen. In addition to the complexity of the printing technology and effect, the printing production process is lengthy, taking more than 4 months from production to sales. The production of the screen plate will take 1–2 months, and the production process requires more workforce, time, and energy. Furthermore, the production of the screen and equipment cleaning also requires a considerable amount of water resources. If the screen is not used further, then it becomes a waste. Such a production method significantly impacts the ecological environment and does not meet clean production requirements. On the other hand, digital inkjet printing can directly print on textiles without needing expensive and time-consuming screens, as shown in Fig. 1.4.

### 1.5.2 Breakthrough the bottleneck of traditional textile printing

Digital printing technology breakthrough the bottleneck of traditional textile printing, that is, the combination of image processing software, printing machine,

**Figure 1.4** Inkjet printing on the textiles.
*Source*: Atexco, Hangzhou, China, provided the photos.

printing ink, and printing substrate processing. The actual image or pattern digitally stored can be directly printed on the fabric, with the diversity of pattern design and color change, and is widely used in the popular design and fashion apparel industry. This technology is particularly suitable for small, diversified, and customized production processes, which significantly reduces the cost of online operation by 50% and the timeline by 60% and dramatically shortens the overall production schedule, which can quickly respond to customer needs. Moreover, this technology reduces the wastewater output caused by screen cleaning in the printing process and reduces waste by 80%, in line with the clean production requirements. Digital textile printing technology makes the dyeing and finishing industry more technological, environmentally friendly, faster, and diversified.

### 1.5.3 Digital printing technology has a vast space for development in the international application

Now, digital printing products in Italy have accounted for more than 30% of its total domestic printing volume. The growth rate of digital printing depends on the industrial structure and cost. Italy is a famous printing design market, producing most of the world's high-end printed textiles. Texprint, a well-known Italian digital textile printing industry, has recently purchased eight sets of DT32 from Atexco with cotton and rayon printing production capability of 50,000 m per day. Some of the images of their facilities are shown in Fig. 1.5.

### 1.5.4 Products meet market demand

In terms of the production cost of Italian printing, the price per square meter of the production of 400 m small batches of products is nearly 2 euros, while the same batch of products in Turkey and China production costs less than 1 euro; if it is a small batch production of 800−1200 m, it is also nearly 1 euro/m$^2$. Such a cost difference makes digital printing mainstream. Digital printing is in line with its market

Introduction to digital textile printing 15

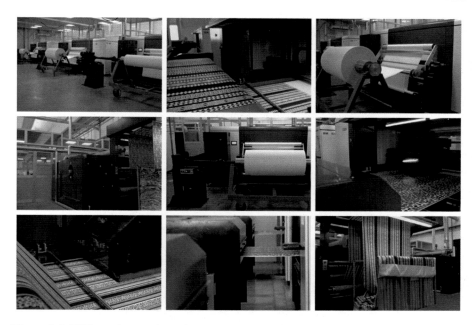

**Figure 1.5** Different images from the printing facility at Texprint using Atexco.
*Source*: Atexco, Hangzhou, China, provided the photos.

demand. According to relevant statistics, the world's traditional printing is mainly concentrated in Asia, including China, Southeast Asia, and India, accounting for 66% of the total printing volume. China accounts for nearly 30% of the world's production in traditional printing; however, digital printing is less than 1% of domestic printing, and there is enormous room for growth. The digital printing floor of Zhejiang Huzhou Puxin Home Textile Co., Ltd. is shown in Fig. 1.6. It can be seen that Atexco Model X plus is installed for production.

### 1.5.5 Production costs are relatively low

Traditional printing needs to prepare screen plates in small batch production, which makes the initial cost of traditional printing relatively high. The cost of mass production is, however, relatively low due to the cost dilution caused by increased printing volume. The main cost of digital printing is the nozzle and ink, regardless of whether the cost of large or small batch production is similar. The management cost will be relatively low as long as the printing speed is fast enough, which is also why the equipment manufacturer develops the concept of a high-speed machine. If the ink quality is good, the wear and tear on the nozzle will be relatively reduced, and the service life of the nozzle will naturally increase. If the ink quality is not good, it will hurt the nozzle and halts the printing. The ink composition includes pigments, moisturizers, surfactants, fungicides, and other additives.

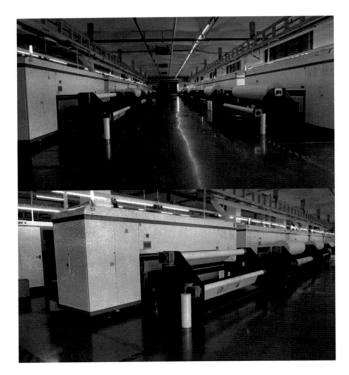

**Figure 1.6** Printing floor of Zhejiang Huzhou Puxin Home Textile Co., Ltd.
*Source*: Atexco, Hangzhou, China, provided the photos.

Pigments are the key to the achievement of printing color. The ink quality is half successful when the pigment is good.

### 1.5.6 Attach importance to the concept of technological innovation

It is an integrated technology that requires nozzles to be matched with inkjet printing equipment, inks, printing software, and fabric preprocessing and postprocessing. The warranty period of the industrial nozzle at this stage is generally half a year to 1 year. The drive engine (nozzle) in the printing industry is expensive. The use of multiple nozzle design equipment also has potential risks in case one of the nozzles in the printing process of the problem will cause losses. At present, industrial nozzle manufacturers want to extract the nozzle warranty fee from the equipment manufacturer and the ink dealer, which also invisibly increases the cost of ink consumed. Given the continuous introduction of new nozzles and the application of different fabrics, ink manufacturers must develop special inks suitable for various nozzles and fabrics. Most of today's ink pigments are derived from existing traditional pigments and may not be able to meet the demand.[31] The product must also consider diversification and coloring, which are more complex and challenging for end users.

Furthermore, today's software is expensive for most end users. The ink and fabric must be remade to the coloring curve for the different printing equipment, which is more complicated than traditional printing. Digital printing differs from traditional paper printing because it requires fabric before and after processing to create the fabric color. Although the fabric before and after treatment is familiar to the traditional printing factory, digital printing is the key as it can only print out the print that customers need to show its value. The previous work will be abandoned if a problem occurs in the pretreatment and steaming. Digital printing has good prospects for future development. Nowadays, the popular elements of textile printing are rapidly changing, requiring more design. The number of repeated orders is gradually declining, the design is becoming highly complex, the design requirements are higher than before, and the environmental awareness to the public is improved.

Moreover, the tolerance of color variation and fabric specifications is becoming stricter; Fig. 1.7 clearly shows that such detailed printing is only possible through digital inkjet printing with an Atexco vision machine, which was otherwise impossible by other means of digital textile printing. Moreover, one can print on the embroidered fabrics using this patented technology, as shown in Fig. 1.8. The deviation of print is only ≤ 0.2 mm, achieved by precise positioning by taking 10,000 photos/second, and can be used in printing for lace, embroidery, jacquard, or double-sided prints.

To this end, digital printing must meet the above-related needs. More nozzles and equipment manufacturers, ink merchants, software vendors, and printing

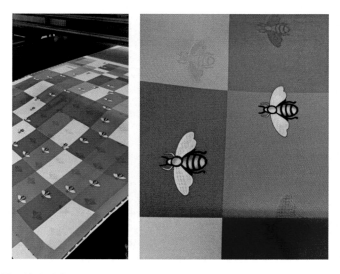

**Figure 1.7** The high delicacy of designs is possible at the fingertips using digital inkjet printing machines.
*Source*: Atexco, Hangzhou, China, provided the photos.

**Figure 1.8** The high delicacy of printing over the embroidery of the laces is now possible. *Source*: Atexco, Hangzhou, China, provided the photos.

manufacturers will need to invest funds in the future. Furthermore, software vendors must develop user-friendly software for printing device manufacturers in the future. Printing manufacturers must continue to cultivate talents and improve their professional skills in digital printing to master the ability to control digital printing equipment and ink chemistries independently. Only in this way can digital printing technology advance faster and more effectively.

# References

1. Hada, J. S.; Meena, C. R. Dabu, The Sustainable Resist Printed Fabric of Rajasthan. In *Sustainable Approaches in Textiles and Fashion: Manufacturing Processes and Chemicals;* Muthu, S. S., Ed.; Springer Singapore: Singapore, 2022; pp 69−91.
2. Hann, M. Symmetry in Regular Repeating Patterns: Case Studies From Various Cultural Settings. *J. Text. Inst.* **1992**, *83*, 579−590.
3. Giele, E. Using Early Chinese Manuscripts as Historical Source Materials. *Monumenta Serica* **2003**, *51*, 409−438.
4. Schäfer, D.; Riello, G.; Molà, L. *Introduction. Seri-Technics: Historical Silk Technologies. Seri-Technics: Historical Silk Technologies;* Max Planck Institute for the History of Science, 20205−11.
5. Merrill, J. *The Showgirl Costume: An Illustrated History;* McFarland, 2018.
6. Tierney, J. Design Quality, Mechanization and Taste in the British Textile Printing Industry, 1839−1899. *J. Des. History* **2017**, *30*, 249−264.
7. Rattee, I.D. Transfer Printing. In *Textile Printing*, 2nd ed.; Miles, L.W.C., Ed.; Society of Dyers and Colourists: Perkin House, 82 Grattan Road, Bradford, West Yorkshire BD1 2JB, England, 1994; p. 58.
8. Sweet, R.G. Fluid Droplet Recorder, 1971.
9. Sweet, R.G.; Cumming, R.C. Fluid Droplet Recorder With a Plurality of Jets, 1968.
10. Schneider, J.; Hendricks, C. Source of Uniform-Sized Liquid Droplets. *Rev. Sci. Instrum.* **1964**, *35*, 1349−1350.

11. Rayleigh, L. The Influence of Electricity on Colliding Water Drops. *Proc. R. Soc. Lond.* **1878**, *28*, 405–409.
12. Folch, A. *Hidden in Plain Sight: The History, Science, and Engineering of Microfluidic Technology;* MIT Press, 2022.
13. Ren, G. Review of Piezoelectric Material Power Supply. In Proceedings of the 2021 International Conference on Electronics, Circuits and Information Engineering (ECIE), 22–24 Jan. 2021, 2021; pp. 136–139.
14. Cie, C. *Ink Jet Textile Printing;* Elsevier, 2015.
15. Elmqvist, R. Measuring Instrument of the Recording Type, 1951.
16. Zoltan, S.I. Pulsed Droplet Ejecting System, 1972.
17. Uchino, K. *The Development of Piezoelectric Materials and the New Perspective. Advanced Piezoelectric Materials;* Elsevier, 20171–92.
18. Stemme, N. Arrangement of Writing Mechanisms for Writing on Paper With a Coloredliquid, 1973.
19. Kyser, E.L.; Sears, S.B. Method and Apparatus for Recording With Writing Fluids and Drop Projection Means Therefor, 1976.
20. Cummins, G.; Desmulliez, M. P. Inkjet Printing of Conductive Materials: A Review. *Circuit. World* **2012**, *38*, 193–213.
21. Heinzl, J. Printing With Ink Droplets From a Multi-Nozzle Device. *Adv. Non-Impact Print. Technol. Computer Office Appl.* **1981**, 1191–1201.
22. Mark, N. Sudden Steam Printer, 1965.
23. Ho, M.F.; Keefe, B. Challenges in the Development of High-Speed True 600-dpi Thermal Ink-Jet Printing. In Proceedings of the Color Imaging: Device-Independent Color, Color Hard Copy, and Graphic Arts, 1996; pp. 51–57.
24. Aden, J. S.; Bohórquez, J. H.; Collins, D. M.; Douglas Crook, M. The Third-Generation HP Thermal Inkjet Printhead. *Hewlett Packard J.* **1994**, *45*, 41.
25. Endo, I.; Ohno, S.; Sato, Y.; Saito, S.; Nakagiri, T. Liquid Jet Recording Process and Apparatus Therefor, 1982.
26. Le, H. P. Progress and Trends in Ink-Jet Printing Technology. *J. Imaging Sci. Technol.* **1998**, *42*, 49–62.
27. Winston, C.R. Method and Apparatus for Transferring Ink, 1962.
28. Wieselman, I. L.; Tomash, E. Marks on Paper: Part 2. A Historical Survey of Computer Output Printing. *Ann. History Comput.* **1991**, *13*, 203–222.
29. Xu, L.; Huang, Y.; Xu, Q. Precise Digital Printing Method for Fabric by Utilizing Digital Printing Machine and Fabric, 2018.
30. Wang, M. Digital Inkjet Textile Printing. Savonia University of Applied Sciences Finland, 2017.
31. Memon, H.; Khoso, N. A.; Memon, S.; Wang, N. N.; Zhu, C. Y. Formulation of Eco-Friendly Inks for Ink-Jet Printing of Polyester and Cotton Blended Fabric. *Key Eng. Mater.* **2016**, *671*, 109–114. Available from: http://www.scientific.net/KEM.671.109.

# Digital printing mechanisms

**2**

Hanur Meku Yesuf[1,2], Abdul Khalique Jhatial[3],
Pardeep Kumar Gianchandani[3], Amna Siddique[4] and Altaf Ahmed Simair[5]
[1]Ethiopian Institute of Textile and Fashion Technology, Bahir Dar University, Bahir Dar, Ethiopia, [2]Key Laboratory of Textile Science and Technology, Ministry of Education, College of Textiles, Donghua University, Shanghai, P.R. China, [3]Department of Textile Engineering, Mehran University of Engineering and Technology, Jamshoro, Sindh, Pakistan, [4]School of Engineering and Technology, National Textile University, Faisalabad, Pakistan, [5]Department of Botany, Government College University, Hyderabad, Sindh, Pakistan

## 2.1 Introduction

Inkjet printing is computer printing that recreates a digital image by propelling ink droplets onto paper, plastic, fabric, and other substrates.[1,2] Digital printing adds colorful designs to various substrates using digital inkjet technology.[3] This can be accomplished mechanically via a push or electrically.[4] Drops can be pulled out of a nozzle by a substantial applied voltage, fluid can be pushed out of a nozzle by sound waves, or a chamber deformation can expel a drop. Although sometimes, a pretreatment process might be needed to enhance the properties of the printed fabric.[5] Digital printing technologies use bitmap images or computer-generated patterns to deposit ink onto target substrates, and digital printing has been widely used for decades for printing graphics and documents.[1,2,6]

Textiles and artificial intelligence are working hand in hand in the current era[7]; computing power or artificial intelligence is essential for spinning,[8] weaving,[9] knitting,[10] nonwoven,[11] or testing and quality control.[12] In digital printing, computing power is used in three phases.[13] First, skilled print designers must create the design; second, the picture must be processed so it can be printed; third, the printer must be controlled so that each pixel is put down in the correct location with the correct mix of inks. Digital technology allows for significantly better resolution printing, the use of as many colors as the designer desires, and no constraints on how patterns are repeated.[14]

There is enormous potential for textiles, and many countries (i.e., universities) are coming forward as prominent places for the textile discipline, particularly textile printing.[15,16] Digital printers can print on various materials, including thicker cardboard, heavyweight papers, folded carton, cloth, plastics, and manmade surfaces. T-shirt printing is a common digital printing application, and some digital printers can also print on linen and polyester in addition to T-shirts. Microdroplets of colored liquid ink are sprayed to the surface of exact spots using small nozzles in digital (inkjet) printing. The color of ink jetting, the volume of ink, and the position of the tiny droplets are all controlled by computers. The four primary colors used by digital printers (yellow, magenta, cyan, and black) present distinct issues in color mixing for textiles.[17]

Digital printing companies follow the following steps during the printing processes[18–26]:

Digital printing firms use customers' specifications and instructions to generate unique designs. The customer will be notified when the design is complete, and any suggested modifications will be implemented before the final version is created.

- Whenever the finalized design is accepted, it is stored in suitable formats and resolutions so that the printer can easily recognize the product and print it without errors.
- The print heads are first cleaned with a specific solution to prevent them from drying out and becoming damaged. This operation is done after every 100 prints; the heads may need to be cleaned even before the 100th print depending on the color of the image. A range of inspections is carried out during this period.
- The printer produces waste ink, with each print deposited in a drum. This drum must be checked and emptied regularly to guarantee that no ink leaks out.
- The printer has a cleaner container and is replaced when the cleaner levels become low. Because this cleanser is essential to the printing operation, care must be taken to ensure that the printer never runs out of it. An inadequate cleaner might harm the heads. The printer uses a set regimen of cleaning; as a result, it stops and begins every time.
- The printer should function at a temperature of 20°C–25°C. Checking ink temperature is necessary to avoid damage to printer heads if the temperature falls outside this range.
- The proper-sized pallet is mounted to the printer, and the item is placed on it, depending on what is printed.
- The printed substrate is set entirely flat on the board with no wrinkles. The design will be altered if there are any wrinkles.
- The inkjet printer starts publishing by spraying the graphic on the substrate by shifting the print heads from edge to edge.
- The materials or products are carefully taken off the pallet when the printing is finished. The print is baked or sticks permanently to the substrate by passing it through a large drier at the proper temperature.
- The final stage is to pack and prepare the items for shipping after being convinced of their quality.

## 2.2 Digital printing methods

### 2.2.1 Fine art inkjet printing

Fine art is visual art that has been created primarily for esthetic and intellectual purposes and judged for its beauty and meaningfulness rather than its functional value, precisely, printmaking, painting, sculpture, drawing, watercolor, graphics, and architecture.[27] The digital fine art gallery is an elevated collection of unique fine art inkjet sheets. Bright colors, deep blacks, high contrast, remarkable detail reproduction, and startling visual depth are guaranteed with pure artist papers, specific surface textures, and a comprehensive selection of superior inkjet coatings. The inkjet substrates in the digital fine art collection have an aging resistance of more than 100 years,[28] making them ideal for fine art applications such as fine art prints, unique editions, protracted exhibits, and art collections.[29,30] Conventional fine-art media, such as arches of watercolor paper, treated and untreated canvas, experimental substrates (like metal and plastic), and fabric, are used in fine-art inkjet prints. Publishing from a computerized picture source to an inkjet printer as an outcome is known as fine art inkjet printing. It emerged from Kodak, 3M, and other big manufacturers' computerized proofing technologies, with artists and other printers attempting to convert these specialized prepress proofing tools to fine-art printing.[31]

## 2.2.2 Notable digital laser exposure

Lasers expose digital pictures onto real, light-sensitive photographic paper, which is then developed and fixed.[32] Photographic paper, like photographic film, is a paper laminated with light-sensitive chemical components used to make photo-realistic images. When photographic paper is treated to light, a latent picture is captured and later developed into a visible image.[33] These prints are real pictures with a consistent tone across the image. The print's archival quality equals the manufacturer's grade for the photo paper. The significant advantage of large format prints is that there is no difference between direct or detail distortion in the corners because no lens is used.[34]

Electrostatic digital printing is what laser printing is all about. By continually passing a laser beam backward and forth on a negatively charged cylinder known as a drum to create a differently charged picture, it generates high-quality letters and pictures and moderate-quality images. After that, the drum accumulates electrically charged powdered ink (toner) and transfers the picture to paper, which is baked to securely bond the text, images, or both to the paper. Laser printers use the same xerographic printing technique that digital photocopiers use. As implemented in analog photocopiers, the picture is created by reflecting light off an existing paper onto the exposed disc in classic xerography.[35,36]

An image of the page to be printed is projected onto an electrically charged, selenium-coated, revolving, cylindrical disc by a laser beam, which is generally an aluminum gallium arsenide (AlGaAs) semiconductor laser that can generate red or infrared light. Charged electrons can fall away from light-exposed regions due to photoconductivity. Electrostatic attraction attracts powdered ink particles to the charged parts of the drum that have not been laser-beamed. The picture is subsequently transferred to paper by the drum, then transported through the machine through physical touch. Finally, the paper is sent through a finisher, which employs heat to instantly fuse the toner that represents the picture onto the paper.[37]

There are typically seven steps involved in the process. After the raster image production is completed, all phases of the printing process can happen one after the other in quick succession. This allows for the employment of a very tiny and compact unit in which the photoreceptor is charged, rotated a few degrees and is scanned, rotated a few more degrees and developed, and so on. The entire procedure may be performed in less than one drum revolution. Raster image processing, charging, exposing, developing, transferring, fusing, cleaning and recharging are distinctive stages in the procedure.

## 2.2.3 Digital cylinder printing

Equipment that directly deposits ink onto a curved surface, commonly the wall of an item with a circular cross-section and a constant, tapered, or variable diameter,[38,39] is known as digital cylinder printing. Digital cylinder printing uses digital imaging technologies to reproduce black-and-white or full-color images and text onto cylindrical objects, mainly promotional merchandise. Multiple design strategies (mirror prints, tone-on-tone, contouring, and etched)[40–43] are enabled by the capabilities of digital cylinder printing machines to print full color in one pass, including primers, varnishes, and specialized inks.

Seamless edges with no visual overlapping are another advantage of full-wrap cylindrical printing. Original design artwork should be able to be printed on cylinders and tapered things without needing to be manipulated or distorted for ease of print file preparation. Flat pictures will print to scale on a curved surface, and adjustment is made by the software automatically. These requirements can be met by the most modern systems on the market. Different imaging techniques are used by digital cylinder printing machines as shown in Fig. 2.1.

Inputting a cylinder-shaped object, or part, into a framework, which firmly retains it in position, is the first step in the digital cylindrical printing operation. The component is then passed through a print head mechanism, which releases tiny droplets of CMYK (cyan, magenta, yellow, and black)[46] inks in a precise pattern to create an image. Typically, one portion is printed at a time, and depending on the intricacy and sharpness of the design, it can take anywhere from 8 to 45 seconds to finish. After that, a UV coating is applied to give it a glossy surface and protect it from damage. Cups, tumblers, thermos bottles, cosmetics cases, machine components, carrier tubes, markers, pipes, jars, and other objects can be produced using digital cylindrical techniques.

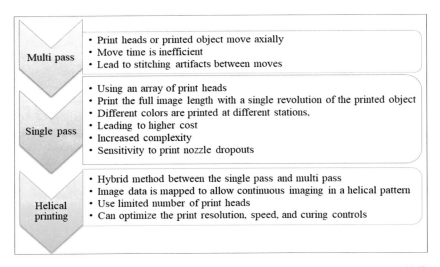

**Figure 2.1** Different imaging techniques used by digital cylinder printing machines.[44,45]

## 2.3 Generation of droplets

Continuous inkjet (CIJ) printing and drop-on-demand (DOD) printing are the two methods used by inkjet printers to create droplets as shown below in Fig. 2.2.[47,48] Both drops generating systems may create fluid droplets with diameters ranging from 10 to 150 μm[49,50]: DOD printing is dominant in graphics and text printing with a lower drop diameter, generally 20−50 μm. CIJ printing is mainly used for coding and marking applications with a drop diameter of roughly 100 μm.

### 2.3.1 Continuous inkjet

Continuous inkjet refers to a pressured and continuous flow of ink. Commercially, the continuous inkjet (CIJ) technology[51] is used to label and code items and packaging. CIJ technology uses a high-pressure pump to push liquid ink from a reservoir via a gun body and a small nozzle (typically 0.003-in. diameter), resulting in a continuous spray of ink droplets due to Plateau−Rayleigh instability. A piezoelectric crystal[45,52] vibrates inside the gun body, creating sound waves that lead the jet of ink to shatter into irregular-sized ink droplets at regular intervals. 64,000−165,000[53,54] irregular-sized ink droplets per second may be obtained.

An electrostatic field formed by a charging electrode or a magnetic flux field is applied to the ink droplets. The field changes depending on how much drop deflection is required. The electrostatic charge causes each droplet to deflect in a regulated manner. One or more uncharged guard droplets may isolate charged droplets to reduce electrostatic repulsion among adjoining droplets. The droplets travel across another electrostatic or magnetic field. They are driven to publish on the receptor surface by electrostatic deflection plates or flux fields or permitted to continue redirected to a collecting gutter for reuse by electrostatic deflection plates or flux fields. Droplets that are more strongly charged are deflected largely. Only a tiny percentage of the droplets are used for printing, with the rest being recycled.[55,56]

The toner droplets' extremely high velocity ($\approx 20$ m/s), which permits for a reasonably extended separation between print head and substrate, and the very high drop ejection frequency, which provides for faster printing, are the main benefits of CIJ.[57]

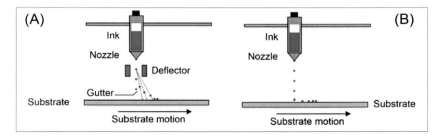

**Figure 2.2** Schematic diagrams of printing methods used in Inkjet printing: (A) continuous inkjet (CIJ); (B) drop-on-demand (DOD).[48]

An additional benefit is the lack of nozzle blocking because the jet is continually active, enabling volatile solvents like ketones and alcohols, allowing the ink to bite into the substrate and dry fast. The ink system demands active solvent regulation to prevent solvent evaporation during the moment of flight, which is the time between nozzle ejection and gutter recycling. The venting process involves air being drawn into the gutter, and the unused drops are vented from the reservoir. The viscosity is measured to compensate for solvent loss, and a solvent (or solvent blend) is applied.[49]

## 2.3.2 Drop-on-demand

Drop-on-demand indicates that the liquid is ejected one drop at a time from a single jet nozzle or hundreds of nozzles.[50,58] A drop-on-demand (DOD) inkjet can be made in various methods. Thermal DOD and piezoelectric DOD are common approaches to increasing the frequency of drops.[6,59] One DOD method use software that instructs the heads to apply zero to eight droplets of ink per dot just where they are required. Pastes, epoxies, hot-melt inks, biological fluids, and other materials are now used in inkjet fluids. The original way was mechanical DOD, followed by electrical means like piezoelectric devices and thermal or heat expansion methods.

### 2.3.2.1 Thermal DOD printing

The print cartridges in the thermal inkjet method are made of small compartments, each incorporating a heater built using photolithography.[59] A pulse of current is sent across the heater to expel a droplet from every chamber, producing quick vaporization of the ink in the chamber and the formation of a bubble, which generates a considerable pressure rise, pushing a droplet of ink onto the paper. Since thermal print heads lack the power of piezo DOD or continuous inkjet, the distance between the head's face and the substrate is crucial.

The surface tension of the ink, as well as the condensation and eventual collapse of the vapor bubble, draws more ink into the compartment through a small channel connected to an ink reservoir. The inks used are typically water-based and contain pigments or dyes as colorants. The inks must have a volatile component; if not, droplet ejection will not be possible to generate the vapor bubble. The print head is often less expensive to manufacture than inkjet technologies that need unique components.[49]

### 2.3.2.2 Piezoelectric DOD printing

Piezos are polarized ceramic gadgets that are electrically polarized, similar to how a magnet is polarized. Instead of a heating element, many industrial and commercial inkjet printing machines, and maybe even some household printers, use a piezoelectric material in an ink-filled chamber behind each nozzle.[60] When a voltage is induced, the piezoelectric material changes form, causing a pressure pulse in the fluid and pushing an ink droplet through the nozzle.[61] Sound waves in the ink reservoir discharge the droplets from single tubular inkjets and fluid resonator compartments.

Since there is no precondition for volatile constituents and no problem with ink residue buildup, piezoelectric inkjet enables a broader range of toners than thermal inkjet. Piezoelectric print heads are more costly due to the need for a piezoelectric element, typically zirconium titanate (PZT). On the other hand, the ink cartridges may be detached from the head and changed separately as needed, which might result in cheaper operating expenses.

At equivalent drop quantities, piezo heads are believed to produce quicker firing rates than thermal heads. The Piezo inkjet technique is frequently used to label items on production lines. The "use-before" date, for example, is frequently affixed to items using this approach; in this case, the head is stationary as the product travels past. This application demands a significant distance between the print head and the substrate, producing fast speeds, prolonged operation lifetime, and minimal operational costs.[49]

## 2.4 Printing parameters and ink formulation

Color/effect, density, viscosity, and surface tension qualities[62–65] describe the chosen ink, ensuring that the liquid is retained throughout a span under the pressure with which it is traveling in a bridge or stream. The force supplied to the toner in the production of the stream must be adequate to create a jet and apply adequate energy to transport the jet as a continuous liquid mass.

Among the essential features that ink must have is viscosity.[64,66] Since the nozzles on print heads, firing chambers, and conducts are small and tight, only ink with the appropriate viscosity will flow smoothly and meet the best drop generation efficiency.[67] The ink flow speed within the print heads is heavily influenced by viscosity. The ink will move readily within print heads if the viscosity is too low,[62] but droplets may break while jetting. Other issues may arise, such as the presence of satellites, drop positioning variation, or nozzle plate wetness. It may also cause the ink to flow beyond the nozzles, resulting in significant flaws in the printed substrates. If the viscosity of the ink is too high,[62] it will have a more challenging time flowing within the print heads and may even prevent ink jetting.

The production of drops is significantly linked to surface tension.[66] It will make the best drop-shape generation.[67] If this parameter is beyond tolerance, the drop may break or have deviation issues during the jetting operation. The liquid percentage determines the surface tension's standard operating limit of 28–32 dyn/cm.[62] The temperature also affects it, although it is unusual to fall out of range. This parameter likewise determines the contact angle with the receptor surface. As a result, the way the ink wets the surface will change.

Density is affected by solid and liquid fractions[62,68] and fluctuates according to temperature. Because of the usage of ingredients of varying densities, the density of the solid fraction in ink can be highly varied. Drop volume is generally calculated using density.[69] Inks of varying densities will produce drops of varying weights, but it is crucial to remember that the print head determines the drop volume and is independent of the ink.[68]

Moreover, the ink and carrier colors must have similar hues to provide adequate visual contrast. Toner and ink for digital printing include cyan, magenta, yellow, and black (CMYK),[46] other color gamut inks like orange, blue, green, and specialty dry inks for metallic, white, and clear finishes. A hot-melt ink with a solid phase at carrier temperature and a liquid phase at a higher temperature is preferable.

The fundamental issue with inkjet inks is the contradiction between a coloring agent that will stay on the surface and the need for a quick bleed through the fluid medium.[70] Depending on a combination of water, glycol, and dyes or pigments, aqueous ink is often used in desktop digital printing in businesses and households.[71] Some inks might need some posttreatments, such as in the case of reactive printing, it might be needed to postcure the reactive dyes on the cotton.[72,73] These inks are cheap to make yet difficult to manage on the media's surface, necessitating the use of unique coating material. Sulfonated poly azo black dye, nitrates, and other chemicals are in HP inks.

## 2.4.1 Aqueous inks

Aqueous inks use water as their primary solvent and are used for printing in every imaginable application. Aqueous inks can use pigment, dye, or combinations of dyes and pigment as the colorant[74]—printers with thermal inkjet heads[75] demand water to accomplish their ink-expelling activity. As a result, they use aqueous inks primarily. While aqueous inks provide the broadest and most vibrant color spectrum, they are not waterproof unless coated or laminated after printing.[76] Despite being the least costly, most dye-based inks quickly fade when subjected to light or ozone. Pigment-based aqueous inks are often more expensive but have higher long-term durability and UV resistance.[77] Pigment-based inks are typically promoted as an archival grade. Aqueous inks are used by certain professional wide format printers, although the majority use a far broader spectrum of inks, the bulk of which need piezo inkjet heads and substantial maintenance.

## 2.4.2 Solvent inks

Volatile organic compounds (VOCs) with high vapor pressures are the primary constituents in solvent inks. Pigments are used instead of dyes to create color for excellent fading resistance. Solvent inks are affordable and allow printing on flexible, uncoated vinyl substrates, such as those used in automobile graphics, posters, placards, and adhesive stickers.[67] Potential downsides are the solvent's vapor and the necessity to discard spent solvent. Unlike most aqueous inks, graphics generated with solvent-based inks are usually waterproof and UV-resistant for outdoor usage without additional over-coatings.

Because of their fast print speeds, several solvent printing machines require specific curing technology. They generally consist of a mix of heaters and blowers. The substrate is usually heated before and after the print heads apply ink. Hard solvent inks are the most durable without purpose-built over-coatings, but they need customized air movement of the printing zone to prevent exposure to toxic fumes.

Mild or eco-solvent inks, not as safe as aqueous inks, are intended for use in confined areas without special-purpose airflow of the printing area. Mild solvent inks have exploded in popularity in recent decades due to improved color quality and durability at a lower cost.

## 2.4.3 UV-curable inks

Acrylic monomers[78] with an initiator package make up the majority of UV-curable inks. The ink is cured after printing by exposing it to intense UV light. When ink is subjected to UV light, photoinitiators induce the ink constituents to cross link into a solid, resulting in a chemical reaction. The drying procedure is usually done using a shuttered mercury-vapor lamp or a UV LED. Curing inks on thermally sensitive surfaces is possible because of high-power, short-duration (microsecond) curing methods. UV inks do not evaporate; instead, this chemical reaction causes them to cure or set. Because no material is evaporated or removed, about 100% of the provided volume is used for coloring.

This process occurs rapidly, resulting in fast drying and a fully hardened image in seconds. This enables a lightning-fast printing process. Because of this fast chemical bond, no solvents enter the substrate after it comes off the printing machine, resulting in high-resolution prints. UV-curable inks benefit from drying immediately after curing, being able to be transferred to a variety of uncoated surfaces, and producing a highly durable picture. They are costly, necessitate costly hardening chambers in the printer, and the cured ink has a large volume, resulting in a modest relief on the substrate.[13] Although technological advancements are being achieved, UV-curable inks are subject to breaking when placed on a flexible substrate due to their quantity. As a result, they are frequently found in big flatbed printers that print directly on stiff materials like plastic, wood, or metal, where flexibility is not an issue.

## 2.4.4 Dye sublimation inks

The word "sublimation" was given to the dye because it had been supposed to change from a solid to a gas without passing into a liquid form. These inks employ sublimation dyes to publish directly or indirectly on textiles with a high percentage of polyester fibers. The dyes sublimate into the fibers during the heating, creating a vibrant picture with a long-lasting hue. These inks are often employed to decorate clothing, signage, posters, and novelty products like mobile phone covers, plaques, coffee mugs, and other sublimation-friendly substrates.

Dispersed dyes include sublimation colorants, which are employed in inkjet applications. These are organic colors insoluble in water and other organic solvents, as their name implies. Sublimation dye molecules can be equally dispersed in the polymer matrix at the molecular level rather than lying on the exterior of the substrate when hydrophobic synthetic polymeric materials used with sublimation imaging are transparent or translucent. It achieves a pleasing color impression or vibrancy without needing a gloss coat or other surface treatment. Light diffracts from inside the polymer matrix and the polymer surface/air barrier.[79]

### 2.4.5 Solid ink

Solid inks, also known as hot melt inks, are primarily of paraffin components. They are melted above their melting point to permit printing and solidify when they contact a cooled substrate. In graphic printing, hot-melt inks are often used for masking operations. Rather than the fluid ink or toner powder commonly used in printers, solid ink technology employs solid ink sticks, crayons, pearls, or granular solid substances. Small solid ink spheres or pucks are held in a hopper before being transported to the printing head by a worm gear or melted as necessary in some solid ink printers. Once the solid ink is put into the printer, it is melted and used to print pictures on the poster or any other substrate in a process that resembles offset or conventional printing.

### 2.4.6 Metal nanoparticle ink

Metal Nanoparticles Inks, also known as functional Inks, are used in printed electronics manufacturing.[80] Electrical conductivity, semi-conductivity, electroluminescence, and insulating or dielectric properties of functional inks enable the additive manufacture of electronic circuits, electrodes, and connectors. Industrial applications need high-capacity inks that can print continuously for many hours.[81,82]

Metal nanoparticle conductive inks are composed of dispersion and organic or inorganic solvents (e.g., tetradecane, alcohols, water, etc.) that stabilize metal nanoparticles (such as silver, copper, and gold). Water is a polar solvent that aids in the reduction of environmental impact. Because of the minimal viscosity of a nonpolar solvent, it is simple to raise the metal percentage.[82]

For functional inks, nanoparticles with at minimum one dimension smaller than 100 nm are often used. The dimensions of submicron particles are more significant than 100 nm, and at least one dimension is less than 1 μm. Microparticles range in size from 1 to 1000 μm across all dimensions. Nanoparticles can be made in two ways: top-down (by breaking up bigger particles into nanoparticles) or bottom-up (by growing the particles).[81–83]

## 2.5 Printing heads

### 2.5.1 Fixed head

The fixed-head concept includes an unavoidable print head, also known as a gaiter-head, meant to endure the printer's lifetime. Consumable expenses may be reduced, and the head itself can be more precise than a cheap disposable one because there is no need to change the head every time the ink runs out. If a fixed head is destroyed and assuming that removing and replacing the head is even feasible, procuring a spare head might be costly. If the head of the printer cannot be changed, the printer must be avoided.

Consumer devices with fixed heads are available, although high-end industrial printers and big format plotters are more likely to have them. Fixed-head printers are predominantly produced by Epson and Canon in the consumer market; however, many more modern Hewlett-Packard[84] machines, such as the Officejet Pro 8620 and HP's Pagewide series, employ a fixed-head.

### 2.5.2 Disposable head

The throwaway head concept uses a print head as part of a changeable ink cartridge. When a cartridge is empty, it is changed with a fresh one, including the print head. This raises the cost of replacement parts and makes it more challenging to produce a low-cost, high-precision head, but it also means that a broken or clogged print head is a minor issue; the user can replace the cartridge.

A replaceable ink container is coupled to a disposable head, which only changes each tenth ink tank. This configuration is used on most high-volume HP inkjet printers, using disposable print heads on lower-volume versions. Kodak has an identical technique, with a low-cost print head that the customer can update. Canon uses changeable print heads in most models, which are intended to last the printer's life but may be changed by the owner if they are clogged.

Manufacturing with additives is a relatively new concept. Failures due to internal clogs, orifice breakage from impacting impediments on the print table, calibration failures from overstressed, piezoelectric bond life failures, and other unanticipated factors will cause failures in print heads. Most long-life 3D printers have spare print heads on their component lists. Kyocera is one of the well-known manufacturers of print heads, and it can be seen in Fig. 2.3; it is also installed on the VGEA 3180DT.

## 2.6 Cleaning mechanisms

Clogging is a common issue with inkjet nozzles since they are so small. Cleaning them consumes ink, either during user-initiated cleaning or, in many cases, regular cleaning done automatically by the printer.[21] Ink drying on the print head's nozzles causes the pigments and dyes to dry out and create a solid block of hardened material that jams the tiny ink passages, which is the most common cause of inkjet printing issues. Most printers employ a rubber cap to keep the print head nozzles from drying out when the printer is not in use. The print head can be left uncapped due to unexpected power outages or disconnecting the printer before it has finished capping the print head.

Although the head is covered, the gasket is imperfect, and humidity or other solvents can leak out for many weeks, allowing the ink to dry and solidify. Once the ink accumulates and hardens, the drop volume, drop trajectory, and even the nozzle's ability to jet ink can all be compromised. For water-soluble inks, specialist solvents can be used to clean inkjet printing head nozzles, or they can be soaked in warm distilled water for a brief time.

Almost all inkjet printers have a system to reintroduce humidity to the print head to prevent it from drying out. There is no separate supply of pure ink-free solvent

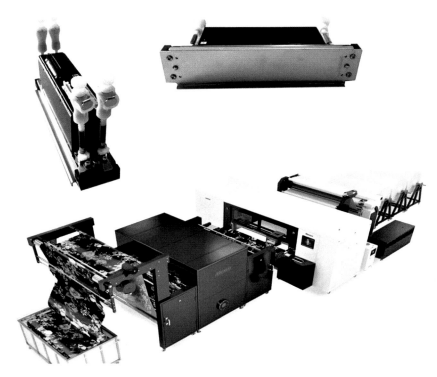

**Figure 2.3** Kyocera print heads are installed on the VGEA 3180DT.
*Source*: The image was produced with the permission of Atexco, Hangzhou, China.

on hand in most cases to do this task; therefore, the inks are used to moisten the print head. Some of it flows over the print head to the dry channels as the ink jets out, partially softening the solid ink. After spraying, a rubber wiper blade is swiped across the print head to distribute the moisture uniformly, and the jets are fired all at once to clear any ink clumps that have clogged the channels.

Some printers use a secondary air-suction pump, sucking ink through a badly blocked cartridge through the rubber capping station. The page feed stepper motor, attached to the end of the shaft, is usually used to power the suction pump mechanism. The pump only activates when the shaft rotates backward, which causes the rollers to reverse while the head is being cleaned. The suction pump is also required to prime the ink channels inside a new printer and reprime the channels between ink tank changes due to the built-in head design.

A manual clean mode is typically included in professional solvent and UV-curable wide-format inkjet printers. This allows the operator to clean the print heads manually and change wiper blades and other elements used in the automatic cleaning operations. The large amount of ink used in these printers frequently causes overspray and results in dried ink accumulation in many locations that automated systems cannot remove.

The ink used in the cleaning operation must be collected to prevent ink from leaking into the printer. The spittoon, an open plastic tray beneath Hewlett-Packard printers' cleaning/wiping unit, is the collection area. A large absorption pad is usually found in a pan beneath the paper feed plate in Epson printers. It is typical for dried ink in the spittoon to develop a mound that can build up and touch the print heads, clogging the printer, especially in older printers. Some more large professional printers who use solvent inks may use a changeable plastic receptacle to collect waste ink and solvent, which must be emptied or replaced as it fills up.

The majority of printers cannot avoid the second kind of ink drying. For the ink to jet from the cartridge, air must pass to expel the ink that has been withdrawn. The air enters through a long, thin labyrinth tube that wraps back and forth across the ink tank, measuring up to 10 cm in length. Although the channel is lengthy and narrow to prevent moisture evaporation through the vent tube, some evaporation does occur, and the ink cartridge gradually dries up from the inside out. Many wide-format printer cartridge designs include the ink in an airtight, foldable bag that does not require a vent to fight this problem, which is especially acute with professional fast-drying solvent inks. Until the cartridge is empty, the bag simply shrinks.

Some printers' routine cleaning consumes a significant amount of ink and significantly influences expense calculations. A typical test pattern can be printed on the page to discover blocked nozzles. Specific software workaround approaches are known for re-routing printing information from a blocked nozzle to a functional nozzle.

## 2.7 Advantages and disadvantages of inkjet printing

The advancement of digital printing onto substrates is shifting printing techniques and reducing limitations that designers have conventionally experienced. Designers can deal with thousands of colors and generate designs with a high level of detail. Redundant patterns and color variations are critical aspects of screen and roller printing.[14,85]

### 2.7.1 Advantages of digital printing

- Superior image quality can be obtained.[1,82,85–89]
- There is no need to balance water and ink because the print is consistent.
- Digital printing allows any quantity to be printed.
- Human power reductions and high-quality finish can be achieved In digital printing. Digital printing is environmentally sustainable and user-friendly.
- Digital printing reduces the number of unwanted prints.
- Low-volume prints are cheap in digital printing.
- Large-format print exceeding 10 ft in diameter is possible in digital printing.
- Images may be changed as many times as needed because no printing plates are required.
- Modification may be done during printing without affecting the printing process.
- Print turnaround time is substantially faster than offset printing.
- Versatile: Inks compatible with any selected surface may usually be generated. Digital printing is possible on various surfaces, including paper, marble, glass, and metal.

- The input data is not restricted by size and does not necessary to repeat because the data is retrieved from a computer file.
- Ink does not come into touch with the substrate allowing for printing on flat or curved, smooth or rough, and delicate or rigid substrates.
- Multi-color: Dozens of colors may be printed without using a color kitchen using the cyan−magenta−yellow−black color range.
- High speed: Printing speeds are, of course, affected by resolution, the type of printing required, head technology, and other factors.
- No moving components: The only movable components in the printer are the mechanism to move the heads concerning the substrate. Only the ink, not a mechanical instrument, moves during printing.
- Inkjet printers are quieter than impact dot matrix or daisywheel printers in operation.
- There is almost no warm-up period.

### 2.7.2 Disadvantages of digital printing

- Digital printing machine is pricey.[2,89]
- Digital printing is better for low-volume prints.
- Digital printouts are prone to scuffing and marking.
- The print is not as delicate in gradients, tints, and huge solid sections.
- Since digital print colors are fully absorbed, fractures may occur in edges that have been folded after printing has ended.
- Color interpretation, paper flatness, graphic edge sharpness, back sticks, unclean ink spots, and white lines are all inkjet digital printing problems.
- Many intelligent ink cartridges have a microchip that detects the estimated ink level in the printer; this might lead the printer to generate an error message or wrongly warn the operator that the ink cartridge is empty.
- Many printer suppliers advise users against using third-party inks, claiming that they might harm print heads due to different formulations, cause leaks, and poor print quality.

## 2.8 Solidification mechanisms

The final phase in the printing process is transforming from a liquid deposit to the required solid substance. Almost usually, a loss in volume will accompany this change. Because only dilute solutions of long-chain polymers and small quantities of suspended particles are generally printed, the volume shift may be massive when solidification occurs via solute evaporation. Inkjet printing of conducting and semi-conducting polymers and printing of metal and ceramic nanoparticle suspensions, the principles of solid creation by evaporation from a solution, are of particular relevance.

The well-known coffee stain effect, in which solute firmly segregates to the first contact line, can significantly impact the form of a printed drop upon drying. The solvent evaporation rate is highest along the contact line because of the facile vapor movement in the surrounding dry substrate. As a result, precipitation begins at the contact line. Because this deposit will pin the contact line, fluid will flow from the drop's center to replace the evaporating fluid from the contact line's vicinity. This peripheral flow continually feeds the solidification at the interface region, and the last deposit will have a distinctive ring where the solute has separated during the drying process. This phenomenon is significant for regulating the form of inkjet-printed droplets and drying liquid beads, resulting in a distinctive dual-ridged line profile following solidification.[49]

## 2.9 Applications

Some applications of Inkjet printing include[51,67,74,81,82,88,90–93]:

- Desktop printing—Digital techniques that eliminate the need for printing plates allow for low-cost home and business printing.
- Business Stationery—This includes business cards and letterheads.
- Variable data printing—bulk customizing of printed items using database-driven print files.
- Fine art—archival digital printing techniques include exposure prints on actual photo paper and pigment-based ink prints on watercolor paper.
- Print-on-Demand—digital printing is used for personalized printing, such as children's books, photo books (such as wedding picture books), and others.
- Advertising—outdoor vinyl banners and event signs, trade exhibitions, point of sale or purchase in the retail industry, and tailored direct mail campaigns are all examples.
- Photos—digital printing has revolutionized photo printing in terms of the ability to retouch and color correct a photograph before printing.
- Photos—the capacity to repair and color correct an image before printing has been revolutionized by digital printing.
- Architectural Design—Interior and outside areas may be altered with digitally printed murals and floor graphics thanks to innovative material that adapts to a range of surfaces.
- Sleeking—The technique of using digital ink adhesion to create foil, holographic effects, or even glossy and dull surfaces. This is accomplished by digitally printing a rich black area where the sleeking is desired. The machine registers this, and it solely conforms to this specific region.

## References

1. Kwon, K.-S.; Rahman, M. K.; Phung, T. H.; Hoath, S.; Jeong, S.; Kim, J. S. Review of Digital Printing Technologies for Electronic Materials. *Flex. Print. Electron.* **2020**. Available from: https://doi.org/10.1088/2058-8585/abc8ca.
2. Li, X.; Luo, S.; Li, S.; Zhao, Y.; Deng, G.; Cao, G. The Solutions to the Quality Defects of Inkjet Printing. In *Proceedings of the Applied Sciences in Graphic Communication and Packaging*, Singapore, 2018, **2018**; pp. 411–416.
3. Singh, R.; Shrestha, A. Namuna College of Fashion Technology: Pioneering in Fashion and Textile Education in Nepal. In *Textile and Fashion Education Internationalization: A Promising Discipline from South Asia;* Yan, X., Chen, L., Memon, H., Eds.; Springer Nature Singapore: Singapore, **2022**; pp 103–118.

4. Han, D.; Lee, H. Recent Advances in Multi-Material Additive Manufacturing: Methods and Applications. *Curr. Opin. Chem. Eng.* **2020**, *28*, 158–166.
5. Jhatial, A. K.; Yesuf, H. M.; Wagaye, B. T. Pretreatment of Cotton. In *Cotton Science and Processing Technology: Gene, Ginning, Garment and Green Recycling;* Wang, H., Memon, H., Eds.; Springer Singapore: Singapore, **2020**; pp 333–353.
6. Abdolmaleki, H.; Kidmose, P.; Agarwala, S. Droplet-Based Techniques for Printing of Functional Inks for Flexible Physical Sensors. *Adv. Mater.* **2021**, *33*, 2006792.
7. Halepoto, H.; Gong, T.; Noor, S.; Memon, H. Bibliometric Analysis of Artificial Intelligence in Textiles. *Materials* **2022**, *15*, 2910.
8. Bakhsh, N.; Khan, M. Q.; Ahmad, A.; Hassan, T. Recent Advancements in Cotton Spinning. In *Cotton Science and Processing Technology: Gene, Ginning, Garment and Green Recycling;* Wang, H., Memon, H., Eds.; Springer Singapore: Singapore, **2020**; pp 143–164.
9. Smriti, S. A.; Farha, F. I.; Siddiqa, F.; Jawad Ibn Amin, M.; Farzana, N. Cotton in Weaving Technology. In *Cotton Science and Processing Technology: Gene, Ginning, Garment and Green Recycling;* Wang, H., Memon, H., Eds.; Springer Singapore: Singapore, **2020**; pp 191–246.
10. Khankhadjaeva, N. R. Role of Cotton Fiber in Knitting Industry. In *Cotton Science and Processing Technology: Gene, Ginning, Garment and Green Recycling;* Wang, H., Memon, H., Eds.; Springer Singapore: Singapore, **2020**; pp 247–303.
11. Imran, M. A.; Khan, M. Q.; Salam, A.; Ahmad, A. Cotton in Nonwoven Products. In *Cotton Science and Processing Technology: Gene, Ginning, Garment and Green Recycling;* Wang, H., Memon, H., Eds.; Springer Singapore: Singapore, **2020**; pp 305–332.
12. Siddiqui, M. Q.; Wang, H.; Memon, H. Cotton Fiber Testing. In *Cotton Science and Processing Technology: Gene, Ginning, Garment and Green Recycling;* Wang, H., Memon, H., Eds.; Springer Singapore: Singapore, **2020**; pp 99–119.
13. Sang, R.; Manley, A. J.; Wu, Z.; Feng, X. Digital 3D Wood Texture: UV-Curable Inkjet Printing on Board Surface. *Coatings* **2020**, *10*, 1144.
14. Tyler, D. J. 12 - Digital Printing Technology for Textiles and Apparel. In *Computer Technology for Textiles and Apparel;* Hu, J., Ed.; Woodhead Publishing, **2011**; pp 259–282.
15. Uddin, M. F. Brief Analysis on the Past, Present, and Future of Textile Education in Bangladesh. In *Textile and Fashion Education Internationalization: A Promising Discipline from South Asia;* Yan, X., Chen, L., Memon, H., Eds.; Springer Nature Singapore: Singapore, **2022**; pp 35–57.
16. Ali Hayat, G.; Hussain, M.; Qamar Khan, M.; Javed, Z. Textile Education in Pakistan. In *Textile and Fashion Education Internationalization: A Promising Discipline from South Asia;* Yan, X., Chen, L., Memon, H., Eds.; Springer Nature Singapore: Singapore, **2022**; pp 59–82.
17. El-Kashouti, M.; Elhadad, S.; Abdel-Zaher, K. Printing Technology on Textile Fibers. *J. Textiles Coloration Polym. Sci.* **2019**, *16*, 129–138.
18. Carré, B.; Magnin, L.; Ayala, C. Digital Prints: A Survey of the Various Deinkability Behaviours. In *Proceedings of the Proceedings of the 7th Research Forum on Recycling, PAPTAC,* **2004**.
19. Campbell, J.; Parsons, J. Taking Advantage of the Design Potential of Digital Printing Technology for Apparel. *J. Text. Apparel Technol. Manag.* **2005**, *4*, 1–10.
20. Rokia, W.E.-S.A. Utilizing of Digital Printing Techniques to Create Customize Designs for Textile Floor Covering. **2021** مجلة الفنون والأدب وعلوم الإنسانيات والاجتماع; pp. 261–305.
21. Castrejón-Pita, A. A.; Betton, E. S.; Campbell, N.; Jackson, N.; Morgan, J.; Tuladhar, T. R.; Vadillo, D. C.; Castrejon-Pita, J. R. Formulation, Quality, Cleaning, and Other Advances in Inkjet Printing. *Atomization Sprays* **2021**, *31*.

22. Datta, P.; Mohi, G. K.; Chander, J. New Type Versatile Electric Hand Pallet Jack. *J. Laboratory Phys.* **2018,** *10*, 6.
23. Lahti, J.; Savolainen, A.; Räsänen, J. P.; Suominen, T.; Huhtinen, H. The Role of Surface Modification in Digital Printing on Polymer-Coated Packaging Boards. *Polym. Eng. Sci.* **2004,** *44*, 2052−2060. Available from: https://doi.org/10.1002/pen.20209.
24. Mizes, H.; Spencer, S.; Sjolander, C.; Yeh, A. Active Alignment of Print Heads. In *Proceedings of the NIP & Digital Fabrication Conference*, **2009**; pp. 711−714.
25. Tafoya, R. R.; Secor, E. B. Understanding Effects of Printhead Geometry in Aerosol Jet Printing. *Flex. Print. Electron.* **2020,** *5*, 035004.
26. Wang, J.; Si, Z.-J. The Study of the Quality Test and Control Technology of the Digital Printing. In *Proceedings of the The 21st IAPRI World Conference on Packaging*, **2018**.
27. Nickelson, J. *Fine Art Inkjet Printing: The Craft and Art of the Fine Digital Print;* Rocky Nook, Inc, **2018**.
28. Album, DDDP; Storage, D.; Are, U. Epson Premium Glossy Photo Paper (250) 85 years 98 years 60 years> 300 years> 100 years very high high no Epson Premium Luster Photo Paper (260) 83 years> 200 years 45 years> 200 years> 100 years very high high yes Epson Premium Semimatte Photo Paper (260) 83 years> 200 years 45 years> 200 years> 100 years very high high yes Epson Exhibition Fiber Paper 90 years 150 years 44 years> 200 years> 100 years very high moderate (11) yes, **2007**.
29. Olen, M. K. *The Development of Multi-Channel Inkjet Printing Methodologies for Fine Art Applications;* University of the West of England, **2017**.
30. Parraman, C. *Colour Print Workflow and Methods for Multilayering of Colour and Decorative Inks Using UV Inkjet for Fine Art Printing*, Vol. 8292. SPIE, **2012**.
31. Karthikeyan, A. S. *A Study Using a High-Addressability Inkjet Proofer to Produce AM Halftone Proofs Matching Kodak Approval in Color, Screening, and Subject Moiré;* Rochester Institute of Technology, **2009**.
32. Freeman, S. K.; Strickler, S. R. Imaging the J. Paul Getty Museum's Collection of Ultra Light-sensitive Photographs under Safelight. *J. Am. Inst. Conserv.* **2020,** *59*, 262−270.
33. Langford, M. *Basic Photography;* Routledge, **2013**.
34. Kordecki, A.; Palus, H.; Bal, A. Practical Vignetting Correction Method for Digital Camera With Measurement of Surface Luminance Distribution. *Signal, Image Video Process.* **2016,** *10*, 1417−1424.
35. Harris, T. How Laser Printers Work. *How Stuff Work. Space* **2007,** *200*, 1998−2015.
36. Nakaya, F.; Fukase, Y. 5 Laser Printer. *Color. Deskt. Print. Technol.* **2018**, 157.
37. Lin, L.; Kollipara, P. S.; Zheng, Y. Digital Manufacturing of Advanced Materials: Challenges and Perspective. *Mater. Today* **2019,** *28*, 49−62.
38. Chatow, U.; Samuel, R. Digital Labels Printing. In *Proceedings of the NIP & Digital Fabrication Conference*, **2003**; pp. 476−481.
39. Hall, D. Variable Data Cylinder Printing. In *Proceedings of the NIP & Digital Fabrication Conference*, **2008**; pp. 398−401.
40. Majnarić, I.; Cigula, T. Influence of the Varnishing "Surface" Coverage on Optical Print Characteristics. *Tehnički Glas.* **2020,** *14*, 428−433.
41. Friedman, M.; Walsh, G. High Performance Films: Review of New Materials and Trends. *Polym. Eng. Sci.* **2002,** *42*, 1756−1788.
42. Alexander, G.; Alexander, R. Use of 'Reverse Prints', 'Mirror Image' or 'Horizontal Flip' in Ear Reconstruction. *Indian. J. Plastic Surg.* **2013,** *46*, 591−592.
43. Suganthan, S.; MacDonald, L. Shadow Removal From Image of Stained Glass Windows. *Int. J. Imaging Syst. Technol.* **2010,** *20*, 223−236.

44. Arango, I.; Cifuentes, C. Design to Achieve Accuracy in Ink-Jet Cylindrical Printing Machines. *Machines* **2019**, *7*, 6.
45. Scaccabarozzi, D.; Magni, M.; Saggin, B.; Tarabini, M.; Cioffi, C.; Nasatti, S. Measurement Method for Quality Control of Cylinders in Roll-to-Roll Printing Machines. *Machines* **2020**, *8*, 16.
46. Hiremath, S. S. *A Study of High-Chroma Inks for Expanding CMYK Color Gamut*; Rochester Institute of Technology, **2018**.
47. Wiklund, J.; Karakoç, A.; Palko, T.; Yiğitler, H.; Ruttik, K.; Jäntti, R.; Paltakari, J. A Review on Printed Electronics: Fabrication Methods, Inks, Substrates, Applications and Environmental Impacts. *J. Manuf. Mater. Process.* **2021**, *5*, 89.
48. Solís Pinargote, N. W.; Smirnov, A.; Peretyagin, N.; Seleznev, A.; Peretyagin, P. Direct Ink Writing Technology (3D Printing) of Graphene-Based Ceramic Nanocomposites: A Review. *Nanomaterials* **2020**, *10*, 1300.
49. Derby, B. Inkjet Printing of Functional and Structural Materials: Fluid Property Requirements, Feature Stability, and Resolution. **2010**; 40, pp. 395−414.
50. Liu, Y.; Derby, B. Experimental Study of the Parameters for Stable Drop-On-Demand Inkjet Performance. *Phys. Fluids* **2019**, *31*, 032004.
51. Piatt, M.; Bugner, D.; Chwalek, J.; Katerberg, J. KODAK's Stream Inkjet Technology. In *Handbook of Industrial Inkjet Printing*, **2017**; pp. 351−360.
52. Wang, T.; Lin, J.; Guo, X.; Lei, Y.; Fu, H. A New Method for Producing Uniform Droplets by Continuous-Ink-Jet Technology. *Rev. Sci. Instrum.* **2018**, *89*, 085008.
53. Viluksela, P.; Kariniemi, M.; Nors, M. Environmental Performance of Digital Printing. *VTT Res. Notes* **2010**, 2538.
54. Poozesh, S. *Inkjet Printing: Facing Challenges and Its New Applications in Coating Industry*; University of Kentucky, **2015**.
55. Wheeler, J.; Yeates, S.; Reynolds, S.; Duffy, J. Investigating the Effect of Ink-Jet Printing on Polymer Properties.
56. Wheeler, J. S.; Reynolds, S. W.; Lancaster, S.; Romanguera, V. S.; Yeates, S. G. Polymer Degradation During Continuous Inkjet Printing. *Polym. Degrad. Stab.* **2014**, *105*, 116−121.
57. Ellinger, C.; Xie, Y. Lateral Merging Continuous Inkjet. In *Proceedings of the NIP & Digital Fabrication Conference*, **2011**; pp. 343−346.
58. Beedasy, V.; Smith, P. J. Printed Electronics as Prepared by Inkjet Printing. *Materials* **2020**, *13*, 704.
59. Prapty, T. S. *Thermal Inkjet Printing: Prospects and Applications in the Development of Medicine*; Brac University, **2019**.
60. Kang, S.-H.; Kim, S.; Sohn, D. K.; Ko, H. S. Analysis of Drop-on-Demand Piezo Inkjet Performance. *Phys. Fluids* **2020**, *32*, 022007.
61. Wang, S.; Zhong, Y.; Fang, H. Deformation Characteristics of a Single Droplet Driven by a Piezoelectric Nozzle of the Drop-on-Demand Inkjet System. *J. Fluid Mech.* **2019**, *869*, 634−645.
62. Department, D.C. Ceramic Inkjet Inks. In *Handbook of Industrial Inkjet Printing*; **2017**; pp. 151−162.
63. Huang, B.; Lü, Y.; Ma, L.; Wei, X.; Wang, H. Research on Printing and Prototyping Performance of Three-Dimensional Printing Materials for UV-Curing Inkjet. *Advances in Graphic Communication, Printing and Packaging*; Springer, **2019**, 827−839.
64. Bae, J. Color in Inkjet Printing: Influence of Structural and Optical Characteristics of Textiles, **2008**.
65. Memon, H.; Khoso, N. A.; Memon, S.; Wang, N. N.; Zhu, C. Y. Formulation of Eco-Friendly Inks for Ink-Jet Printing of Polyester and Cotton Blended Fabric. *Key Eng. Mater.* **2016**, *671*, 109−114. Available from: http://www.scientific.net/KEM.671.109.

66. Krainer, S.; Smit, C.; Hirn, U. The Effect of Viscosity and Surface Tension on Inkjet Printed Picoliter Dots. *RSC Adv.* **2019**, *9*, 31708−31719.
67. Bale, M. A System Approach to Develop New Platforms of Industrial Inkjet Inks. In *Handbook of Industrial Inkjet Printing*; **2017**; pp. 23−58.
68. Genina, N.; Fors, D.; Palo, M.; Peltonen, J.; Sandler, N. Behavior of Printable Formulations of Loperamide and Caffeine on Different Substrates—Effect of Print Density in Inkjet Printing. *Int. J. Pharm.* **2013**, *453*, 488−497.
69. Heilmann, J.; Lindqvist, U. Effect of Drop Size on the Print Quality in Continuous Ink Jet Printing. *J. Imaging Sci. Technol.* **2000**, *44*, 491−494.
70. Edinger, M.; Bar-Shalom, D.; Sandler, N.; Rantanen, J.; Genina, N. QR Encoded Smart Oral Dosage Forms by Inkjet Printing. *Int. J. Pharm.* **2018**, *536*, 138−145.
71. Li, B.; Hu, N.; Su, Y.; Yang, Z.; Shao, F.; Li, G.; Zhang, C.; Zhang, Y. Direct Inkjet Printing of Aqueous Inks to Flexible All-Solid-State Graphene Hybrid Micro-Supercapacitors. *ACS Appl. Mater. Interfaces* **2019**, *11*, 46044−46053.
72. Dutta, S.; Bansal, P. Cotton Fiber and Yarn Dyeing. In *Cotton Science and Processing Technology: Gene, Ginning, Garment and Green Recycling;* Wang, H., Memon, H., Eds.; Springer Singapore: Singapore, **2020**; pp 355−375.
73. Memon, H.; Khatri, A.; Ali, N.; Memon, S. Dyeing Recipe Optimization for Eco-Friendly Dyeing and Mechanical Property Analysis of Eco-Friendly Dyed Cotton Fabric: Better Fixation, Strength, and Color Yield by Biodegradable Salts. *J. Nat. Fibers* **2016**, *13*, 749−758. Available from: https://doi.org/10.1080/15440478.2015.1137527.
74. Double, P.; Stoffel, J. Aqueous Inks and Their Application Areas in Industrial Inkjet Printing and Desktop Printing. In *Handbook of Industrial Inkjet Printing*; **2017**; pp. 163−178.
75. Buanz, A.; Saunders, M. H.; Basit, A. W.; Gaisford, S. Preparation of Personalized-Dose Salbutamol Sulphate Oral Films With Thermal Inkjet Printing. *Pharm. Res.* **2011**, *28*, 2386−2392.
76. Tomerlin, R.; Tomiša, M.; Vusić, D. The Influence of Printing, Lamination and High Pressure Processing on Spot Color Characterisation. *Tehnički Glas.* **2019**, *13*, 218−225.
77. Baez, E. *Nanopigment Development and Dispersion in Aqueous Systems;* RMIT University, **2011**.
78. Graindourze, M. UV-Curable Inkjet Inks and Their Applications in Industrial Inkjet Printing, Including Low-Migration Inks for Food Packaging. In *Handbook of Industrial Inkjet Printing*; **2017**; pp. 129−150.
79. Xu, M. Dye Sublimation Inkjet Inks and Applications. In *Handbook of Industrial Inkjet Printing*; **2017**; pp. 179−194.
80. Aliqué, M.; Simão, C. D.; Murillo, G.; Moya, A. Fully-Printed Piezoelectric Devices for Flexible Electronics Applications. *Adv. Mater. Technol.* **2021**, *6*, 2001020.
81. Schauer, C.; Rösch, A. A Full-System Approach to Formulation of Metal Nanoparticle Inks for Industrial Inkjet Printing. *Handb. Ind. Inkjet Print.* **2017**, 195−214.
82. Saito, H.; Nakajo, H. Metal Nanoparticle Conductive Inks for Industrial Inkjet Printing Applications. *Handb. Ind. Inkjet Print.* **2017**, 215−224.
83. Jing, S.; Meifei, Z.; Tingfang, M.; Houyong, Y.; Juming, Y. Preparation and Properties of Digital Printing Inks Based on the Cellulose Nanoparticles as Dispersant. *Adv. Text. Technol.* **2019**, *27*, 69−72. Available from: https://doi.org/10.19398/j.att.201803020.
84. Simske, S. J. Hewlett Packard's Inkjet Printhead Technology. *Handb. Ind. Inkjet Print.* **2017**, 313−334.
85. Kapoor, B. Developing Designs for Teenage Girls Using Digital Textile Printing.
86. Tyler, D. J. Textile Digital Printing Technologies. *Text. Prog.* **2005**, *37*, 1−65. Available from: https://doi.org/10.1533/tepr.2005.0004.

87. Mejtoft, T. Strategies for Successful Digital Printing. *J. Media Bus. Stud.* **2006**, *3*, 53–74. Available from: https://doi.org/10.1080/16522354.2006.11073469.
88. Puyot, M. Memjet's Inkjet Printhead Technology and Associated Printer Components. In *Handbook of Industrial Inkjet Printing*; **2017**; pp. 335–350.
89. Zapka, W. Pros and Cons of Inkjet Technology in Industrial Inkjet Printing. In *Handbook of Industrial Inkjet Printing*; **2017**; pp. 1–6.
90. Brünahl, J.; Condie, A.; Crankshaw, M.; Cruz-Uribe, T.; Zapka, W. Xaar's Inkjet Printing Technology and Applications. In *Handbook of Industrial Inkjet Printing*; **2017**; pp. 285–312.
91. Corrall, J. Konica Minolta's Inkjet Printhead Technology. In *Handbook of Industrial Inkjet Printing*; **2017**; pp. 253–284.
92. N. Rosario, T. Concepts and Strategies to Adapt Inkjet Printing to Industrial Application Requirements. In *Handbook of Industrial Inkjet Printing*; **2017**; pp. 239–252.
93. Lange, A.; Wedel, A. Organic Light-Emitting Diode (OLED) and Quantum Dot (QD) Inks and Application. In *Handbook of Industrial Inkjet Printing*; **2017**; pp. 225–238.

# Overview of different digital textile printing machines

Aijaz Ahmed Babar, Pardeep Kumar Gianchandani and Abdul Khalique Jhatial
Department of Textile Engineering, Mehran University of Engineering and Technology, Jamshoro, Sindh, Pakistan

## 3.1 Introduction

In recent decades, digital printing technology (DPT) has received massive attention for printing textile fabrics because of its unique characteristic features, such as high design accuracy, color consistency, easy design repeatability, stain-free printed designs, better image quality, and ability to print with both reactive, as well as disperse inks, widely used class of colorants for most commonly used textile materials, that is, cotton and polyester.[1] Additionally, DPT is reported to have nearly 35% less water and 45% less electricity requirement compared to its counterparts. Sustainability is of vital importance for the textile and garment sector worldwide.[2] DPT also tends to produce relatively lower waste compared to traditional printing techniques.[3–5] These characteristic features of the DPT process give it an edge over traditional textile screen-printing methods. Additionally, the superfast changeover time between inks, i.e., only 10 minutes, in DPT gives it an edge over the traditional process, which requires ~ 120 minutes to change inks, making DPT an ideal choice for the textile printing industry, especially for small lots and samples. DPT is generally based on inkjet technologies. Inkjet technology delivers the ink droplets in such a way that only ink droplets encounter with medium.[6] Thus, it is a nonimpact printing technique, unlike conventional printing techniques, which are impact printing methods.[7]

The principle of DPT can be comprehended by witnessing a dripping water tap (Fig. 3.1A). Drops continue to fall on the substrate placed below the tap. If the movement of the tap, and ultimately the movement of the drop, is precisely controlled, every drop will then fall on the designated area of the substrate. Furthermore, if the movement of the droplet is controlled to the extent that the drop falls on the designated area of substrate only when it is needed, then this approach echoes the digital printing process fairly well.[8]

The concept of primary inkjet printing dates back to the early 19th century. In 1833, it was Savart who introduced the idea of uniform droplet formation from a liquid jet using an orifice, which was later scientifically explained by Lord Rayleigh and Weber[9]; however, there was no major development until Siemens Elema developed the first inkjet printer in 1948 using Rayleigh's concept.[9,10]

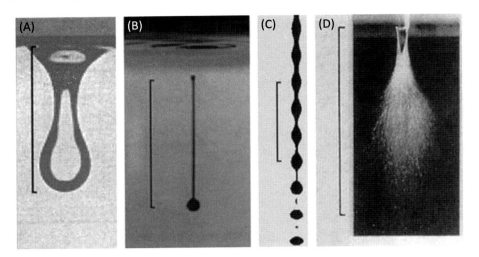

**Figure 3.1** Fluids emerging from a nozzle, illustrating the effect of increasing flow rate: (A) dripping faucet; (B) drop-on-demand (DOD); (C) continuous inkjet (CIJ); (D) atomization.[8]
*Source*: Courtesy of Xennia Technology Ltd.

Afterward, the theory of employing pressure waves to break inkjet into ink droplets was established in the early 1960s.[9,11] Later, the drop-on-demand method was reported in the 1970s. It was the 1980s when DPT first reached the consumer market. At that time it was only Canon, Epson, and Hewlett Packard who developed and produced digital printing systems commercially.[9] Since then, several other companies have entered the market, and there has been consistent progress in digital printing and print heads to achieve the desired quality of print and high throughput to meet industrial production needs.[12–16] It is the combination of the print head, the capability of inks, and the nature of substrates that determines the quality of the print. To that end, major developments in overcoming the bottleneck of controlling the flow of ink and reducing its droplet size to a very fine scale of (1–2 pL) have been achieved to improve print quality and enhance the production rate.[17] Massive research, however, is still being carried out to increase the production rate of digital printing and to make the process more affordable in terms of running costs.

## 3.2 Classification of digital printing machines

Digital printers use three approaches, drop-on-demand, continuous inkjet, and atomization, depending on the scale and speed of printing (Fig. 3.1B–D).[8] In the drop-on-demand approach, a characteristic tiny ink droplets with a long tail is formed. Regulating this tail of such a tiny drop to ensure it does not smear the print design is quite a challenging task. Whereas, in the continuous inkjet approach, a continuous jet that changes into a series of tiny ink droplets is formed. Ink droplets that are not needed on the substrate are diverted to the gutter to recirculate for reuse avoid ink wastage and prevent printed images from being stained (Fig. 3.2).

Similarly, manipulating this jet of tiny droplets to ensure they only fall on the desired and specified areas is a critical task; however, individual droplet movement cannot be controlled in the atomization case. Since the atomization technique does offer control over droplet movement, it is not preferred for the digital printing of textiles. Detailed classification of drop-on-demand and continuous inkjet technology is given in Fig. 3.3. Therefore, in this chapter, mainly continuous inkjet and drop-on-demand techniques have been discussed.

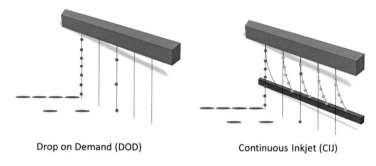

**Figure 3.2** Inkjet printing techniques.

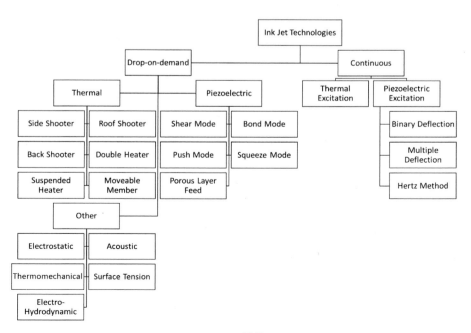

**Figure 3.3** Classification of printing techniques.[18,19]

## 3.3 Drop-on-demand inkjet technology

In this technique, drops of ink are only ejected when they are needed to develop an image. For this end, there are mainly two mechanisms used to eject/generate drop, i.e., thermal inkjet and piezoelectric inkjet. The thermal inkjet technique employs an electrical heater placed inside the faucet to enhance the ink temperature to the level of bubble nucleation, and the fiery expansion of the vapor bubble drives the ink out of the faucet. In comparison, the piezoelectric inkjet technique uses an actuator to regulate the amount of ink in the nozzle and squeezes the ink droplet out of the nozzle.

Since the thermal inkjet technique uses an electric heater inside the faucet, a current pulse is used to rapidly increase the temperature of ink present in the faucet up to 300°C or more. This leads to vicious nucleation and expansion of vapor bubbles, resulting in the ejection of ink droplets out of the faucet vent. Moreover, owing to the explosive nature, it is difficult to control the process apart from the power applied and pulse length. Sometimes, a short prepulse or train of prepulses is used to prewarm the ink in the faucet, which offers limited control to regulate the ejection volume of ink. Water tends to produce more fiery bubble growth than other solvents, and that is why water-based inks are always preferred in thermal inkjet printing.[17]

Additionally, several types of ejectors are used to eject the ink droplets in the thermal inkjet technique. Among these ejector types, side-shooter, where the heater plane is parallel to the faucet plane, and roof-shooter, where the heater plane is perpendicular to the faucet planes, are the most common type of ejectors (Fig. 3.4). Another famous type of ink ejector used in thermal inkjet is the "back-shooter," where the heater is located on the back of the ink bath.[20] Mega companies like Canon and Sony introduced side-shooter ejectors having multiple heaters and roof-shooter ejectors having two independently driven side-by-side heaters, respectively, to enable drop modulation and regulate the movement of the ejected droplets. To develop energy-efficient ink drop ejectors, ejector designs with suspended heaters have also been proposed, where the heater is embedded in the ink[17]; thus, most of the heat generated by the heater is instantly transferred to the ink resulting in better energy efficiency than their counterparts. Furthermore, a novel design of a drop ejector with an adjustable member inhibiting ink movement back into the ink trough to make the process more energy efficient has been disclosed by Canon in a

**Figure 3.4** Thermal inkjet printheads, (A) roof-shooter, (B) side-shooter, and (C) suspended heater.[16]

series of patents.[17] The movement of the adjustable member is auto-regulated during the vapor bubble expansion.

Additionally, thermal inkjet printing heads are manufactured using same techniques that are used in semiconductor and integrated circuit (IC) industries; thus, printing heads can have decent electronic control systems to regulate the movement and behavior of printing heads. That is why numerous thermal inkjet printing heads are precisely controlled. Well-established technology of IC and semiconductor industries makes the design and fabrication of thermal inkjet printing heads a relatively low-cost process and enables the designing of multifaucet printing head arrays.[17,21]

On the other hand, a typical piezoelectric element is used to generate the ink droplets in the piezoelectric inkjet. These piezoelectric elements are generally made of typical lead zirconate titanate, whereas their placement varies depending on the design of the printing head. In piezoelectric inkjet, the applied voltage is used to reduce the volume of the chamber, which leads to squeezing the ink droplet out of the nozzle. Contrary to the thermal inkjet print heads, which are classified based on the position of the heating element, piezoelectric inkjet print heads are categorized according to the operating principle and geometry drop ejector. As displayed in Fig. 3.5, the famous types of drop ejectors used in piezoelectric inkjet are "shear mode," "bend mode," "push mode," "squeeze mode," "nozzle excitation," and "porous layer feed." In shear mode, the electric field is perpendicular to the poling direction, whereas the electric field is parallel to the poling direction in bend mode. Likewise, push mode also uses an electric field parallel to the poling direction; however, in this case, the membrane is placed in the expanding direction of the piezoelectric material. A hollow tube is used as a drop ejector in the squeeze mode.

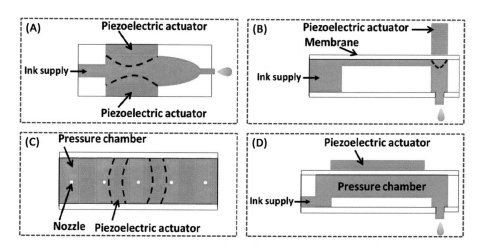

**Figure 3.5** Piezo-driven inkjet printheads (A) squeeze mode, (B) push mode, (C) shear mode, (D) bend mode.[19]

Another design of the piezoelectric print head uses a porous metal layer-based actuator where ink is supplied to the chamber using this porous layer.[18]

Lately, piezoelectric drop ejectors were also mounted on the nozzle plate. The novel design offered substantial cost advantages and significantly reduced the effect of air bubbles present in the ink trough. Moreover, the ink ejection process in the piezoelectric inkjet method can be regulated by waveforms, and controlled short prepulses can help in regulating the ink droplet sizes.

Besides thermal inkjet and piezoelectric inkjet, there are some other drop-on-demand printing techniques too. For instance, an electrostatic inkjet drop ejector system directly employs an electric field to move the membrane of an ink chamber to eject the ink droplet.[22] Another example of drop-on-demand inkjet printing is thermomechanical technology. This technology uses an electric heater to induce varying thermal expansion coefficients, which results in the sudden movement of composite structure. Moreover, faucet-free ink droplet ejection by employing an acoustic excitation focused on the surface of ink has also been reported. It has also been asserted that drop ejection can be initiated by moving the paddle behind the faucet, whereas faucet movement can also initiate drop ejection.[17]

Besides electrostatic ink droplet ejection, the electrohydrodynamic method has also been reported for ink droplet ejection (Fig. 3.6). In this method, two electrodes are installed on the front end of the nozzle and another at the back of the medium. The electrode present before the nozzle ejects the ink droplets in the presence of high applied voltage, whereas the electrode installed at the back of the nozzle guides the ink droplets to the medium.[23] Another concept disclosed is "Surface Tension Driven Inkjet," which produces ink droplets by generating equilibrium between the surface tension of the ink and the positive driving force. When the surface tension is lowered by heating the ink, and high voltage is applied, ink droplets are ejected through the faucet.[17]

## 3.4 Continuous inkjet technology

Contrary to drop-on-demand technology, as the name implies, the continuous inkjet technique generates a jet of ink. A pump is used to direct ink from a reservoir to

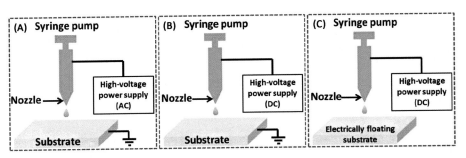

**Figure 3.6** Electrohydrodynamic (EHD) jet printing systems with (A) AC, (B) pulsed DC, and (C) single potential AC.[24]

the nozzle, producing an endless stream of drops at high frequency. This ink stream then breaks into variable-sized droplets after leaving the nozzle. The nozzle is periodically excited in a way that translates into a spatial perturbation in the fluid jet to regulate the size of these droplets in traditional continuous inkjets. The accuracy of the droplet size can be controlled by manipulating the excitation frequency and jet velocity. Generally, this periodic excitation is received from mechanical oscillations *via* a piezoelectric transducer attached to the print head. Afterward, the charged ink droplets pass through an electric field, which regulates their movement and directs them to the desired location of the substrate, and unused droplets are collected in a collector for recirculation and reuse (Fig. 3.7).[25]

Generally, two methods are used for controlling the movement of these inkjet droplets in the piezoelectric continuous inkjet technique. The first one is the binary deflection method, in which droplets are directed to a single pixel in the substrate or towards the reservoir, whereas another one is the multiple deflection method, in which ink droplets are diverted to several pixels simultaneously (Fig. 3.8). Furthermore, the Hertz method is another variant of continuous inkjet used when a variable amount of ink per pixel is required. To that end, very fine drops of ink are generated by a high excitation frequency (1 MHz). To avoid their merger during flight, these tiny ink droplets are given smaller charges.[17]

Another famous approach employed in continuous inkjet technology is using an annular heater with each nozzle, which produces thermal pulses to disintegrate the stream of ink. This way, the temperature of the ink is raised in the vicinity of the faucet, and the viscosity of the ink lowers in the surroundings of the nozzle. This combination of constant ink stream velocity and intermittent thermal pulse breaks the inkjet into uniform droplets. Standard electric field-driven, air deflection, and differential thermal deflection (i.e., based on differential thermal energy of two independent heaters placed on the diametrically opposite sides of the faucet) methods are generally used to regulate the direction and size of ink droplets in the continuous inkjet technology.

Although, continuous inkjet technology offers relatively fast printing capability compared to its counterparts. Highly volatile solvents give continuous inkjet

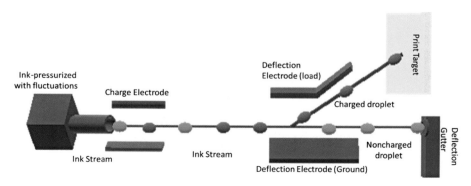

**Figure 3.7** Schematic demonstration of principles for continuous inkjet technology.[25]

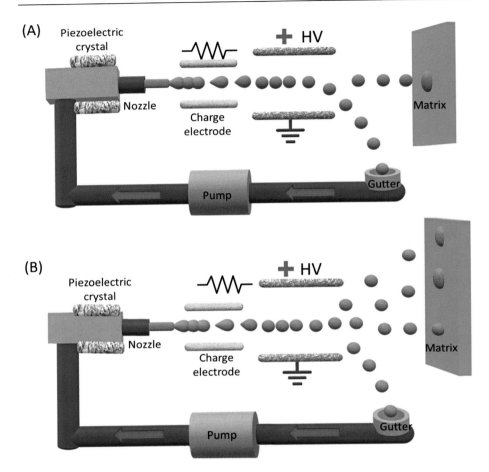

**Figure 3.8** Schematic demonstration (A) Binary deflection method and (B) multiple deflection method of continuous inkjet printing.[23]

technology an edge over other techniques; however, lower print quality, especially in terms of low resolution, costly print heads, and high maintenance cost, limit its acceptability. Additionally, using highly volatile solvents also makes it an environmentally unfriendly process. Volatile organic compounds are always considered a threat to health and thus need to be minimized.[26]

## 3.5 Print head selection

The print head plays a critical role in determining the quality of print images and the production rate of the printing process. Therefore, making the right choice about the technology and version of the print head for any specific application is extremely important. To that end, there are multiple factors (i.e., cost, image quality, and production rate of the printer) that must be kept in sight while making the selection of the appropriate print head. Cost is one of the primary factors for deciding the manufacturing or processing route to be followed for any product. The cost of the print head and its running cost mainly depend on the inkprint head interaction and minimum working life of the print head, because in absence of recommended ink, the minimum working life of the print head is not guaranteed. Since the print head is a key component in digital printing technology and its replacement is costly, the minimum life of the print head plays a vital role in calculating the running cost of the process. The minimum life of the print head is measured using Weibull statistics in the presence of prescribed ink and operating conditions. Several factors, such as loss of hydrophilicity of nozzle plate and contamination ink delivery system, etc., can affect the lifespan of the print head. This way, the cost of the print head ultimately impacts not only the cost of the equipment but also affects the cost of the product. Furthermore, running the cost of the print head is also regulated by the number of nozzles and operating frequency of the print head. Therefore, the cost of the print head is generally calculated in terms of cost/nozzle to standardize the cost of the print head.

Another extremely important factor considered for selecting an appropriate print head is image quality. The lower the volume of the drop, the finer are the specifics of the image, as it is the volume of the drop that establishes the size of the printed dot. With the evolution of digital printing, the volume of drop ejected by print heads of commercial digital printing machines has substantially reduced from $>100\,\mu L$ to as low as 1.5 pL. Smaller ink volume drops can drastically improve the resolution of the resultant printed image. Image resolution indicates the scale of the grid over which dots are placed, and it is conventionally measured in "dots per inch" and measured in both longitudinal and latitudinal directions. Additionally, image quality can also be characterized by the ability of the print head capable of producing variable sizes of ink droplets, also known as grayscale. Since print heads with the capability of grayscale add more to productivity than to image quality, thus, drop volume is yet the dominant metric to define image quality even for those print heads with grayscale capabilities.

Moreover, the production rate of the print head is defined as the amount of ink delivered per unit of time by the print head. It can be calculated by multiplying the number of nozzles, operating frequency, and drop volume.[17] In contrast, the productivity of the nozzle can be measured by multiplying the operating frequency and volume of the drop. Another way to describe the productivity cost of a print head is the ratio of print head cost and its productivity; it is calculated as the dollar cost of one liter per hour of productivity per nozzle.

## 3.6 Companies active in print head technology

Since digital printing has been there for some time, therefore, several manufacturers of print heads and digital printers for textiles have been reaping the benefits of digital printing from the market. There is huge potential for digital textile printing in countries like India, China, Pakistan, and Bangladesh, etc.[27–29] The inkjet print head market has been dominated by two major players, Canon and Hewlett-Packard, followed by Lexmark, etc.[9,30] Other famous companies involved in the manufacturing of different printhead technology are Xaar, Kodak, Memjet, Konica Mintola, etc.[12–15]

Xaar develops a variety of print heads targeting a wide range of applications from ceramic tile decoration to precision jetting for graphical, labeling, advanced manufacturing, etc. The key feature of Xaar's print heads is the highly reliable jetting of numerous fluids using specialized waveforms with ink recirculation and ink Throughflow. Additionally, Xaar's new print heads 5601, developed using Micro-Electro-Mechanical Systems (MEMS) technology, are suitable for printing textiles, laminates, wide-format graphics, different labels, etc. In this print head, Xaar has used four dies placed in an alternating offset manner, and it used a vernier system to ensure that nozzles are in close alignment. Ink droplets volume on each die in the vernier regions is controlled by a special waveform that trims the excessive volume of the ink droplets to maintain uniform density. The temperature of the print head is observed with a sensor and is regulated by circulating the temperature-controlled ink. Bubble-free ink flow is properly maintained to cool the Application Specific Integrated Circuit (ASIC), and this sustained flow of ink also prohibits the thickening and skin forming on the surface of ink to safeguard the efficient working of nozzles. A regulated drive pulse is used to overcome the small variations coming from nozzles in the form of drop speed, volume, and trajectory angle in the way of media movement. To that end, Xaar uses a precisely designed special pulse shape to edge the waveform delivered to the printer from the amplifier. In short, Xaar printhead reaps the benefits of advances in fluid design, waveforms, drive electronics, and advanced materials to precisely control ink's speed and drop volume in the inkjet printing process.

Kodak Enterprise Inkjet Systems Division is the successor to the Medium Extended Air Defense (Mead) Advanced SystemsTM, which first introduced continuous inkjet systems back in the 1970s for commercial-scale printing. Since then, this technology has seen rapid development in print quality because of reduced dot size, and increased speed has improved productivity resulting in lower cost per print. In Kodak's stream continuous inkjet technology, thermal drop stimulation is used in combination with air deflection, which differs from traditional inkjet technology, where acoustic drop stimulation is used for drop formation while electrostatic deflection is employed for print drop selection. Thermal drop stimulation and air deflection enable Kodak's stream continuous inkjet technology to operate at a relatively much higher speed while maintaining the print image quality. Compared to drop-on-demand print heads, in continuous inkjet technology, the nozzles are not locally energized to eject fluids. Drop volume, velocity, and trajectory are thermally managed using very low energy ($\sim 10$ nJ) by using small thermal perturbations to fluid jets in the form of heat bands from heaters that surround the faucets. There is a little surface temperature rise in each succeeding band along the jet length, whereas each band is a fraction of the jet radius. The presence of these perturbations leads to the reduced surface tension of the fluid jet and commences collapse or pinch-off, a hydrodynamic process of the jet. The lowered surface tension of the jet enables the pinch-off process with relatively low energy. This way, regulated pinch-off timing can easily and precisely control ink droplets volume. Moreover, an integrated complementary metal-oxide-semiconductor (CMOS)-MEMS device using very large-scale integration (VLSI) processing produces precise geometries for jet exit orifice and heaters, which guarantees accurate drop direction and homogeneous drop volume. Furthermore, the printhead life of stream continuous inkjet technology is independent of coverage and can print up to 20 million linear meters of images. As stream continuous inkjet technology is relatively new, these unique features and printing characteristics of Kodak's printing technology demonstrate its high potential to scale up image quality and resolution, printing speed, and printing width.

Memjet introduced a novel class of inkjet printing by employing disruptive "waterfall" printing technology in the previous decade. Memjet printhead technology has developed rapidly and is well documented and protected with hundreds of global patents signifying innovation and the value of this solution. Current Memjet printheads are developed using a combination of CMOS and MEMS techniques; however, the uniqueness of Memjet printheads is that their fabrication involves up to 15 times less use of silicon than other thermal inkjet printers. At first, CMOS layers are manufactured by developing CMOS chips on a silicon wafer, which later helps regulate the firing of the nozzles for each line of ink droplets. Afterward, benefiting from MEMS processes, ink chambers and heaters are built, which regulate the temperature of ink during the printing process, and collectively they are named as MEMS layer. Further, using the MEMS process, holes are crafted in CMOS chips, which act as pathways for ink supply to the ink firing chamber. Finally, the printhead assembly is completed by precisely interconnecting the CMOS layer and MEMS layer, and jetting array subassembly within a frame.

Moreover, the present Memjet thermal inkjet printhead system has robust and efficient ink recirculation through the printhead by moving it through one end of manifold structures and moving out from another. This robust ink recirculation and the presence of air bubble removal and filtration of the ink particulate system substantially advance the overall printing efficiency of Memjet technology. Memjet printhead technology offers even velocity, and trajectory homogenous drop sizes, by ensuring the distribution of ink from one nozzle to another. The Memjet thermal inkjet printhead system is based on 70,400 nozzles with five independent channels, and each channel comprises two linear nozzle arrays.

Owing to the robustness of water-based inks and their relatively high viscosity requirements for textile industrial applications, piezoelectric inkjet technology is preferred over thermal inkjet printing technology because of its better water compatibility. Additionally, drop-on-demand inkjet is generally used in industrial printers for digital printing of textiles over continuous inkjet approach because of the high cost and constraints of continuous inkjet technology, such as lower time availability for dye fixation to textile. Dye fixation is crucial for textile materials to achieve acceptable color fastness.[31] Some major players that use drop-on-demand

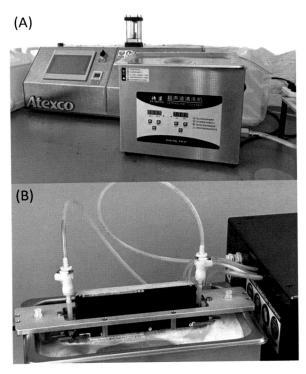

**Figure 3.9** Cleaning equipment (after-sales equipment) (A) Advanced Yumeng ultrasonic cleaner by Atexco (B) the print head mounted on ultrasonic cleaner.

or continuous inkjet technology for producing industrial printers are SPG Prints, Epson, Canon, Aprion, Seiko, Colorjet, Konika Minolta, etc. Digital images of their commercial, industrial printers used for digital textile printing can be viewed on their respective websites for more reference.

In addition, there has been significant improvement in head cleaning devices to ensure that a single head can be used many times; for instance, the ultrasonic cleaner may be used for cleaning print heads, as shown in Fig. 3.9.

## 3.7 Outlook

This chapter illustrated a basic understanding of major digital printing machines and discussed different types of digital printing machines. There is emerging potential to use artificial intelligence tools to support and improve the quality of textile materials and their manufacturing performance.[32,33] Fast printing speed, design accuracy, and the ability to print with water-based inks, including reactive and dispersed inks, are some of the major reasons why the process has gained massive attraction in recent years. Additionally, the process can realize almost any design with much superior image resolution, which gives it an edge over traditional printing techniques. Despite having these benefits, DPT is still in the development phase, and intensive research is being carried out to explore the full potential of the technology. Moreover, reducing the running cost of the process by increasing the production rate and nozzle choking are the major hurdles; overcoming them will pave pathways for its acceptability on the mega scale and consumer market.

## 3.8 Conclusion

In summary, this chapter briefly overviews digital printing machines for textiles, discusses their possible mechanisms, and elaborates on the technological edge of DPT over conventional screen printing of textiles. Some of the key advantages that DPT brings to the industry are high design accuracy, marvelous printing resolutions ability to print with colorants acceptable to a wide range of textile materials, and considerably fast printing speed, etc., making it an ideal printing technology capable of overcoming the limitations faced by conventional screen printing. Superfast change time between inks in DPT makes it the choice of the textile printing industry, especially for small orders and printing of samples. Furthermore, the chapter briefly elaborates on the selection criteria for opting for an appropriate printhead and the effect of various parameters of the printhead selection. Finally, it concludes with the future outlook of DPT, discussing limitations that need to be addressed urgently to enhance the scope of DPT and make it readily acceptable to the textile printing industry for mega-scale production.

## References

1. Koseoglu, A. U.; Gungor, Y.; Arik, Y. Innovations and Analysis of Textile Digital Printing Technology. *Int. J. Sci. Technol. Soc.* **2019**, *7* (2), 38–43.
2. Memon, H.; Jin, X.; Tian, W.; Zhu, C. Sustainable Textile Marketing—Editorial. *Sustainability* **2022**, *14* (19), 11860. Available from: https://www.mdpi.com/2071-1050/14/19/11860.
3. Briggs-Goode, A.; Russell, A. 5 - Printed Textile Design. In *Textile Design;* Briggs-Goode, A., Townsend, K., Eds.; Woodhead Publishing, 2011; pp 105–129e.
4. Haverinen, H. M.; Myllyla, R. A.; Jabbour, G. E. Inkjet Printed RGB Quantum Dot-Hybrid LED. *J. Disp. Technol.* **2010**, *6* (3), 87–89. Available from: https://doi.org/10.1109/JDT.2009.2039019.
5. Wen, D., et al. Inkjet Printing Transparent and Conductive MXene (Ti3C2Tx) Films: A Strategy for Flexible Energy Storage Devices. *ACS Appl. Mater. Interfaces* **2021**, *13* (15), 17766–17780. Available from: https://doi.org/10.1021/acsami.1c00724.
6. Memon, H.; Khoso, N. A.; Memon, S.; Wang, N. N.; Zhu, C. Y. Formulation of Eco-Friendly Inks for Ink-Jet Printing of Polyester and Cotton Blended Fabric. *Key Eng. Mater.* **2016**, *671*, 109–114. Available from: http://www.scientific.net/KEM.671.109.

7. Bates, I.; Plazonić, I.; Petric Maretić, K.; Rudolf, M.; Radić Seleš, V. Assessment of the UV Ink-Jet Ink Penetration Into Laboratory Papers Within Triticale Pulp and Its Influence on Print Quality. *Coloration Technol.* **2022**, *138* (1), 16−27. Available from: https://doi.org/10.1111/cote.12563.
8. Tyler, D. J. 12 - Digital Printing Technology for Textiles and Apparel. In *Computer Technology for Textiles and Apparel;* Hu, J., Ed.; Woodhead Publishing, 2011; pp 259−282.
9. Soleimani-Gorgani, A. 14 - Inkjet Printing. In *Printing on Polymers;* Izdebska, J., Thomas, S., Eds.; William Andrew Publishing, 2016; pp 231−246.
10. Rune, E. Measuring Instrument of the Recording Type, ed. Google Patents, 1951.
11. Sweet, R. G. Fluid Droplet Recorder. ed. Google Patents, 1971.
12. Brünahl, J.; Condie, A.; Crankshaw, M.; Cruz-Uribe, T.; Zapka, W. Xaar's Inkjet Printing Technology and Applications. In *Handbook of Industrial Inkjet Printing*, 2017; pp. 285−312.
13. Puyot, M. Memjet's Inkjet Printhead Technology and Associated Printer Components. In *Handbook of Industrial Inkjet Printing*, 2017; pp. 335−350.
14. Corrall, J. Konica Minolta's Inkjet Printhead Technology. In *Handbook of Industrial Inkjet Printing*, 2017; pp. 253−284.
15. Piatt, M.; Bugner, D., Chwalek, J.; Katerberg, J. KODAK's Stream Inkjet Technology. In *Handbook of Industrial Inkjet Printing*, 2017; pp. 351−360.
16. Simske, S. J. Hewlett Packard's Inkjet Printhead Technology. In *Handbook of Industrial Inkjet Printing*, 2017; pp. 313−334.
17. Mariano Freire, E. 3 - Ink Jet Printing Technology (CIJ/DOD). In *Digital Printing of Textiles;* Ujiie, H., Ed.; Woodhead Publishing, 2006; pp 29−52.
18. Shah, M. A.; Lee, D. G.; Lee, B. Y.; Hur, S. Classifications and Applications of Inkjet Printing Technology: A Review. *IEEE Access.* **2021**, *9*, 140079−140102. Available from: https://doi.org/10.1109/ACCESS.2021.3119219.
19. Wijshoff, H. The Dynamics of the Piezo Ink-Jet Printhead Operation. *Phys. Rep.* **2010**, *491* (4), 77−177. Available from: https://doi.org/10.1016/j.physrep.2010.03.003.
20. Lee, C.-s.; Na, K.-w.; Lee, S.-W.; Kim, H.-c; Oh, Y.-s. Bubble-Jet Type Ink-Jet Printhead and Manufacturing Method Thereof. ed. Google Patents, 2004.
21. Frits Dijksman, J. Ed. Introduction. In *Design of Piezo Inkjet Print Heads: From Acoustics to Applications*, 1st ed.; Wiley: Hoboken, NJ, USA, 2019.
22. Suzuki, Y.; Takagishi, K.; Umezu, S. Development of a High-Precision Viscous Chocolate Printer Utilizing Electrostatic Ink-Jet Printing. *J. Food Process. Eng.* **2019**, *42* (1), e12934. Available from: https://doi.org/10.1111/jfpe.12934.
23. Li, J.; Rossignol, F.; Macdonald, J. Inkjet Printing for Biosensor Fabrication: Combining Chemistry and Technology for Advanced Manufacturing. *Lab. Chip* **2015**, *15* (12), 2538−2558. Available from: https://doi.org/10.1039/C5LC00235D.
24. Yudistira, H. T.; Nguyen, V. D.; Dutta, P.; Byun, D. Flight Behavior of Charged Droplets in Electrohydrodynamic Ink-Jet Printing. *Appl. Phys. Lett.* **2010**, *96* (2), 023503.
25. Ikegawa, M.; Ishikawa, M.; Ishii, E.; Harada, N.; Takagishi, T. Ink-Particle Simulation for Continuous Ink-jet Type Printer. *NIP & Digital Fabrication Conference*, 2015, 01/01 2015.
26. Memon, H.; Yasin, S.; Ali Khoso, N.; Hussain, M. Indoor Decontamination Textiles by Photocatalytic Oxidation: A Review. *J. Nanotechnol.* **2015**, *2015*, 104142. Available from: https://doi.org/10.1155/2015/104142.

27. Dutta, S.; Bansal, P. Textile Academics in India—An Overview. In *Textile and Fashion Education Internationalization: A Promising Discipline from South Asia;* Yan, X., Chen, L., Memon, H., Eds.; Springer Nature Singapore: Singapore, 2022; pp 13–34.
28. Uddin, M. F. Brief Analysis on the Past, Present, and Future of Textile Education in Bangladesh. In *Textile and Fashion Education Internationalization: A Promising Discipline from South Asia,* X. Yan, L. Chen, H. Memon, Eds.; Singapore: Springer Nature Singapore, 2022; pp. 35–57.
29. Ali Hayat, G.; Hussain, M.; Qamar Khan, M.; Javed, Z. Textile Education in Pakistan. In *Textile and Fashion Education Internationalization: A Promising Discipline from South Asia;* Yan, X., Chen, L., Memon, H., Eds.; Springer Nature Singapore: Singapore, 2022; pp 59–82.
30. Keeling, M. R. Ink Jet Printing. *Phys. Technol.* **1981,** *12* (5), 196–203. Available from: https://doi.org/10.1088/0305-4624/12/5/302.
31. Dutta, S.; Bansal, P. Cotton Fiber and Yarn Dyeing. In *Cotton Science and Processing Technology: Gene, Ginning, Garment and Green Recycling;* Wang, H., Memon, H., Eds.; Springer Singapore: Singapore, 2020; pp 355–375.
32. Halepoto, H.; Gong, T.; Noor, S.; Memon, H. Bibliometric Analysis of Artificial Intelligence in Textiles. *Materials* **2022,** *15* (8), 2910. Available from: https://www.mdpi.com/1996-1944/15/8/2910.
33. Wang, H.; Halepoto, H.; Hussain, M. A. I.; Noor, S. Cotton Melange Yarn and Image Processing. In *Cotton Science and Processing Technology: Gene, Ginning, Garment and Green Recycling;* Wang, H., Memon, H., Eds.; Springer Singapore: Singapore, 2020; pp 547–565.

# Color management and design software for textiles

Pardeep Kumar Gianchandani[1], Abdul Khalique Jhatial[1], Aijaz Ahmed Babar[1] and Hanur Meku Yesuf[2,3]
[1]Department of Textile Engineering, Mehran University of Engineering and Technology, Jamshoro, Sindh, Pakistan, [2]Key Laboratory of Textile Science and Technology, Ministry of Education, College of Textiles, Donghua University, Shanghai, P.R. China, [3]Ethiopian Institute of Textile and Fashion Technology, Bahir Dar University, Bahir Dar, Ethiopia

## 4.1 Introduction

Textiles are purchased because of their appearance, which includes color ink and design.[1] In the last two decades, there has been an increasing interest in color management and design by producers as demanded by consumers. Color management was not a primary business of many suppliers and producers then. Thus, the industry lacks many ways to manage color. Color is a very complex phenomenon of the eye-brain visual system. The fundamental challenge for color managers is that it has many nuances to take part in color, such as appearance, likes, and dislikes.[2,3] The potential challenge for color managers is to satisfy or meet the requirements of producers, that is, the textile industry and end users. A number of color management systems (CMSs) have been developed so far, but none of them have an exact correlation with human vision.[4] Thus, to make things work, several easy systems, instruments, and software have been designed and developed so far, which will be discussed later in this chapter.

The color we see is the interaction of light, objects, and our eye-mind system, as shown in Fig. 4.1. We humans are limited in vision to 360–780 nm wavelength. Color managers play with basic colors (RED, GREEN, and BLUE in additive color mixing) and (CYAN, MAGENTA, and YELLOW in substantive color mixing) to produce a variety of colors. Color for casting and measurement has become an integral part of the textile industry (from fiber to finished garment).[5] Color systems have evolved to numerical information of colors and become the indispensable, integrated part of the entire CMS.[6]

The eye-mind system is superior to any machine or instrument; however, the instruments are developed to replicate light, objects, and eye-mind systems that can precisely process information and identify color.[7] Thus, instrumental analysis became an invaluable tool for observing the colors at a commercial scale. Eye-mind system evaluation varies from person to person.[8]

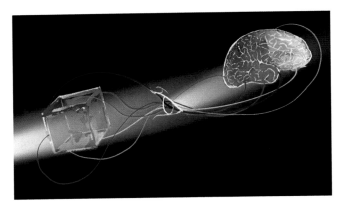

**Figure 4.1** The light, object, and eye-mind system interaction.
*Source*: Image is courtesy of Miss XinRui Hui @ College of Textile Science and Engineering, Zhejiang Sci-Tech University, China.

The textile industry has advanced a lot, particularly in the color sector (dyeing and printing), that is, the digitalization of colors started a couple of decades ago. The textile printing industry needs to understand and adopt the developments in color measuring, managing, and designing systems. After the industrial revolution and the digitalization of the globe, advanced technology is available to every corner of the globe. Industries are shifting from conventional approaches to advanced solutions. Digital technology has opened numerous creative opportunities for artists, artisans, designers, and colorists. Until 1970, before the invention of the first CAD (computer-aided design), the designs were translated using graph paper and punch cards for jacquard designing, which consumes a lot of paper material and requires ample space for usage and storage. Artificial intelligence and computer science have recently established great importance in all the fields of textiles.[9] In particular, the tools and software that are helpful for image processing are greatly acknowledged for this contribution.[10] The difference made by digital technology is notable; designs and patterns are completely controlled and reviewed before production. The success rate of designs is almost 100% compared to 15%−20% before digitalization. Traditional color measuring and management were done by producing numerous samples and their visual inspection by experienced personnel in the industry and then sending them for approval to the customer, which is time-consuming and a lot of materials waste. Moreover, the redesign was also a time taking process and required effort as you are starting a new job and dependency on designers is high.

It is important and necessary to understand the potential of new technologies used for color measurement, management, and design. Since the digital inkjet printing technology might be applied to any substrate, that is, cotton, polyester, or their blends.[11] Modern tools and technology give freedom and customization to producers and customers. These tools can reproduce the design in no time and create a design database for future needs. Most importantly, color rendering creates a realistic appearance that does not require producing samples for samples' physical final appearance. The digital design and color patterns can be viewed under different light resources to minimize the effect of metamerism. The modern system can communicate with printers directly. The objective of this chapter is to review the basics of color, color mixing, appearance, color measurement systems, CMSs, and color designing systems and software used for digital printing.

## 4.2  Elements of color management

The natural CMS has three elements (1) light, (2) Object, and (3) Eye-mind, which are known as the triplet of colors. Whereas modern commercial CMSs are based on two additional elements, along with these three are (4) Computers/Hardware and (5) software.[4,5,12,13]

1. Light
2. Object
3. Eye-mind system
4. Computer/Hardware
5. Software

The color we humans see is the reflectance of selected wavelength sensed by the eye-mind system. The human eye has Photoreceptors (i.e., cones and rods). Rods sense the intensity of light, and codes are of three types (i.e., red, green, and blue) responsible for actual color. Light sources play an important role in identifying and sensing the actual color.[6,14] With full illumination, such as in daylight, the rods and cones work together, and humans can identify the exact details of colors. Only rods work well as the light intensity reduces, and we can only see shadows and shapes. Color vision and identification are subjective. It varies from person to person; thus, for proper color management and measurement, scientists have developed various systems and software's will be discussed later in this chapter.

## 4.3  Classification of colors

Color science has advanced a lot, and color scientists have developed standard methods of classifying and defining colors. To answer the question, "How are colors defined and organized"? Scientists have fixed three attributes of colors (1) Hue, (2) Chroma, and (3) light and various systems to classify the colors.

Hue is the actual color that the eye-mind system perceives, such as yellow, green, red, etc., represented in a circle in Fig. 4.2, known as a color wheel. Chroma is color purity and intensity represented along the radius of the wheel, the higher the chroma brighter will be the color, and the lower the chroma duller will be the appearance of the color.[15] Lightness is the reflection of only black and white light from the object. All these three-color attributes are independent of each other.

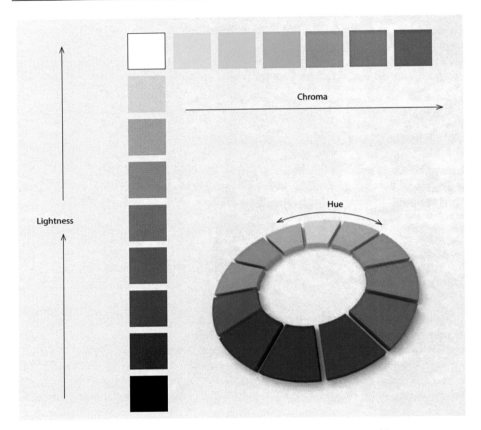

**Figure 4.2** Natural classification of color based on hue, chroma, and light.[15]

Fig. 4.2 represents the natural classification of color based on three color attributes. There are various other systems that have been developed so far, such as

1. Munsell atlas
2. The DIN standard table
3. The OSA (Optical Society of America) color system
4. The NCS (Natural Color System) atlas
5. The RAL Design system
6. The CIE (International Color Consortium) color system

## 4.4 Color measurement

Color measurement is a visualization of color understandard elements of color measurement, such as light source and observer with standard observing conditions, that is, angle of observation.[4] The International Commission on Illumination CIE

("Commission Internationale de l'Éclairage") has defined standard conditions for color measurement (observation and visualization). CIE proposed the standard color stimulus, standard observer, and standard light resources.

CIE has fixed the three colors (1) Red 700 nm, (2) Green 546 (3) blue 435 as a primary color. The sensitivity of cones is angle-dependent; thus, CIE also fixed the standard angle of observation, that is, 2 and 10 degrees, along with standard illumination for numerical (colorimetric) calculations.[6,14] There are spectral differences among the colors by which we may distinguish and discriminate the textile materials.[16] The human eye's spectral luminous efficiency is considered a standard as the eye perceives sample spectral radiance at different lightness values, that is, in day and night vision for instrumental measurement. CIE has developed a system that uses three primary light sources (name RGB) as essential color stimuli. The desired color can be achieved by varying the amount (numeric value) of individual primaries. Different colors would require different numbers of primaries. Color is three-dimensional, and it is easy to accurately measure and reproduce particular wavelengths.

Color measurement is a measurement of the primary light source, that is, additive color mixing, which is easy to manage than the color mixing of dyes and pigments, that is, substantive color mixing. CIE color measuring system is based on additive color mixing of primaries.[1,17,18] The particular quantity of each primary is selected/measured for a specific color, known as tristimulus values. The limitation of this system is that single primary cannot be produced by using the two other primaries; negative values are not desired and thus ignored, and primaries are considered imaginary primaries (X Y Z) to avoid negative values. Thus, CIE knowing the fact and in 1931, decided that the system should calculate the tristimulus values from the reflectance of a color measuring surface. The measurement must be made with/understandard primaries, standard light resources, standard observers, and standard observing and viewing conditions.[4]

The real color is measured using the positive tristimulus values from the following equations.

$$C\lambda 1 \equiv 0.73467(X) + 0.26533(Y) + 0.00000(Z) \qquad (1)$$

$$C\lambda 2 \equiv 0.27376(X) + 0.71741(Y) + 0.00883(Z) \qquad (2)$$

$$C\lambda 3 \equiv 0.16658(X) + 0.00886(Y) + 0.82456(Z) \qquad (3)$$

$$SE \equiv 0.33333(X) + 0.33333(Y) + 0.33333(Z)$$

Where $\lambda 11 = 700$ nm, $\lambda 12 = 546.1$ nm, $\lambda 13 = 435.8$ nm, and SE = equal energy, tristimulus values have equal energy at all wavelengths throughout the visible spectrum. The amount of reflected light is a product of reflected light, and the amount of incident is the appearance of the color. Thus, the appearance of color depends on the source of light. Various light sources have various phases, mainly daylight, fluorescent, and tungsten tubes. For proper measurement, the reflectance of light, that is, the appearance of the color.[2,18] It is essential to specify the light resources. Thus, CIE has standardized the three standard illuminates, A, B, and C, for the measurement of color based on their spectral energy distribution, and color temperature varies according to different phases of daylight. The standard luminant A has an absolute temperature of 2856K, Illuminate B represents the sunlight with a color temperature of 4874K, and Illuminate C represents the average daylight with a color temperature of 6774K.[14]

CIE has set standard illumination and viewing conditions. Illumination should be at 45 degrees, viewing should be normal, and vice versa to avoid measurement variation.

### 4.4.1 Instrumental color measurement

Color matching and measurement by humans are subjective. It varies from person to person. Thus, color scientists have developed instruments for measuring and managing colors. Instrumental color measurement measures the reflectance of incident light under standard illuminates, observer, and standard viewing conditions.[17] Instruments can calculate the tristimulus values of incident light. Instrumental color measurement can be divided into reflectance and transmittance. Both cases can be a diffused (some amount of light transmitted) or regular (light reflected/scattered like a mirror) measurement. Color measurement by instruments has many advantages overhuman measurement or assessment, such as defining the spectral range, data interval, bandwidth, colorimetric values, geometry, measurement area, accuracy, precision, references, user-friendly software having data export options, and many more.[2,7] Color measurement goes through all these parameters, not all in the same or one measurement. Instrumental measurement is like a modular measurement capable of measuring according to international standards (ISO, ASTM CIE) and can take many factors that are not possible for humans.

### 4.4.2 Measurement instruments

Two types of instruments are used for color measurement (1) tristimulus colorimeter (2) Spectrophotometer and Based on geometric and color attributes. Both use different instruments, methodologies, and color descriptors. The recent trend in instrumental process control has resulted in the use of online instruments. Most colorimetric instruments are, however, offline and used chiefly in laboratories. Laboratory instruments should be highly accurate and standardized, while online instruments should be rugged under various environments and have good precision and firmness.

The classification of color-measuring instruments is given in Fig. 4.3.

#### 4.4.2.1 Tristimulus colorimeter

This meter uses filters to record the reflection of incident light in different wavelengths across the visible spectrum. Tristimulus Colorimeter uses red, green, and

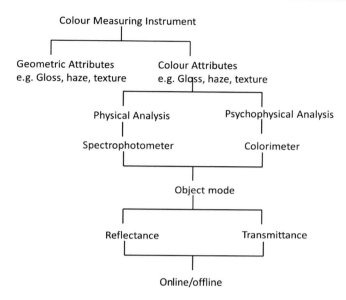

**Figure 4.3** Classification of measuring instruments.

blue filters corresponding to CIE 2 degrees standard observing and viewing conditions. Color measured by filters depends on the light resource used. This meter produces the tristimulus values for one light source and one observer. Tristimulus Colorimeter is very quick in measurements, easy to operate, and inexpensive. The disadvantage of this instrument is that it cannot detect the metamerism nor be used for color formula calculations. Thus, it can be adopted in quality for finding color differences and tolerances.

### 4.4.2.2 Spectrophotometer

The spectrophotometer measures the reflectance or transmittance of specimens at different wavelengths along with reference measurements in the visible spectrum. This instrument is installed with different illuminates; therefore, it can measure the metamerism or change in change underdifferent light sources. The spectrophotometer can measure the fraction of reflectance and transmittance of the specimen. This instrument is capable of measuring a single measurement for a variety of illuminates and observers. Spectrophotometers are simple and fast to operate.

Both colorimeters and spectrophotometers provide transmission or reflectance data obtained overthe same range of visible wavelengths (about 400–700 nm), but they may treat these data differently. The differences between these two types of instruments are listed in Table 4.1 (Hunterlab, 2008).

**Table 4.1** The difference between a calorimeter and a spectrophotometer.

| Colorimeter | Spectrophotometer |
|---|---|
| It provides the measurement related to mind-eye observation. It provides the data of colorimeter values (XYZ) | It provides the wavelength-by-wavelength spectral reflectance or transmittance with the help of a computer and |
| It consists of a sensor and a simple data processor | |
| Defined illuminates and a standard observer condition, usually C/2 degrees are used | Many illuminates/observer combinations can be used for calculating the tristimulus values and metamerism index |
| Tristimulus filters are used for isolating the broad bands of wavelength. | It isolates the narrow band of wavelength by using the prism. |
| It is a rugged and straightforward instrument than a spectrophotometer | It is versatile and works for the color formation and measurement of metamerism. |

## 4.5 Color management

Color management is managing color through the entire process of the supply chain in the textile industry, starting from the raw material to the finished product. Whereas Color measurement is the measurement of color attributes (i.e., sample and reference specimen). The entire process is known as a CMS, starting from acquiring, displaying, and printing. CMS is a complete control system by which color is measured consistently and transferred accurately to the printing device.

CMS draws upon at least three main disciplines: color science, device modeling, and human-computer interface design. Color science supplies the interchange of color spaces (such as the CIE XYZ system). A color management system (or CMS) is a collection of software and data application programs to achieve improved cross-device color rendering. Color management software is composed of a number of elements, including device modeling and gamut mapping.

Nowadays, online and digital color management is a vital part of today's industrial services and survival. Digital color measurement and management were first realized by photographic companies (Kodak, Agfa, Fuji), computer manufacturers (Microsoft, IBM, Apple Macintosh), Printing machine manufacturers (Heidelberg, LinoColor, Scitex), and color measuring instrument manufacturers (Barco, X-Rite, GretagMacbeth, and Datacolor). In the recent past, CMSs have advanced; each industry has its own customized CMS systems, which are not limited to dyeing or printing only but for whole wet processing units; it has become an essential part of communicating with buyers nationally and internationally.

CMSs can measure color values (tristimulus values), reflection, transmittance color difference, and metamerism under standard illumination and color design possibilities. It should be noted that color space is central to all the devices related to images and printing, that is, vega3180di, as shown in Fig. 4.4. Customization of

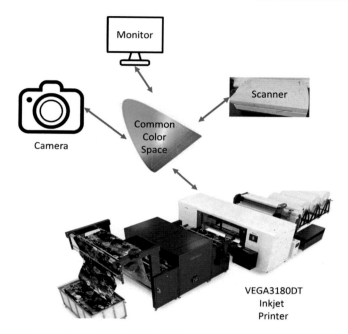

**Figure 4.4** Color space is central to all the devices.

CMS systems is highly demanded, and several customized CMS systems (i.e., software) are available in the market listed in Table 4.2.

### 4.5.1 Agfa

Color Tune is designed to create profiles for different solutions, such as ICC profiles for industrial inkjet digital proofers. It can produce multiple color profiles, for example, Gray and RGB. The color tune is the newest version of Agfa integrated with the advanced Agfa library.

### 4.5.2 Aleph

Aleph specializes in developing and manufacturing advanced direct-to-fabric and direct-to-paper inkjet printers for the textile and visual communication industries and provides its customers with total, integrated solutions. Aleph has developed a wide range of customized software named (1) Smart color, (2) Smart print, (3) Newton, and (4) Design collection. All these modules are compatible to work with spectrophotometer and colorimeter.

Smart color is designed for color matching and realistic textile and screen-printing simulation. It can process multiple color tasks and generate customized color atlases through the use of color charts and is capable of storing data in a single file.

Table 4.2 Color measurement and design systems.

| S. No | Manufacturer | Product name |
|---|---|---|
| 1 | Agfa (http://www.agfa.com/graphics) | ColorTune CMM |
| 2 | Aleph (http://www.alephteam.com) | Smart print |
| | | Smart color |
| | | Newton |
| | | Design collection |
| 3 | Chromix (http://www.chromix.com) | ColorValet Pro |
| | | ColorThink Pro |
| 4 | Color savvy (http://www.colorsavvy.com) | SavvyProfile suite |
| 5 | Color solutions (http://www.color.com) | ColorBlind Prove-It |
| 6 | Colorburst systems (http://www.colorburstip.com) | SpectraCore |
| 7 | Colorvision/datacolor Ms (http://www.colorcal.com) | ProfilerPRO, DoctorPRO |
| 8 | Ergosoft (http://www.ergosoftus.com) | ErgoSoft RIP 16 |
| 9 | Fujifilm (http://www.colorprofiling.com) | ColorKit profiler |
| 10 | GretagMacbeth (http://www.gretagmacbeth.com/il) | NetProfiler, profilemaker |
| 11 | Heidelberg (http://www.heidelbergusa.com) | Prinect calibrator/profiler |
| 12 | Kodak polychrome (http://www.kpgraphics.com) | Matchprint |
| 13 | Pantone (http://www.pantone.com) | ColorVision and colorplus |
| 14 | X-rite (http://www.xrite.com) | MonacoEZcolor & rofiler |
| 15 | Datacolor http://www.datacolor.com | SpyderX |
| | | Spyder5 |
| 16 | Atexco http://www.atexco.com/en/ | Ajet RIP |
| | | Vega print |

Smart print is designed for the needs of digital textile prints. It is flexible. Smart print can interface with all industrial printers to produce prints on textiles and paper. This module can also estimate the cost of printing jobs and keep records of printing tasks.

Aleph's newton software is designed to measure and modify the color profiles. Newton uses the spectrophotometer/colorimeter to read the reference colors and perform the color profile.

Aleph's design collection is a solution for managing the design database and bringing the designs with a few clicks. It can generate design catalogs.

### 4.5.3 Chromix

Chromix was founded in 1998 for color management and image fidelity. Chromix's latest products are ColorThink Pro and ColorValvetPro

1. ColorThink is a pro equipped with color worksheets smartcolor guide (acts as a consultant), and a profile inspector; these are its revolutionary features.

2. ColorValvetPro is color management software for printers. It can profile the color multiple times. One can have a color database with access to measurements done for years.[19,20]

### 4.5.4 Aquario design

Aquario design is a leading provider of fashion, textile, and print design applications for Adobe, photoshop, and illustrator. It was founded in 2009 by a team of fashion designers. Aquario design deals in apparel, graphics, textiles, printing, and color management. It has two commercial products (1) Textile Design and (2) Aquario suits.[21]

#### 4.5.4.1 Aquario textile design

Aquario Textile Design is a plug-in that can easily be integrated with adobe photoshop for the familiar working environment; it is equipped with an (1) AqurioTD repeat for easy, quick, and error-free repeat patterns and (2) AquarioTD Colorist for recoloring the artwork in Adobe and photoshop.[22]

#### 4.5.4.2 Aquario suites

Aquario suites are composed of (1) graphics suites (geometric design and Palette management), (2) Fashion suites (sketch, color management, fashion utility, auto exporting), and (3) print suites (raster, color reduction, repeat color management).[23]

### 4.5.5 Textile print by adobe portable document format print engine

Adobe has developed an engine that can integrate with Portable Document Format (PDF) for the growing needs of the inkjet printing market. Pdf is the most trusted file format for the printing industry as it preserves the real appearance of color and design. PDF Print Engine can reproduce prints of any color, size, or grade, particularly when paired with Adobe PDF, because it employs the same core technologies as Adobe Photoshop, Illustrator, and Acrobat.[24]

### 4.5.6 Colorburst systems—spectracore

Spectracore is designed with custom LED phosphors to provide accurate color capture. It has a high color rendering index (CRI), that is, 98, that has an excellent response overa visible spectrum. Spectracore 180 is a single source LED at 5000K. it offers a very balanced neither warm nor light. The daylight temperature of this natural light of red and blue are accurate. Spectracore can also operate in low temperatures, from 150 to 210 °F. It offers very uniform light without any hot spots. It has smooth, flicker-free dimming measurements in the range of 5%–100% to maintain the CRI.

## 4.5.7 Ergosoft

Ergosoft is based on "Raster Image Processing" (RIP) Software used for the digital printing industry. It processes the image via a color management routine to ensure consistent, correct, and high-quality output for the printing machine. Ergosoft 16 is an advanced version of RIP software equipped with many advanced features to increase productivity and printing. RIP software is a production essential that controls and monitors the complete printing workflow. Ergosoft 16 is designed for four areas of workflow starting from (1) Job editing or designing the layout, (2) Color profiling, (3) Processing, and (4) production management.[25]

## 4.5.8 X-rite

X-rite was Founded in 1958 and is a complete solution for textile color management. X-rite provides the solution for color development and formulation of dyes and improves overall color quality. X-Rite has solutions to support everyone in the textile industry to ensure color quality and consistency in two categories (1) Apparel brands and (2) Digital Textile Printing.

### 4.5.8.1 Apparel brands

This has been developed to provide an innovative solution for apparel color from concept to market. This category deals in color specification and communication, such as Ci7860, is a spectrophotometer (benchtop) ideal for digital color. Color iQC is quality software that ensures color consistency and is adaptable and configurable. Pantora is effective in communicating digital standard color effectively. MetaVue VS3200 is one of the noncontact spectrophotometers used for multicolor measurement and quantification of colors.

### 4.5.8.2 Digital textile printing

Measuring and managing color for textile Printing is a more difficult job. Inconsistency in color measurement for textile print is the most common problem. Digital printing requires consistent and accurate color measurement tools and technology to match the color. X-rite solutions are customizable for textile digital designers and printers. X-rite is a general supplier of color measurement and color management tools, which can be used for color management in the textile industry.

1. i1Pro 3 Plus is designed for a wide variety of textiles materials; i1Pro 3 Plus has built-in color management software that allows one to calibrate and create customized profiles.
2. i1iO Automated Measurement Table is another tool that can be integrated with i1Pro 3 Plus for automatic color scanning/profiling to speed upthe process and reduce the risks of improper measurement.

X-rite has developed inline measuring systems for many industries to measure the color without stopping production. The carpet measurement system and ERX30 spectrophotometer are two products for inline measurement in the textile industry for apparel, dye houses, textile vendors, and textile printers.

### 4.5.9  MatchPrint II by kodak polychrome

It has been developed to meet the needs of inkjet printing. Matchprint II has been designed to exploit, at best, the functions and performances of the various printers with an image motor rendering that allows printing immediately without waiting (on-the-fly ripping), also at the maximum speed provided today by the Ms printers.[26]

### 4.5.10  Datacolor

Datacolor, founded in the 1970s, has developed color measuring tools and software for better management of color. Datacolor has developed various color measuring instruments, that is, spectrophotometers (both benchtop and portable), along with color management solutions. Management solutions are for the paint and textile industry. Dyer color formulation software: Match Textile is one of the advanced and comprehensive color matching software. Match Textile is based on a modular approach feature to be added to the package according to specific customer needs. The basic level serves all general color matching needs and comes bundled with Datacolor tool (spectrophotometer) color and QC software. The key features of multiilluminate matching, customer-specific matching parameters, smart recipe matching, dye compatibility management, and 3D graphics are key features.

### 4.5.11  Atexco digital

Atexco Digital has been deeply engaged in the textile industry for 30 years and is a complete solution supplier for textile digital printing. Atexco is aimed at the characteristics of the textile digital printing production process, such as changeable ink and substrate, the large amount of CMYK + N (spot color) color mixing, complex color rendering rules, and long color processing process, after years of research, Atexco has gradually built a CMS framework in line with the textile digital printing industry, and created an application scenario integrating color measurement technology and color processing scheme. Atexco can provide effective solutions in color measurement, measurement error processing, color calculation, soft proofing, color tracking, machine color feature consistency, etc. The Ajet RIP or vega print are two well-known Atexco printing software tools within the module of color management function. Besides, there are some other software, such as ATsoftproof and ATmatcher.

Analog printing software ATsoftproof, also known as soft proofing software, can simulate printing and preview the effect of printing, whereas ATmatcher is used for color consistency across multiple machines. There are different regions of ATsoftproof, as shown in Fig. 4.5; the menu toolbar is one of the common sections as any software in daily life; besides, there is the simulation parameter settings

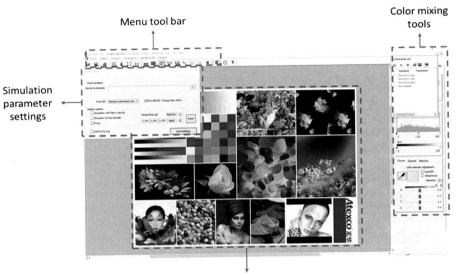

**Figure 4.5** Different regions of windows of ATsoftproof (Softproofing).

section, image palette, preview of the main screen, and color mixing tools. It is used for color palettes and proofing previews. The different regions of ATmatcher are shown in Fig. 4.6; they include data measurement and loading, integrated chromatic aberration assessment, data analysis interface, and color difference distribution. It is a special color-tracking software, and it is used for processing data between different printers.

## 4.6 Future trends

Color can surprisingly be challenging to be right as human perception varies from person to person. Moreover, fast fashion's changing needs require many practical tools and technologies to meet the market speed. Moreover, recently there has been a great demand for green purchases,[27,28] the software that may provide comparisons of the eco-friendliness of textile materials would be acknowledged. Also, there has been an interest to use recycled textiles, or textile production from waste,[29,30] thus, algorithms are needed to meet the industrial requirements of effective usage of textile waste. Thus, conventional tools could delay product launch and add to the business cost. Effective color measurement and management requires the right technology and tool for businesses to remain competitive in the market. Pandemic scenarios have emphasized the need for digitalization of color management. Now and onwards, industries must adopt modern technologies and customization to drive the industries. Futures industries have to align with color evaluation technologies

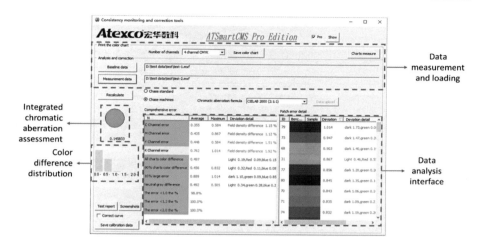

**Figure 4.6** Different regions of windows of ATmatcher (ATSmartCMS Pro Edition).

that use data to minimize the usage of color (ink). Instrument portability is a more significant challenge for color industries. It has advanced a lot, but there is still a need for immediate, accurate color measurement tools that saves many resources, that is, faster turnaround. The most important for future industries is that cloud-based color measuring, managing, and designing systems that could revolutionize the industries to new heights.

# References

1. Ujiie, H. *Digital Printing of Textiles;* Woodhead Publishing, 2006.
2. Javoršek, D.; Javoršek, A. Colour Management in Digital Textile Printing. *Color. Technol.* **2011,** *127*, 235−239.
3. Khatri, Z.; Sheikh, G.Y.; Brohi, K.M.; Uqaili, A.A. Effective Colour Management for Textile Coloration-An Instrumental way Towards Perfection. In *Proceedings of the Proceedings of the 2011 International Conference on Industrial Engineering and Operations Management*, Kuala Lumpur, Malaysia, 2011; pp. 1073−1079.
4. Xin, J. H. *Total Colour Management in Textiles;* Woodhead Publishing, 2006.
5. datacolor. Book Two of Color Management: The Triplet of Color: Light, Object and Observer. 2019, 14.
6. datacolor. Book Three of Color Management: Color Measurement—The CIE Color Space. 2019.
7. Oulton, D. Technology for colour management. In *Proc 1st International Conference on Digital Technologies for the Textile Industries;* University of Manchester, Institute of Science and Technology, 2013; pp 1−10.
8. Williams, S. Practical Colour Management. *Opt. Laser Technol.* **2006,** *38*, 399−404. Available from: https://doi.org/10.1016/j.optlastec.2005.06.001.

9. Halepoto, H.; Gong, T.; Noor, S.; Memon, H. Bibliometric Analysis of Artificial Intelligence in Textiles. *Materials* **2022**, *15*, 2910.
10. Wang, H.; Halepoto, H.; Hussain, M. A. I.; Noor, S. Cotton Melange Yarn and Image Processing. In *Cotton Science and Processing Technology: Gene, Ginning, Garment and Green Recycling*; Wang, H., Memon, H., Eds.; Springer Singapore: Singapore, 2020; pp 547–565.
11. Memon, H.; Khoso, N. A.; Memon, S.; Wang, N. N.; Zhu, C. Y. Formulation of Eco-Friendly Inks for Ink-Jet Printing of Polyester and Cotton Blended Fabric. *Key Eng. Mater.* **2016**, *671*, 109–114. Available from: https://doi.org/10.4028/http://www.scientific.net/KEM.671.109.
12. datacolor. Book Five of Color Management:Measurement Techniques in Colorimetry. 2019, 9.
13. Wong, W. K. *Applications of Computer Vision in Fashion and Textiles*; Woodhead Publishing, 2017.
14. datacolor. Book Four of Color Management: Color Distances, Metamerism and Practical Color Equations. 2019, 15.
15. datacolor. Book One of Color Management: Color Fundamentals and Color Perception. 2019.
16. Jin, X.; Memon, H.; Tian, W.; Yin, Q.; Zhan, X.; Zhu, C. Spectral Characterization and Discrimination of Synthetic Fibers with Near-Infrared Hyperspectral Imaging System. *Appl. Optics* **2017**, *56*, 3570–3576. Available from: https://doi.org/10.1364/ao.56.003570.
17. Broadbent, A. D. *Basic Principles of Textile Coloration*, Vol. 132. Society of Dyers and Colorists: Bradford, UK, 2001.
18. Tyler, D. J. Digital Printing Technology for Textiles and Apparel. *Computer Technology for Textiles and Apparel* **2011**, 259–282.
19. Colorthink Pro. http://www.chromix.com/colorthink/?-session = SessID:B6BF2D8A187e20ED4FjoKqBA0635.
20. Colorvalvet Pro. http://www.chromix.com/colorvalet/pro/?cvpro = true&-session = SessID: B6BF2D8A187e20ED4FjoKqBA0635.
21. Aquario Design. https://aquariodesign.com/.
22. Aquario Textile Design https://aquariodesign.com/aquario-textile-designer/.
23. Aqurio Suits. https://aquariodesign.com/product-suites/.
24. Textile Print by Adobe Pdf Print Engine https://www.adobe.com/products/textiles.html.
25. Ergosoft. https://www.ergosoft.net/rip-software/.
26. MatchPrint II. https://www.msitaly.com/00/w00000001/software.html.
27. Chen, L.; Qie, K.; Memon, H.; Yesuf, H. M. The Empirical Analysis of Green Innovation for Fashion Brands, Perceived Value and Green Purchase Intention—Mediating and Moderating Effects. *Sustainability* **2021**, *13*, 4238.
28. Memon, H.; Jin, X.; Tian, W.; Zhu, C. Sustainable Textile Marketing—Editorial. *Sustainability* **2022**, *14*, 11860.
29. Memon, H.; Ayele, H. S.; Yesuf, H. M.; Sun, L. Investigation of the Physical Properties of Yarn Produced from Textile Waste by Optimizing Their Proportions. *Sustainability* **2022**, *14*, 9453.
30. Wang, H.; Memon, H.; Abro, R.; Shah, A. Sustainable Approach for Mélange Yarn Manufacturers by Recycling Dyed Fibre Waste. *Fibres Text. East. Eur.* **2020**, *28*, 18–22.

# Digital image design and creation of printed images on textile fabrics

Bewuket Teshome Wagaye[1,2], Degu Melaku Kumelachew[1,2] and Biruk Fentahun Adamu[1,2]
[1]Textile Engineering Department, Ethiopian Institute of Textile and Fashion Technology, Bahir Dar University, Bahir Dar, Ethiopia, [2]Key Laboratory of Textile Science and Technology, Ministry of Education, College of Textiles, Donghua University, Shanghai, P.R. China

## 5.1 Introduction

Siemens AG, a German firm, developed the first inkjet printer in 1951. Then, in the early 1960s, the United States, Europe, and Japan competed in an advancement race for an ink-delivering regulator mechanism in a charge-controlled system. This race resulted in the development of high-speed large-format printers for large-scale applications. During the 1970s, a printhead advancement competition amongst precision machinery manufacturers worldwide became intense, and numerous types of inkjet printers were introduced one after the other. In the early 1990s, inkjet printhead manufacturing technology advanced quickly, and low-priced, as well as small-sized printers started to be marketed in street-side electrical appliance stores.[1]

These days Hewlett-Packard Co., Canon Inc., Seiko Epson Corp., and Lexmark International Inc. are familiar and active producers of consumer-use printers.[1] Active manufacturers of textile inkjet printers are shown in Table 5.1.

The method of printing a design or an image directly onto a fabric surface by applying graphic design software is called digital textile printing (DTP).[2] It is derived from the advancement of inkjet technology. The design-colored pattern is made using inkjet technology by spraying tiny drops of various color inks into predefined microarrays on the fabric/garment.[3] Each of these arrays is presented by the smallest image element of the design, known as a pixel. DTP involves printing pixels in which each pixel makes up of different ink colors. In inkjet technology, the delivered liquid ink drops only contact the substrate. That is why it is a nonimpact printing method.[4]

Digital printing offers a number of advantages over traditional printing, including speed, shorter lead time, customization options, cost reduction flexibility, and cleanliness in the production of cotton products.[5,6] The digital inkjet printing method requires particular ink cleanliness and conductivity. Because of this, none

**Table 5.1** Textile inkjet printhead manufacturers. (The specifications mentioned are only for the specified direct to fabric printer models).

| S. No. | Manufacturer | Effective print width (mm) | Max. print width (cm) | Total nozzles (within effective width) | Printing resolution (DPI) | Media fabric type | Max. print speed | Drop volume (pL) | RIP software |
|---|---|---|---|---|---|---|---|---|---|
| 1 | Kyocera Corporation, Japan (KJ4B-0300) | 112.35 | 220 | 1,328 × 2 Channels (1328 × 2) | 300 × 600 × 2 Colors | Flag, Fabric banner, polyester, silk bolting cloth; BackLit fabric | 75–120 m/min | 5, 7, 12, and 18 | Fiery |
| 2 | EFI Reggiani, Italy (efi Reggiani BOLT) | – | 180–340 | – | Upto 600 × 4800 | Home textile, Sportswear, and footwear | 90 m/min | 5 upto 30 | Fiery |
| 3 | EFI, USA (VUTEk FabriVU 180) | – | 180 | – | Up to 2400 | flags, banners, backlit displays, etc. | 7 m²/min | 4 – 18 | Fiery |
| 4 | Konica Minolta IJ Technologies, Inc., Japan (NASSENGER SP-1) | 72 | 185 | 1024 | 720 × 900 | Cotton, silk, flax, wool, nylon, polyester, blended fabrics, etc. | 58 m/min | 3.5 upto 35 | IJ-Manager |
| 5 | Kornit Digital, USA (Presto S) | – | 180 | – | 1000 × 800 | Cotton, polyester, cotton / polyester blends, Lycra, viscose, silk, leather, denim, linen, and wool... | 65 m/min | – | Compatible with any RIP software |
| 6 | Epson, Japan (ML-8000) | – | 184.4 | – | 1200 x 1200 | vinyl, banner, vehicle wrap, and other traditional signage prints | 5 m/min | – | – |
| 7 | Mimaki, USA (JV300 Plus series) | – | 161 | – | 1440 | Tapestry, flag, sports apparel, fashion textile, poster, wallpaper, and interior fabrics | 32.7 m²/h | 7 upto 35 | TxLink4 Lite or RasterLink6Plus |

*Source:* Extracted from manufacturers brochure.

of the conventional chemicals are directly used during ink production.[6] In conventional textile printing methods, such as screen printing, the designer is limited to the colors and repeat size that the screen setup allows. With digital printing, any number of color-produced designs can be printed digitally to achieve the same esthetic value.[4] Merely replacing the traditional set of screens allows the technology to be integrated as a substitute method, although it may not yet realize that the capabilities of DTP have conceptual possibilities that reach far beyond the technical. DTP has a virtually unlimited color choice and repeats size, thus allowing for vastly alternative esthetic decisions.

DTP cuts the production stages hence the processing time (Fig. 5.1). Fast and easy design corrections are possible since the whole process is digitized. In addition, the amount of wastewater discharged to the environment is reduced. Since the textile industry is considered one of the most polluting industry, thus, any effort to make it environmentally friendly is worthy.[7] Textile factories that use DTP machines can minimize water consumption by upto 90% and electric consumption by 30%. It enables unlimited color and design options, and DTP's final printed fabric quality (clarity and color) exceeds conventional ones. Furthermore, it is perfect for product differentiation, individuation, short runs, and customized designs.[2] Since DTP eliminates color differentiation and screen preparation, it has lower emission rates than screen printing.[8]

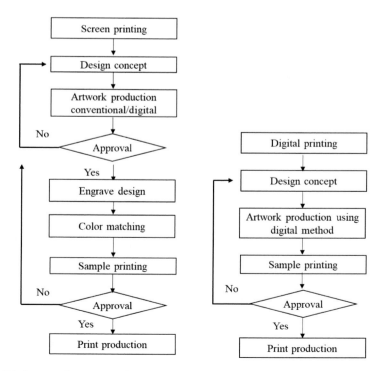

**Figure 5.1** Process flow comparison between screen printing and digital printing.

## 5.2 Image capture and display

The rapid development of computer-aided design (CAD) systems for print design has significantly impacted the quality and reproducibility.[9] It is possible to prepare design patterns for DTP on several standard graphics-based programs. Nevertheless, patterns need distinct editing necessities. Therefore, it is a usual practice to use proprietary software. The software creates an opportunity to edit the textile design. In addition, it can also be augmented with additional characteristics to create an integrated color management system (CMS). The CMS integrates an image scanner, monitor display, and an inkjet printer, which enables precise and consistent replication of the colors of the original design. Recently additional features have been added to the design systems. In addition to recording the color information of the printed design, the feature allows recording the 3D surface texture properties of knitted, woven, and nonwoven fabrics. Most design systems allow using of all operating platforms for professional reprographics. The most popular operating system is Microsoft Windows. Table 5.2 shows an example of existing design software and operating systems. New design systems include 3D modeling elements like examining the consequence of the design as a garment or even draped on a figurine with rendered shading effects.[10]

Table 5.2 CAD systems applicable for editing textile print patterns.

| Design software | Operating system | Developer (website) |
|---|---|---|
| Adobe lightroom | MS windows, macOS, iOS, Android, and tvOS (Apple TV) | Adobe (USA) (https://www.adobe.com/) |
| Skylum luminar AI | MS windows and macOS | Skylum (USA) (https://www.skylum.com/) |
| Adobe photoshop | Windows and macOS | Adobe (USA) (https://www.adobe.com/) |
| DxO photolab 4 | Microsoft windows and OS X | DxO labs (France) (https://www.dxo.com/) |
| ON1 photo raw | MS windows | ON1 (USA) (https://www.on1.com/) |
| Corel paintshop pro | Windows 7 and later | Corel Corporation (Canada) (https://www.corel.com) |
| Serif affinity photo | MS windows | Serif (United Kingdom) (https://www.serif.com) |
| ACDSee photo studio ultimate | Windows, macOS, and iOS | ACDSee (USA) (https://acdsee.com) |
| GIMP | Linux, macOS, and microsoft windows | GIMP (USA) (https://www.gimp.org) |
| Canva | Windows and Mac OSX | Canva (Australia) (https://www.canva.com) |

*Source*: Extracted from developers' websites.

Digital image design and creation of printed images on textile fabrics

The design image can be drawn manually onto transparent film for photographic development of diapositive images, or CAD systems are applied with the help of a design input scanner.[11]

A digital image is a 2D rectilinear array of pixels. CAD systems to generate pattern and colorway displays have become a standard procedure in printed textile production. Fig. 5.2 shows present and future developments in pattern design, screen production, and recipe formulation. These systems typically allow the capture of the designer's work by a flatbed scanner or, for larger patterns, a rotary scanning device. Even if it is less common color cameras are also used for image capture. The data thus attained represents the location and color of each individual pixel of the design, and the number of these will depend on the definition at which the artwork is scanned (usually a few hundred dots per inch (dpi)).[12] The lowest scan definition needed for a textile design is usually 300 ± 600 dpi, but modern scanners capture up to 2000 dpi.[10]

An inkjet printer needs the appropriate software for image scanning, color separation, and color management.[14] Most printers require a flattened tiff file in either RGB or Lab color with a resolution of no more than 300 dpi at the size the design is to be printed.[15]

### 5.2.1 Image capture

There are three types of images. The easiest one is called a binary image, which takes only black and white or 0 and 1. It consists of a 1-bit image, and a pixel is represented using 1 binary digit. They are usually used for general shapes or

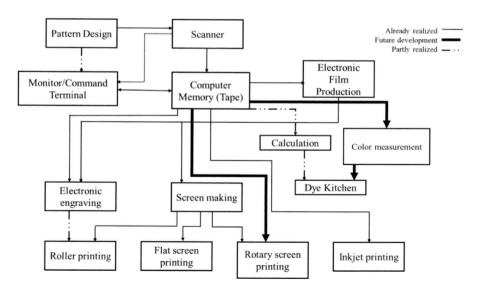

**Figure 5.2** Scheme of present and future developments in pattern design/screen production/recipe formulation.[13]

outlines. Threshold operation is used to generate binary images. As soon as a pixel is above the threshold value, it is turned white ('1') while pixels below the threshold value are turned black ('0'). The second one is a grayscale image. They have only one color (monochrome). Any information about the color is not included in grayscale images. Each pixel determines the existing various gray levels. A standard grayscale image comprises 8 bits in each pixel and 256 different gray levels. The third one is color images. 3-band monochrome images have a different color in each band; the real data preserved in the digital image are called color images. Gray-level data is contained in each spectral band in the color images. The images are represented by the colors red, green, and blue (RGB images). The bit depth of each color image is 24 bits per pixel, corresponding to 8 bits for each of the three-color bands.

Digital image files for DTP should be compatible with a variety of operating systems, including Mac and Windows, and formatted with the right resolution, color mode, size, and scale. With particular image editing methods, CAD technologies became easier and cost-efficient, opening up textile design technology to a wider audience and allowing creativity and efficient design possibilities. Users may digitally produce patterns within the application using a mouse or stylus pen or scanning existing artwork into the computer. Most textile designs are digital until they are printed, enabling easier editing and engineering textile patterns for final 3D items and producing distinctive and new textiles.[16]

The graphics sector recently developed a uniform communication protocol known as TWAIN, regardless of the type of scanner used. TWAIN is an application programming interface and communication protocol that control communication among software and digital imaging devices. The first release was in 1992 and was last updated in 2015.

Modern flatbed scanners use a contact image sensor. The scan head, a light sensory array, is passed over the pattern design. The lamp is also illuminated in the design. Light is bounced back at the light sensory array (charge-coupled device, CCD) using a series of mirrors and lenses. The brighter the light reflected, the greater the electrical charge. The lens splits the image into three colors, that is, R, G, and B, and an electrical charge is measured. These measurements are changed into binary code. The binary code represents a digital version of the design.

The photosensor array of digital cameras comprises a mosaic of red, green, and blue cells commonly organized in a so-called "Bayer pattern," with two G cells for every R or B. Most cameras execute "onboard" demosaicing calculations, using algorithms of varying difficulty to interpolate the two missing values (e.g., the R and B values at a G cell location) for each pixel. The combined data for the individual sensors (known as RAW format) can, however, be stored and processed on an external computer. Foveon and Texas Instruments recently developed a camera chip where each individual cell produces an RGB response, eliminating the requirement for demosaicing algorithms.[10] The Foveon X3 digital camera image sensor is produced by Dongbu Electronics and engineered by Foveon, Inc., which is currently part of Sigma Corporation. It employs a photodiode array having three vertically stacked photodiodes. Each photodiode has a different spectral sensitivity, enabling response to different wavelengths since each can show a different spectrum.[17] After that, the signals from the three photodiodes are converted to additive color data and subsequently to a standard RGB color space. The X3 sensor was initially used in the Sigma SD9 DSLR camera in 2002 and was later adopted by the SD10, SD14, SD15, SD1 (also SD1 Merrill), the original mirrorless compact Sigma DP1, and Sigma DP2 in 2008 and 2009, and the Sigma DP2 Quattro series in 2014, and the Sigma SD Quattro series from 2016.[18]

## 5.2.2 Pattern data encoding compression and storage

An image is the visualization of a BMP. A BMP is a collection of pixels. A pixel is represented with a single color. The color depth of an image denotes the number of bits used to store each pixel's color. Different systems of digital image encoding have been developed. Many keep track of the RGB values of all pixels in a particular fashion. For example, all MS Windows monitors use the BMP display system for pixel information, with RGB values requiring three bytes of data per pixel, which are read line by line from left to right of a displayed image beginning at the lower most point. This leads to big files for high definition and large pattern repeats, which, in addition to requiring large memory, results in slower access, display, or transmission times.[10] The alternative method of encoding comprises Run-length encoding, which is lossless data compression in which the same data value occurs in many consecutive data elements are stored as a single data value and count (RGB values, number of consecutive pixels with these values), rather than as the original run. It is suitable for encoding traditional textile designs with big blocks of consistent color. Other encoding methods allow data compression; nevertheless, when the complete file is restored, the recovered data should match the original exactly.

Image compression minimizes the data required to represent a digital image by eliminating repeated data. Before storage or transmission, this modification is performed. In digital format, images can be represented in a variety of methods. Encoding the contents of a 2D image in raw BMP (raster) format is typically inefficient and might consequence in huge files. Most image file formats use compression because raw image representations typically need substantial storage space and correspondingly long transmission times during file uploads/downloads. Table 5.3 shows some common image encoding formats. The compressed image is decompressed later to reassemble the original image information or a close approximation of it (lossy techniques).[19]

Data encoding formats bring additional information, so-called "metadata." It is a collection of information that specifies and gives details of the rights and management of an image. It enables data conveyance using an image file; hence other software and humans can comprehend it. Pixels of image files are produced by automatic capture using cameras or scanners. Internally, in formats like JPEG, DNG, PNG, TIFF, and others, metadata is embedded in the image file; externally, outside the image file in a digital asset management system or in a "sidecar" file (such as for XMP data) or an external news exchange format document as

Table 5.3 Some common encoding systems (file formats).

| Acronym | File format | File extension(s) | Brief explanation |
|---|---|---|---|
| APNG | Animated portable network graphics | .apng | A better system for lossless animation sequences. GIF has a lower performance. |
| AVIF | AV1 image file format | .avif | Due to its excellent capability and royalty-free picture format are a great option for both images and animated images. It compresses data significantly more efficiently than PNG or JPEG. |
| GIF | Graphics interchange format | .gif | For easy visuals and animations, the best option. For lossless and indexed still images, use PNG; for animation sequences, use WebP, AVIF, or APNG. |
| JPEG | Joint photographic expert group image | .jpg, .jpeg, .jfif, .pjpeg, . | For lossy, compression of still images is the most common. For precise reproduction of images, PNG is preferable; for better reproduction and higher compression, WebP/AVIF is better. |
| PNG | Portable network graphics | .png | For an exact replica of source pictures, or when transparency is needed, PNG is preferable over JPEG. |
| SVG | Scalable vector graphics | .svg | Perfect for user interface elements, icons, infographics, and items that need to be rendered precisely |
| WebP | Web picture format | .webp | A fantastic alternative for both images and animated images. With support for deeper color depths, dynamic frames, transparency, and other features, WebP enables better compression than PNG or JPEG. AVIF has a better advantage in terms of compression. |
| BMP | Bitmap file | .bmp | Bitmap. All Windows-based graphic displays use this format for digital images. File sizes are large. |
| ICO | Mircosoft icon | .ico, .cur | Microsoft created this icon set for Windows desktop icons. The data for each icon can either be a BMP image without the file header or a full PNG image based on BMP. Mostly applicable in textile design systems. |
| TIFF | Tagged image file format | .tif, .tiff | A universal file format with data tag based on BMP. Mostly applicable in textile design systems. |

mentioned by the International Press Telecommunications Council. Descriptive information about the visual material is included in the metadata. This can include the headline, the caption, and keywords.

Additionally, people, places, businesses, artwork, or objects are visible in the image. It also includes the creator's name, credits, and the visual content's underlying rights, such as model and property rights: additional rights usage terms, and extra information for licensing the image's usage. The date and place of production, user instructions, and more descriptions are also included. Metadata associated with an image file must remain intact. It is also important for simplifying workflow, quickly discovering digital files through online search or offline, and tracking image usage.

### 5.2.3 Digital color management systems and color communication

Color management is a controlled translation between different color devices. It allows the printer to predict the color output of their particular printing processes accurately. Device color variances are managed by the International Color Consortium (ICC) profile format. It is a standard describing devices linking their color to a mathematical model of human vision. Color profiling is a fantastic way to get close color matches, but it can be challenging to set up.[20] ICC profiles use an industry-standard file format for both Windows and Macintosh operating systems to capture all devices' color reproduction features (Fig. 5.3). ICC-compliant picture editing, illustration, page layout, web design, web browsing applications, and the drivers and raster image processors (RIPs) that monitor printers can read these profiles.

Photoshop 5 introduced standard working spaces during the 1990s. In Photoshop 4, a "standard" working space called Apple RGB was used; however, it could not be altered by the user. ICC profiles define standard working environment spaces. sRGB, ColorMatch RGB, and Adobe RGB are common graphic arts standard spaces (in order of increasing size). A gamma (contrast), color temperature (white point), and color gamut size are all part of a standard RGB working space profile. The gamut refers to the conceivable colors that a given system may represent. In its most basic form, a color space can be considered the combination of three fundamental colors (red, green, and blue) in a color additive system. When one mixes any of these hues, one can get a wide range of colors (this is the base of the RGB color model). Naturally, the colors the human eye can perceive are what one is interested in. Most existing color space models have limitations in covering the entire range.

Color management's primary objective is to achieve the accurate and reliable color of the actual footage, photographic print, or transparency to the final displayed or printed output.

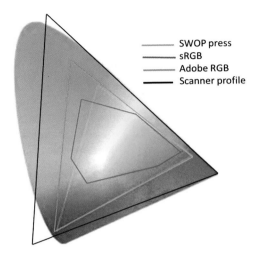

**Figure 5.3** Standard working spaces.

### 5.2.3.1 The source-destination-simulation model

There are four types of ICC profiles.

**Source profile;** It is a scanner or camera profile that defines the color space in which the image was made or now resides. It may also be a typical workplace profile. Preferably, all images having a camera or scanner profile would be transformed into a standard working space, and the standard working space profile would be included in files so that users could recognize it later. Unless what standard working space used was known, color matching is impossible.

**Standard working space profile**; It serves as a repository for data. Adobe Photoshop and other software support a variety of standard RGB working spaces, including Adobe RGB, ColorMatch RGB, and sRGB (in order of size). The largest, Adobe RGB, is now used by the majority of content providers. Standard CMYK working spaces from Adobe, such as U.S. Sheetfed Coated and U.S. Web Coated, can be used by users with CMYK workflows.

**Destination profile**; The device or process to which the file will be output is described in this profile. The destination could be a color monitor, an inkjet printer (for proofing), or an industrial printer (for final output).

**Simulation profile**; It outlines the equipment or procedure that the image should seem like while proofed. For example, the simulation profile would be the same as the production printer when proofed on a smaller inkjet printer.

The procedure depicted in Fig. 5.4 explains the role of the four profile categories. Consider color image files as "on the move" from a source to a destination to track the workflow (monitor, printer, proofer). A profile is used to depict each stage. "Assigning" profiles to a file identifies the device from which it was created, whereas "converting" modifies the color to match the output device.

Digital image design and creation of printed images on textile fabrics

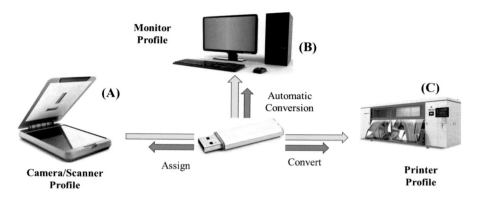

**Figure 5.4** The color model of source-destination simulation. A camera file (A) transformed from a digital camera profile to a standard working space profile (shown using the flash disk) and then exported to a monitor profile (B), a proofer with final output simulation, and a Model X Plus Atexco printer (C).

## 5.2.4  Input and output devices color calibration

To accomplish reliable and high-fidelity color measurement or reproduction in a color imaging system, the imaging device must be calibrated, which often entails two steps denoted by the words calibration and characterization. Calibration is the setting up of a process such that it can generate reproducible results. Calibration can entail maintaining the device's internal controls at their nominal settings (as digital cameras often do).

The link between the device coordinate (typically RGB or CMYK) and some device-independent color spaces, such as CIE XYZ, is defined by characterization.

Many flatbed and film scanners come with a calibration target and software allowing calibrating (bringing your scanner to a known value). The industry often refers to this as "linearizing" the device, essentially the same as "calibrating" it. If this option is offered, make use of it. Starting with a known value or standard is always an innovative approach. One could use the same scanner profile if calibrated regularly because it can always restore the scanner to its original configuration.

Targets for calibrating and profiling scanners are frequently similar or identical. Custom targets are included with some devices and profiling software applications, while others use common targets like Kodak, Agfa, Fuji, and others. A "Targe" measurement table estimates a scanner's precise color recognition capabilities. These targets are covered by ISO standard 12641, updated by the ISO committee with Part 2 in December 2019.

The "IT" target from Kodak is a well-known and widely used target. The LAB values for each color patch are also included in reference files (typically essential text files) provided by the producers of these targets. When the target is scanned, these values serve as a reference point for the color values obtained. A reference point, if one will, for creating a calibrated state or generating a customized profile.

In color calibration, the main goal is to accurately map and predict the printing process results. Calibration can take other device characteristics into account. The main factors affecting the calibrating of digital printing of textiles are a printer, RIP software, substrate, type and volume of ink/dye, and post finishing operations.

The first step in printer calibration is to know the volume of ink/dye absorbed by the fabric since it is mandatory to eject the optimum amount of ink on the fabric, followed by linearization, which decides the output levels of each color requirement to create linear ink coverage at levels ranging from 0% to 100%. The last step is ICC generation. The calibration software has the theoretical yield of what the printer will output. Color change may happen depending on dyes of different colors, the post finishing operations, pretreatment operations, and the change of substrate.[21]

## 5.3 Pixel and image creation using inkjet printers

The creation of tiny ink drops formed using mechanical pulses or thermal jet printers follows a striking pattern. In thermal bubble inkjet printers, ink is projected from the nozzle by heating. Tiny resistors in the print head heat up when an electric current passes through them. The ink vaporizes as a result of the heat. The ink forms a bubble, which forces a small amount of ink through the nozzle and onto the substrate. A vacuum is created when a bubble pops, pulling more ink into place for the next bubble. The drop breaks off to happen in microseconds. The drop tails are driven by the surface tension when the drop separates from the nozzle, which pulls it downward. Break-off is driven by the ink's outward motion and the nozzle's retraction during bubble collapse. Drop ejection typically produces the main drop and smaller trailing drops called satellites. On the substrate surface, the droplets usually create a roughly round spot. The volume of the droplets varies between 5 and 10 picolitres, based on the printhead feature and the type of the electrical driving pulse(s). Ink quality is essential; ink filters are positioned in each nozzle to prevent the passage of foreign matter to the heating resistor. Precision is necessary because it is related to the resolution of the finished product. Resolution is measured in dots per inch (dpi). In general, the higher the number, the better the resolution.

Analog or continuous-tone printing is a method of printing that may produce a full range of color levels (usually 256) at any point. Halftoning is required to reproduce pictorial images because binary or halftone printing applies a procedure with only two levels, ink or no ink. Grayscale printing refers to systems that can generate a few varying degrees of gray with each spot, employing many inks with varied pigment concentrations yet needing some spatial halftoning to replicate the entire picture density and color range. Each pixel should be produced by filling in only part of the pixel area in binary or grayscale digital printing, which requires halftoning.

The halftone pixel is sectioned into an array of N × N subpixels; then, subpixels are individually colored, not the whole pixels. Most graphic art scanners, for example, split pixels into a 12 by 12 array of subpixels. Then the pixel can be exposed (ink deposited) from 0 to 144, allowing it to choose from 145 different average reflectance levels. Unfortunately, the exposing patch should be very small, just 1/12 of pixel width, resulting in higher writer resolution. A printer resolution of 200 × 12 = 2400 subpixels per inch is required to write a 200-line screen image. This is why graphic arts digital plate setters and halftone proofers have 1800−3000 subpixels per inch resolutions. It is required to match the tone reproduction capacity of digital silver halide prints created at a resolution of 200 pixels per inch. A resolution of 16 × 500 (8000 subpixels per inch) would be required to match the quality of 500 pixels per inch and 256-level (8 bits per color) photographic prints. Certainly, such high-resolution writers can produce excellent text and graphics, far superior to photographic systems.

The fact that binary printing techniques require substantially more data to be transmitted to the printer is a second drawback. The visual arts system should send 1 bit per subpixel rather than 8 bits per pixel per color; for reaching similar 256 levels of density, 256 bits per pixel per color is required, a factor of 256/8 = 32 times more. This is why digital graphic arts printers typically have a high-speed RIP and apply several laser beams to produce many subpixels simultaneously.

If a digital printer can write a few distinct levels of density instead of only two of a binary system, then fewer subpixels are required to achieve the required number of pixel densities. As a result, there are significant benefits in lowering the needed writer resolution and the data rate. Hence, inkjet printing requires multiple-density inks or, better yet, multiple drop size capability.[22]

Reprographic printers can print at resolutions of upto 1200/1440 dpi. Models with 300/360 or 600/720 dpi are commonly used for textile printing. The claimed dpi of a printer and the real definition of the printed pattern are confusing, measured in pixels per inch (PPI), which is dependent on the size of the superpixel matrix. As a result, the printed image definition of a 300-dpi printer with a 4 × 4 matrix is similar to a 600-dpi printer with an 8 × 8 matrix, that is, 75 PPI. The finest screen mesh size (LPI; lines per inch) used for traditional printing is typically 120 LPI (48 rasters), and for an average design, 80 ± 100 LPI (32 40 rasters), which can, in theory, be equated with a jet printer working at 600 dpi and a 6 × 6 superpixel matrix.

The printhead's maximum frequency response determines the speed of textile inkjet printing as they are scanning along the width of the fabric (mostly 12−25 kHz; modern piezo types are faster than bubblejets) and whether the print scanning is single-pass or bidirectional. The width of the fabric strip is exposed in each scan; hence the length of the fabric is carried forward after each printhead pass, and the print definition and superpixel matrix size are additional factors. As a result, higher productivity and optimal quality are improbable. Despite advancements, the speed of smaller jet printers is yet only 0.1 ± 0.3 m/min and 0.5 ± 1.0 m/min for larger, more expensive units, compared to, say, 30 ± 50 m/min for a rotary screen machine. Epson Monna Lisa ML-8000 single pass Direct-to-Fabric printer achieved a maximum print speed of 290 m²/hr with a 300 × 600 dpi.[23] Konica Minolta single-pass textile printer NASSENGER SP-1 model achieved a high speed of 4300 m²/h for the print definition of 720 × 540 dpi.[24] Finally, the time the drive data can be processed and communicated can impact printing speed. As a result, a devoted high-specification PC or workstation unit is typically used to monitor one or more printers, where pattern design and editing take place elsewhere.

## 5.4 UV and latex curing methods

UV-curable inks respond to UV radiation from a light source. Curing lamps or LED curing devices emit ultraviolet light within a limited bandwidth, usually shortly after the ink has been applied to the fabric surface. Curing is a chemical interaction between the ink's monomers and active binders. As a result, the ink is completely dry or "cured" right away.

UV inks can often print faster than aqueous and solvent inks. This is especially critical in high-volume applications. UV inks are also less prone to nozzle clogging because they are cured. There are further methods available to ensure that the thickness of the UV-cured ink applied can be managed to finer tolerances.

A new line of flexible UV inks, such as SuperFlex, has been introduced, providing the requisite flexibility for printing vehicle and fleet graphics. UV-curable inks can also be used to print directly on stiff surfaces without using a self-adhesive coating. Volatile organic compounds (VOCs) are considered very harmful, and thus it is always preferred to minimize them.[25] This newer generation of UV-curable inks has minimal VOCs and the adaptability needed to increase their applications significantly.

Latex wide format printers employ a resin-based ink fixed to the media by an inline heat process. A water-dispersed polymer ink is referred to as "Latex" These are man-made polymers (as opposed to the latex associated with the natural rubber made from trees). Although resin or polymer ink would be a more technically correct title for the ink, the term latex has some marketing advantages, as it is well known that green marketing is used to get competitive advantages by most firms these days.[26,27] Latex inks do not have the strong odor of solvent inks. UV inks have a distinct odor of their own. Latex inks have a low VOC concentration, significantly benefiting printers used in offices and retail print shops.

Heat is required to evaporate the latex ink's water content, allowing it to bond permanently to the fabric's surface. Convection increases airflow while concentrating heat, using less energy to dry the ink. The ability of today's latex printers to evaporate the water content of the ink has been improved, allowing for faster printing. Latex printers have become more stable and user-friendly and need less energy. Latex is considered a green printing method because of its low VOC content and rapid drying time.

## 5.5 Printing machine control

The operator can use a specific user interface to control and manage the printing process, which displays the monitor's real-time status of the various machine functions. A number of these characteristics can be specified at the start of the design process on the CAD station. All of the free printing parameters, such as printing width and length, single or bi-directional printing, and so on, can be changed or chosen by the printer operator. The printed meters are displayed in real-time on three screens, with relative hourly production statistics divided into linear and square meters.

A real-time visual representation of the current situation makes controlling and administering the colors supplied into the printer simple.

In addition to levels, other messages reveal ink status, signaling to the operator not just when cartridges are empty but also any problems caused by loading mistakes.

Monna Lisa digital textile printer uses the proprietary RIP (raster image processor) ".ROB" file format, making it possible to print designs developed using different graphic design systems.

This allows the printing company to immediately print designs created by outside companies or other printing companies and assist those organizations that use other brands of colorways software and printing systems for their internal operations.[28]

## 5.6 Printing head performance monitoring

The precise forecast of an inkjet printer or printhead performance was difficult until the printed output was produced. Accurate data is, however, required far earlier in the process. As a result, major inkjet head manufacturers rely on Quality Engineering Associate's (QEA) IAS products to check jetting straightness, drop position, drop size, satellite dots, missing dots, and other essential quality indicators. IAS is a comprehensive suite of image analysis tools. The IAS from QEA measures prints quality on various substrates, including fabrics, glass, and ceramics. Banding, jitter, and coalescence are metrics IAS systems use to assess overall printer performance.[29]

The built-in ink nozzle is driven by a printing signal in inkjet textile printing, and the ink in the nozzle is ejected by exerting pressure to form a pattern on the fabric surface. The printed image is sharper when the ink drop is smaller. As a result, to ensure the inkjet printer's printing quality, a pressure measuring element with high precision and sensitivity must be used to manage the pressure delivered to the ink to obtain the perfect inkjet quality. Microsensor can provide a number of measuring solutions for inkjet printers' pressure monitoring needs.[30]

Using waveform and voltage-based dynamic drop tuning technology for each nozzle in the printhead helps reduce drop volume and drop speed fluctuation across and from the printhead to the printhead. As a result, color density variations and banding are minimized, and changes in nozzle performance overtime are successfully regulated. Cold switching reduces heat generation within the printhead, reducing the influence of associated fluid viscosity fluctuations on printing performance.[31] Voltage trimming is provided by Xaar AcuChp technology for uniform drop volume and velocity. It reduces installation and setup time by improving drop uniformity within and between printheads. Regarding image quality and substrate flexibility, the XaarDOT (Xaar Drop Optimization Technology) variable drop printhead is very flexible in allowing clients to choose the drop size or sizes to use for a task. Image resolution and actual dots per inch are determined by drop size.

XaarSMART technology is included in the Xaar 2001 + series of printheads, which informs ink temperature and printhead situation in real time so that printer performance may be readily modified to deliver consistent print quality throughout a production run.

The XaarGuard is a nickel alloy nozzle guard that increases the printhead's resilience and protects it from mechanical damage. Rather than using internal electronics, it provides a safe pathway for electrostatic discharge. The ink buildup on the nozzle plate is avoided to reduce maintenance intervals. The nozzles are recessed 250 μm below the nozzle guard surface to protect against impact. This increases nozzle life, uptime, and print quality.

## 5.7 Future prospects

Artificial intelligence has significantly important for all textile applications.[32,33] The incredibly fast speed and small dimensions of microelectromechanical systems make them impossible to detect or monitor, making developing a new inkjet printhead technology problematic. Océ tools used a combination of various computer simulation technologies to understand the impact of inkjet design parameters on the performance of the printhead.

The goal for Océ when designing the new printhead was to double the speed while retaining the print quality and consistency at levels equivalent to or better than the previous generation. They also aimed to keep the driving voltage minimum to save energy and the cost of electronics. Doubling the speed of printhead's meant the driving electronics frequency also increased to several tens of kHz. Maintaining a high drop speed was vital to preserve print quality at this speed. The tail is usually responsible for limiting drop speed because too-long tails break into satellite drops, lowering quality. Another important element in boosting printhead speed is refilling the printhead at a higher frequency without wetting the nozzle to the point where print quality is compromised. The nozzle's free-surface flow and the channel's acoustics must be tailored to fast refill the nozzle to shoot drops at a high rate; however, this causes overfilling of the nozzle, resulting in a moist nozzle plate. A wet nozzle might alter drop formation based on the wet layer thickness and drop size.

5-micron wetting, for example, will have little effect on a 50-micron droplet rather than a significant impact on a 10-micron droplet.[34]

## 5.8 Conclusion

DTP created a huge opportunity for design flexibility, shortening processes, and improved quality. Digital image design using computer vector files helps incorporate unlimited colors into the design. Compared to conventional textile printing, it eliminates the screen preparation step, saving time and energy. The image formation by nozzles helps in producing unlimited intricate patterns. The main limitation of DTP is lower production speed. Currently, researchers are focusing on doubling the speed of nozzle ejection.

## References

1. Abe, T. Present State of Inkjet Printing Technology for Textile. *Adv. Mater. Res.* **2012**, *441*, 23−27. Available from: https://doi.org/10.4028/http://www.scientific.net/AMR.441.23.
2. Choi, S.; Cho, K. H.; Namgoong, J. W.; Kim, J. Y.; Yoo, E. S.; Lee, W.; Jung, J. W.; Choi, J. The Synthesis and Characterisation of the Perylene Acid Dye Inks for Digital Textile Printing. *Dyes Pigm.* **2019**, *163*, 381−392. Available from: https://doi.org/10.1016/j.dyepig.2018.12.002.
3. Memon, H.; Khoso, N. A.; Memon, S.; Wang, N. N.; Zhu, C. Y. Formulation of Eco-Friendly Inks for Ink-Jet Printing of Polyester and Cotton Blended Fabric. *Key Eng. Mater.* **2016**, *671*, 109−114. Available from: https://doi.org/10.4028/http://www.scientific.net/KEM.671.109.
4. Thonggoom, O. Digital Textile Printing. *Srinakharinwirot Sci. J.* **2012**, *28*, 2555.
5. Litvinov, O. Russia's Diverse Digital Textile Printing Market. (accessed Feb 28).
6. Kan, C. W.; Yuen, C. W. M.; Tsoi, W. Y. Using Atmospheric Pressure Plasma for Enhancing the Deposition of Printing Paste on Cotton Fabric for Digital Ink-Jet Printing. *Cellulose* **2011**, *18*, 827−839. Available from: https://doi.org/10.1007/s10570-011-9522-2.
7. Memon, H.; Khatri, A.; Ali, N.; Memon, S. Dyeing Recipe Optimization for Eco-Friendly Dyeing and Mechanical Property Analysis of Eco-Friendly Dyed Cotton Fabric: Better Fixation, Strength, and Color Yield by Biodegradable Salts. *J. Nat. Fibers* **2016**, *13*, 749−758. Available from: https://doi.org/10.1080/15440478.2015.1137527.
8. Ren, J.; Chen, G.; Li, X. A Fine Grained Digital Textile Printing System based on Image Registration. *Comput. Ind.* **2017**, *92−93*, 152−160. Available from: https://doi.org/10.1016/j.compind.2017.08.003.
9. Broadbent, A. D. *Basic Principles of Textile Coloration;* Society of Dyers and Colourists: Bradford, England, 2001.
10. Dawson, T. L. 9 - Digital Image Design, Data Encoding And Formation of Printed Images. In *Digital Printing of Textiles;* Ujiie, H., Ed.; Woodhead Publishing, 2006; pp 147−162.

11. Holme, I. *Handbook of Technical Textiles;* Woodhead Publishing Limited in association with The Textile Institute: Cambridge, England, 2000.
12. Dawson, T. L. *Textile Printing*, 2nd ed.; Society of Dyers and Colourists: Bradford, England, 2003.
13. Rouette, H.-K. *Encyclopedia of Textile Finishing;* Springer: Aachen, Germany, 2000.
14. Shang, S. M. 14 - Process Control in Printing of Textiles. In *Process Control in Textile Manufacturing;* Majumdar, A., Das, A., Alagirusamy, R., Kothari, V. K., Eds.; Woodhead Publishing, 2013; pp 339–362.
15. Bowles, M. *Print, Make, Wear Creative Projects for Digital Textile Design;* Laurence King Publishing Ltd: China, 2015.
16. Polston, K.; Parrillo-Chapman, L.; Moore, M. Print-on-Demand Ink-Jet Digital Textile Printing Technology: An Initial Understanding of User Types and Skill Levels. *International Journal of Fashion Design, Technology and Education* **2014**, *8*, 87–96. Available from: https://doi.org/10.1080/17543266.2014.992050.
17. Jin, X.; Memon, H.; Tian, W.; Yin, Q.; Zhan, X.; Zhu, C. Spectral Characterization and Discrimination of Synthetic Fibers with Near-Infrared Hyperspectral Imaging System. *Appl. Optics* **2017**, *56*, 3570–3576. Available from: https://doi.org/10.1364/AO.56.003570.
18. Foveon X3 sensor. https://en.wikipedia.org/wiki/Foveon_X3_sensor (accessed March 03).
19. Marques, O. Image Compression and Coding. In *Encyclopedia of Multimedia;* Furht, B., Ed.; Springer US: Boston, MA, 2006; pp 299–306.
20. Adams, R.; el Asaleh, R.; Habekost, M.; Lisi, J.; Seto, A. Digital Photography for Graphic Communications.
21. Dutta, S.; Bansal, P. Cotton Fiber and Yarn Dyeing. In *Cotton Science and Processing Technology: Gene, Ginning, Garment and Green Recycling;* Wang, H., Memon, H., Eds.; Springer Singapore: Singapore, 2020; pp 355–375.
22. Owens, J.C. A Tutorial on Printing. https://www.imaging.org/site/IST/Resources/Imaging_Tutorials/A_Tutorial_on_Printing/IST/Resources/Tutorials/Printing.aspx?hkey=c1b33ff1-a402-44dd-ba41-670b7b42d26b&TemplateType=P (accessed March 04).
23. EPSON. Epson Monna *Lisa ML-8000 Direct-to-Fabric Printer* [Brochure]; Epson America, Inc, 2021.
24. KONICA MINOLTA. Konica *Minolta Single-Pass Textile Printer Nassenger sp-1* [Brochure]; KONICA MINOLTA, Inc. Industrial Print Business Unit, 2021.
25. Memon, H.; Yasin, S.; Ali Khoso, N.; Hussain, M. Indoor Decontamination Textiles by Photocatalytic Oxidation: A Review. *J. Nanotechnol.* **2015**, 104142. Available from: https://doi.org/10.1155/2015/104142 *2015.*
26. Memon, H.; Jin, X.; Tian, W.; Zhu, C. Sustainable Textile Marketing—Editorial. *Sustainability* **2022**, *14*, 11860.
27. Chen, L.; Qie, K.; Memon, H.; Yesuf, H. M. The Empirical Analysis of Green Innovation for Fashion Brands, Perceived Value and Green Purchase Intention—Mediating and Moderating Effects. *Sustainability* **2021**, *13*, 4238.
28. MONNA LISA. https://www.monnalisadtp.eu/monnalisa (accessed March 04).
29. Ink-jet print quality. https://www.qea.com/resources/application-notes/ (accessed March 30).
30. Inkjet Printers Pressure Monitoring. https://www.microsensorcorp.com/Details_inkjet-printers-pressure-monitoring-solution.html (accessed March 30).
31. Printhead Performance Technologies. https://www.xaar.com/en/technologies/printhead-performance-technologies/ (accessed March 30).

32. Wang, H.; Halepoto, H.; Hussain, M. A. I.; Noor, S. Cotton Melange Yarn and Image Processing. In *Cotton Science and Processing Technology: Gene, Ginning, Garment and Green Recycling;* Wang, H., Memon, H., Eds.; Springer Singapore: Singapore, 2020; pp 547–565.
33. Halepoto, H.; Gong, T.; Noor, S.; Memon, H. Bibliometric Analysis of Artificial Intelligence in Textiles. *Materials* **2022,** *15*, 2910.
34. Ink-jet Printhead Performance. https://www.flow3d.com/computer-simulation-flow-3d-helps-increase-inkjet-printhead-performance/ (accessed March 30).

# Recent developments in the preparatory processes for the digital printing of textiles

Sharjeel Abid[1], Jawad Naeem[1], Amna Siddique[1], Sonia Javed[2], Sheraz Ahmad[1] and Hanur Meku Yesuf[3,4]

[1]School of Engineering and Technology, National Textile University, Faisalabad, Pakistan, [2]Government College Women University, Faisalabad, Pakistan, [3]Key Laboratory of Textile Science and Technology, Ministry of Education, College of Textiles, Donghua University, Shanghai, P.R. China, [4]Ethiopian Institute of Textile and Fashion Technology, Bahir Dar University, Bahir Dar, Ethiopia

## 6.1 Introduction

Digital textile printing is printing an image-like pattern on fabric via graphic design software. In 1996 it was first jointly introduced by KANEBO textiles, Ltd. and Canon, Inc. There are two types of digital textile printing, that is, transfer printing and direct inkjet printing. Digital inkjet printing is the leading process in textile industries, which is considered an economical process with several environmental benefits like flexibility in design change and color to meet the consumer's demand. Furthermore, digital printing can attain water and energy savings of 90% and 30%, respectively.[1] A small printhead ejects the tiny ink droplets to the fabric's location during digital printing. Several factors influence the digital printing quality, like the type of inks, fabric structure, inkjet printhead, and fabric's preand post treatments.[2]

The textiles and apparel sector of Europe and the United States is confronting a demanding situation in developing countries of the world. To conduct their business efficiently in the 21st century, the Textiles and Apparel industries of the United States have embraced the demand-activated manufacturing architecture approach. This scheme entails a time-built contest, swift response, small-scale lot sizes, larger varieties, and nexus to the supply chain.[3]

All these targets can be fulfilled by using digital printing technologies. This offers a small-scale lot size, swift printing of samples, and reduced lead time for printing. Inkjet technology is primarily used in digital textile printing; however, it was restricted to the production and proofing of samples. Advancements in jet configuration, pre-, and postwet processing necessities, and greater throughput might lead to the expansion of digital printing technology in production mode, mainly employed to develop proofing samples.

Maintaining the quality of digital printing is difficult due to the rough surface of fibers, which affects the interaction between ink and fabric substrate. There are

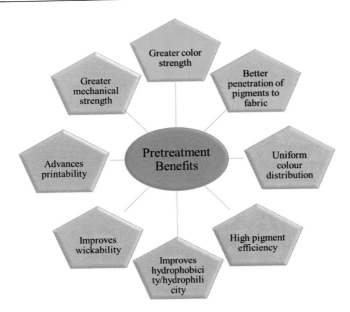

**Figure 6.1** Benefits of pretreatment processes on digital textile printing.

three stages of digital printing, pretreatment, printing, and posttreatment. Different chemicals and techniques are used for the pretreatment of the fabric substrate. Pretreatment of fabric improves the printability of the substrate by inducing charges, providing active sites for the ink substance, and cleaning and smoothing the fabric's surface. Inks having less viscosity usually spread on the fabric's surface, resulting in poor resolution and lower sharpness; therefore, the fabric needs proper pretreatment to avoid such issues. Different types of pretreatment are required for knitted and woven fabrics, dramatically affecting digital textile printing quality.[4] Some of the benefits of pretreatment processes on digital textile printing are mentioned in Fig. 6.1.

## 6.2 The difference between conventional and digital printing pretreatment processes

The initial pretreatment process of conventional and digital printing is the same, including desizing, scouring, bleaching, and mercerization. After that, in conventional printing, fabric goes for printing directly, where a thick paste, including all chemicals and dyes, is applied to the fabric. Another pretreatment process is, however, added before digital printing, containing all the printing chemicals except the ink/dye. In conventional printing for reactive dyes, sodium alginate is most commonly used as a thickener.[5] It has been reported that a high concentration of sodium alginate must be used to enhance the pattern quality of cotton fabric.

**Figure 6.2** Comparison of untreated and pretreated textile substrate for the wicking process.[6]

The manufacturers of digital printing for textiles are cautious in controlling the dropping of ink fired from the head of the printing device; otherwise, they will not be able to achieve effective printing. Because of this reason, the viscosity of ink remains exceptionally low compared to the ink used in other conventional technologies. Therefore, the textile substrate needs to experience a pretreatment process before digital printing. The pretreatment procedure involves using chemicals to control the pH or restrain the transfer of dyes on the surface of the textile substrate.[6]

The values for the viscosity of inkjet fluids range from 3 to 15 mPas. The low viscosity value of inks used in digital printing resulted in ink flow from the print head. Pastes used for conventional printing technologies like screen printing, however, have a high viscosity value of around 5000 mPas. The less viscosity value for digital printing becomes problematic after reaching the surface of textile fabrics when ink is flown away from the destination point due to wicking. This lower viscosity value, however, did not create problems using other substrates like paper. This might be due to its absorbent nature and swift liquid removal from the paper substrate. As a result, thickener padding on textile fabrics becomes essential, resulting in increased absorbent properties of the fabric to impede the wicking process; however, the textile substrate becomes extremely hard in a few circumstances due to thickener application. Because of this reason, the end-use of fabric becomes difficult. Therefore, a scouring procedure is needed to eliminate the thickener or dyes that are not fixed on the textile substrate after printing the textile substrate.

On the other hand, there are some cases where there is no requirement for scouring for the textile substrate. For example, in the case of polyester fabrics used for banners, the pretreatment process improved the drapability of the textile substrate.[6] A comparison of untreated and pretreated textile substrates for wicking the fluid drop is shown in Fig. 6.2.

It is evident from Fig. 6.2 that there was an uncontrolled wicking process on the surface of the untreated textile substrate; however, there was a controlled wicking process on the surface of the pretreated textile substrate. There are various types of pretreatments used in textile printing. Some of them are mentioned below.

## 6.3 Recent developments in pretreatment processes

### 6.3.1 Mercerization

Although mercerization is an additional, unnecessary step, it adds to the fabric characteristics of reactive dyes. Hydroxide compounds like NaOH and KOH are common

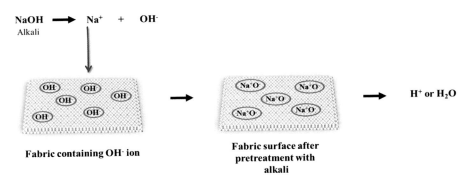

**Figure 6.3** An illustration of the alkaline pretreatment of fabric.

alkali agents.[7] Studies have proved that NaOH treatment elevates the mechanical strength of the fabric, and it can be used for all types of natural fabrics.[8] The mechanism of action of alkali treatment has been shown in Fig. 6.3. Hydroxyl groups of natural fibers and alkali forms $Na^+$ ion and water

Recently, researchers have studied the effect of mercerization on digitally printed fabrics. Zhao and coworkers reported that after caustic mercerization, the penetration of ink droplets between the yarns and fibers decreased; thus, more dye was available on the surface of cotton fabric instead of penetrating as in conventional printing. Using mercerization, the $K/S$ value of the fabric increased from 17.15 to 23.47, which suggested that mercerization was effective. Furthermore, it was noted that mercerization before digital printing reduces wastewater discharge and improves dye uptake, washing, and rubbing fastness.[9]

They have reported that initially, the cotton has higher crystallinity, which reduces the access of hydroxyl groups available for inkjet printing resulting in poor performance. After treating the fabric with caustic soda, the crystallinity, however, changed from 73.9% to 58.5%, with penetration of inks into the fibers.[10]

Among recent developments in the pretreatment processes, the cationization of cotton is popular. Pretreatment of fabric with cationic compounds causes the blocking of active cellulose groups on the surface resulting in uneven coloration after dyeing. Recent development in the cationization process to be performed during the mercerization process, however, resulted in the formation of new cellulose. The change in the crystal lattice structure leads to even distribution and trapping of cationic compounds in cellulose, providing leveling of shades. This technique was introduced in 2003 during the slack mercerization of yarn, and later in 2009, the technique was further developed.[11]

## 6.3.2 Cationization for digital prints

Cationization produces positive charges on the fabric surface, like cotton and viscose. The accumulation of positive charge increases the attraction of negatively charged reactive dyes to the fabrics.[12] The cationization of fabric with amines and quaternary ammonium compounds alters the surface charge of fiber, eliminating the need to use electrolytes in the dyeing process. The cationization process improved the color uptake of fiber by 99%. The cationization agents commonly used consist of short-chained compounds, that is, 3-chloro-2-hydroxypropyltrimethyl ammonium chloride, 2,3-epoxypropyltrimethyl ammonium chloride, and epihalohydrins. Commercially available pretreatment agents with poly ammonium bonds are mostly long-chain compounds. Some commercially available pretreatment agents are Croscolour DRT, Croscolour CF, and Croscolour DRT, which improve color fixation and exhaustion and decrease biological oxygen demand and chemical oxygen demand values. This technique is usually applied by padding and exhaustion, but cationization during mercerization has recently been introduced as a better alternative.[13]

The cationization of cotton for inkjet printing elevates the fabric's printability by inducing a charge to the fiber surface. Recently, researchers have performed the cationization of cotton fabric with a commercially available cationization agent, Resin DWR, which was performed for inkjet printing and compared with untreated fabric and fabric treated with and without the conventional binders. Higher values of $K/S$ indicated greater color strength of cationized fabric compared to other fabrics. Researchers have reported that the higher color strength was due to the fiber's increased ionic attraction of anionic pigments by the cationic charge.

The reaction between the cationic agent and the hydroxyl group of cotton resulted in better penetration of pigments into the fiber, as shown in Fig. 6.4. The microscopic images of treated fabrics with 50% and 100% pigment showed uniform color distribution, high pigment efficiency, and less color penetration for cationized fabric; keeping color components on the fabric's surface preserved the mechanical and physical properties of the fabric.[14]

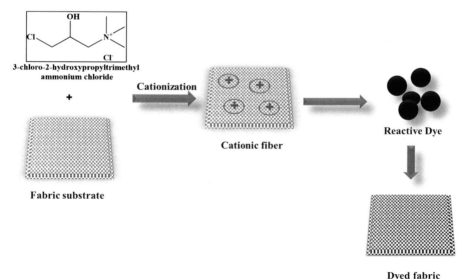

**Figure 6.4** An illustration of the cationization pretreatment of fabric.

One of the drawbacks of the cationic treatment is that it could impart yellowness in the treated/cationized fabrics, and sometimes these are responsible for the staining of nonprint areas in the digitally printed fabrics. Researchers have reported pretreating the cotton fabric with a conventional cationizing agent using the exhaust method and a novel method where the cationic agent (Matexil FC-ER) was printed on the cotton fabric. Then, both fabrics were printed using three different direct dyes. The novel treatment, where the cationic fixer was only applied to the print areas, promoted better color strength, less nonprint area staining, and less fixer consumption than the conventional exhaust pretreatment method. The researchers also reported that in the conventional exhaust treatment, a significant quantity of the cationic fixer remains in the solution instead of coming on the fabric, and the fixer also gets attached to the undesired (print-free) area, which causes cross-staining during washing. Thus, the pretreatment (printing with a cationic agent) before inkjet printing offers a better way to increase the quality of the direct dyes using digital printing.[15]

### 6.3.3 Plasma pretreatment for digital prints

Plasma pretreatment is an environmentally friendly technique for fiber surface modification.[16] Compared to wet processing, plasma treatment only modifies the upper layer of fabric, leaving the other characteristics unaffected. Plasma-activated species alter the fiber surface via cleaning, etching, polymerization, and activation to make additional processing easier[17] (Fig. 6.5). It is one of the preferred methods to pretreat or modifies the textile surface before digital printing.

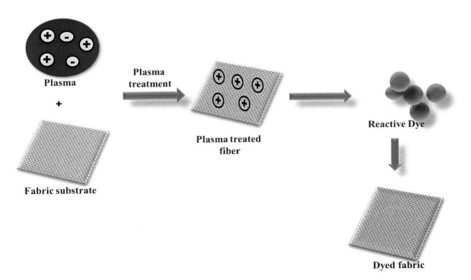

**Figure 6.5** An illustration of the plasma pretreatment of fabric.

Plasma treatment improves hydrophobicity, dyeing, colorfastness, pill resistance, and wicking. In oxygen plasma treatment-free radical and functional groups generates that bombard the fiber surface and increase its roughness. The higher the roughness of the fabric more outstanding the adhesion of deposition substances.[18] In the case of wool fiber, this treatment improves the hydrophilicity by modifying the outer fatty acid layer and protein layer beneath, comprising some disulfide and nitrogen bonds. Plasma treatment improves wool fiber's dyeability, absorbency, and printability.[19]

A comparative analysis was performed on oxygen and argon plasma pretreatment to evaluate the hydrophilicity and wickability of cotton fabric. The results showed higher hydrophilicity with oxygen ($O_2$) than argon (Ar). It is due to the difference in nature of both the gasses; $O_2$ is a chemically reactive gas that functionalizes the fabric surface, while Ar only interacts physically with the fiber. Moreover, $O_2$ induced more hydrophilicity at lower working power of 500 W for 5 minutes, while Ar requires high power and longer treatment duration, 900 W for 10 minutes.[20]

Researchers used the oxygen plasms to treat a polyester nonwoven web before digital printing and reported that treating the fabric with plasma enhanced the color strength due to increased hydrophilicity of the fabric. Furthermore, it was reported that the plasma treatment makes the fabric surface rough and absorbs more coating than untreated fabric.[21]

Zhang and coworkers used plasma treatment to pretreat the silk fabrics and reported a direct link with the aging time of plasma-treated fabric before print. The pretreated fabric was digitally printed at different time intervals after plasma treatment. The experiments reported that the plasma-treated fabric printed within 24 hour of pretreatment provided maximum color yield. After 24 hour of the plasma treatment, the printed fabrics showed a lower color yield than silk fabrics.[22]

### 6.3.4 Enzymatic treatment for digital prints

Enzyme treatment helps improve the quality and appearance of fiber by eliminating short fibers from the fiber surface, bio polishing, stone washing, and removing excess hydrogen peroxide that remains after bleaching. Many environmentally friendly enzymes are used in the textile industry, including lipase, cellulose, pectinase, hydrolases, xylanase, and oxidoreductases. Cellulase is a combination of β-glycosidases, endoglucanases, and exoglucanases, which collectively acts on 1, 4, β- glucosidic bonds on the fiber surface and, as a result of hydrolysis from smaller subunits. Cellulase treatment helps produce a soft, clean, and smooth surface by eliminating tiny fibers, surface fibrils, and other protruding fibers (Fig. 6.6).[23]

Laccase, an essential enzyme from the oxidoreductase group, contains several copper atoms and can oxidize some aromatic compounds and phenols, thus being used widely in industries for environmental purposes. Apart from its use in decolorizing industrial wastewater, it is efficient in removing lignin from flax fibers, increasing the water sorption capability of linen fabric.[24] An enzyme pretreatment of polyethylene terephthalate (PET) fabric was performed for digital inkjet printing. Pretreatment of PET fiber with cutinase enhanced wettability by a substantial drop

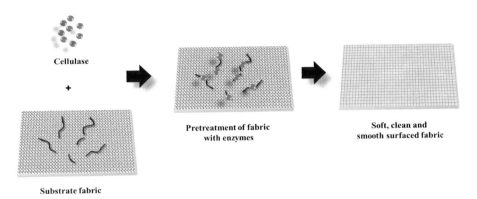

**Figure 6.6** An illustration of the enzymatic pretreatment of fabric.

in contact angle. It is due to the breakage of the ester bond on the PET surface and the formation of new hydrophilic functional groups.[25]

Getting a smooth print of the wool fabric is challenging due to its natural characteristics; however, it is one of the favorite fabrics for countries with extremely cold weather. The protease enzyme can remove the scales and make the fabric hydrophilic;[26,27] thus, it can facilitate uniform inkjet printing on the fabric. Researchers have successfully treated the wool fabrics with protease enzyme and sodium alginate to overcome the drawback. The results showed that the treated fabric had better color strength, good washing, rubbing fastness, and acceptable print sharpness.[28]

### 6.3.5 Other pretreatment processes for digital prints

Several other processes have been used for pretreatment processes, including different chemicals, as mentioned in Table 6.1.[29] These processes have been used for fabric pretreatment before dyeing and have shown beneficial results; however, these are multistep processes that use various chemicals and large consumption of water, which generates tons of hazardous chemical waste, and these methods are very costly.[30]

Researchers have tried to replace and compare the print quality of sodium alginate and pretreated chitosan fabrics. The results showed that the chitosan pretreated fabric could replace the conventional sodium alginate; however, the color strength was lower. On the other hand, the sharpness of digital prints was enhanced.[31] Researchers have tried different pretreatment agents consisting of a thickener or a combination of different thickeners (chitosan, sodium alginate, carboxymethyl cellulose, ludigol (mild oxidizing agent)), urea, and water. For linen inkjet printing, it has, however, been observed that the carboxymethylcellulose chitosan-containing print paste improves printability and colorfastness properties compared to untreated fabric.[32]

There are many differences between the pretreatment of conventional and digital printing. The printing recipe contains all the auxiliaries in conventional printing, including thickener, binder, urea, mild oxidizing agents, wetting agents, fixers, and alkali (depending upon the type of printing). In digital printing,

Table 6.1 Advantages of chemical pretreatment processes and their mechanism of action.

| Pretreatment process | Pretreatment agent | Mechanism of action | Advantages |
|---|---|---|---|
| Acryl treatments | Acrylic acid Acrylonitrile | Performed after alkali treatment of fabric. | Improves fiber strength by reducing the hydrophilicity of the fiber surface. |
| Benzoylation | Benzoyl chloride solution | The HOOH group decomposed to OH radical and then reacted with the $H^+$ group of fiber. | Eliminate unwanted substances and provide smoothness to the fabric. It also increases tensile strength. |
| Peroxidation | Hydrogen peroxide | | Improves mechanical properties and elevates elongation capacity and bending strength. |
| Acetylation | Acetic acid and Acetic anhydride | First, the fabric is treated with alkali followed by acetylation with acidic anhydride, which reacts with the OH group of fiber and makes acetic acid. | Increases tensile strength and thermal stability. |
| Permanganate treatment | $KMnO_4$/acetone solution | Fabric is first treated with alkali and then immersed in $KMnO_4$/acetone solution. Oxygen permanganate combines with the OH group of fabric. | It is a superior process to alkali treatment as it imparts more excellent tensile and flexural strength. |
| Use of coupling agents | Polybutadiene isocyanate, Maleic anhydride-modified Polypropylene, Silanes, and Oligomeric siloxane | Reacts with OH of fiber and forms a strong covalent bond. | It helps to attain greater bonding strength between fiber and matrix. |

however, the printing recipe is only ink.[33] All the other chemicals are previously padded and dried before print.

Recently, researchers have reported using two commercially available pretreatment agents, DP-300 (45% solids) for cotton and DP-302 (15% solids) for cotton and polyester, to check their difference in the final results of digital prints. Shamey and coworkers reported that using pretreatment agents on cotton produced vivid colors and better final effects than polyester fabric under commercial conditions. After pretreatment with these agents, the group also reported a wet and dry crocking fastness loss. The crocking fastness was, however, changed from one color to another, suggesting that the pretreatment agent that had the primary influence and the class and type of color play an essential role in digital printing.[34]

Another aspect of printing good quality fabric is reducing the hydrophilicity of the cotton fabric, which causes excessive absorption of ink, resulting in ink smearing, ultimately leading to poor quality of the final fabric. A group of researchers used different combinations of butyl-acrylate and vinyl-acetate to get optimum print quality on digitally printed fabrics. By choosing an appropriate ratio of these agents, the authors reported improving the K/S of the printed fabrics by 30%.[35] Another group of researchers has reported preparing low-viscosity pretreatment agents based on polyacrylic acid's concentration and molecular weight. They reported that when the viscosity of the combination was less than 50 cps, higher color strength was achieved in the digital printing of fabrics.[36]

## 6.4 Pretreatment processes for pigment digital printing

Polyester fabric is usually used for pigment digital printing because of its stability as outdoor cushions. A commercially available pretreatment agent (DP-302, 15% solids; Lubrizol Corporation) was applied to the scoured polyester fabric at one bar pressure via the Mathis AG padding machine. The pretreatment agent comprises several components, viz. ink coagulants-polyvalent metal salts; acrylic resin for pigment fixation; and other surface tension control and wetting additives such as propylene glycol silicone-based compounds and isopropyl alcohol. 48%–74% increase in % wet pickup was observed after pretreatment. % Wet pickup was calculated by the following formula.

$$\frac{\text{Mass of padded fabric} - \text{Mass of initial fabric}}{\text{Mass of initial fabric}} \times 100$$

After pretreatment, wetting time decreased from 95 to 12.3 seconds, as observed by AATCC Method 79. The spectrophotometric analysis decreased the $L^*$ value, indicating deeper shades after pretreatment than untreated fabric. It was concluded that pretreatment enhanced the overall digital printing quality by providing deeper shades and lowering ink bleeding. It also increases the color range by presenting more saturated hues. Because of their specific nature, pigments are applied in digital printing at the end of the application process. Structures of some of the pigments used in digital printing are mentioned below in Fig. 6.7.[37] Fig. 6.8 shows the main constituents of pigment digital textile printing.

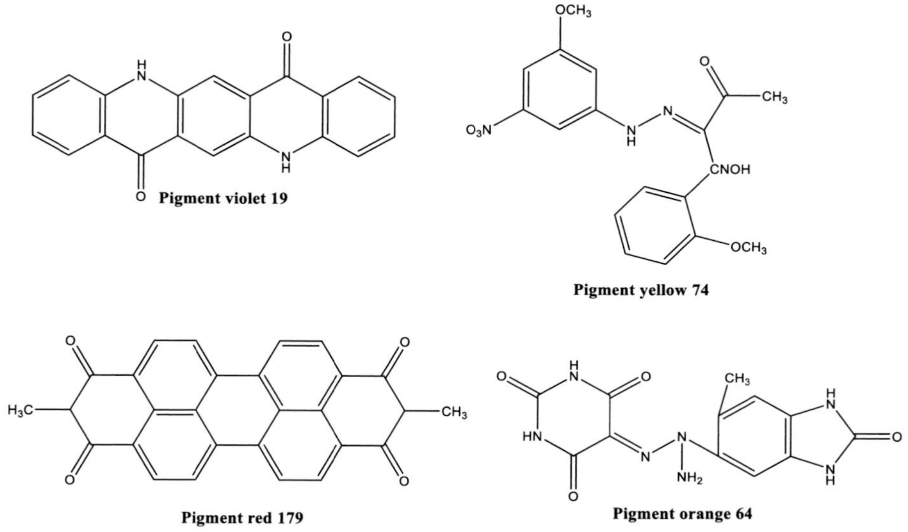

# Recent developments in the preparatory processes for the digital printing of textiles

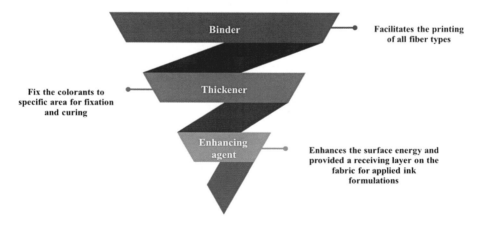

**Figure 6.7** Structures of commonly used pigments in digital textile printing.

**Figure 6.8** Main components in digital pigment textile printing.

## 6.5 Pretreatment of cotton fabrics for digital printing

The purpose of pretreatment and modification of cotton substrate is to increment the reactiveness and compatibility among dyes and cotton fibers.[38] The most efficient methods for the pretreatment of cotton are:

1. Modification with glycidyl trimethylammonium chloride or 3-chloro-2-hydroxypropyl trimethylammonium chloride,[39–41] dendrimer compounds,[6]
2. Modification with chitosan[31,41–44]
3. Pretreatment with gas plasma[45]

These methods showed a specific enhancement in the color yield of printed fabric. For modification of cotton, glycidyl trimethylammonium chloride (GTA) is extremely useful in improving color yield. There is an etherification reaction of GTA with OH groups of cotton fibers, and a hydrolysis side reaction also occurs during cationization. GTA is less expensive and has low toxicity.[46] Furthermore, the molecule of GTA is small and has a low effect on the permeability of dyes and colorfastness. Kanik et al.[39] used GTA and NaOH to alter the cotton substrate at room temperature for 24 hours. Afterward, the specimens were rinsed and dried to remove water from the fabrics. The acquired cationized substrates were employed in inkjet printing of reactive inks to improve color yield. The cationization and pretreatment formations are two separate processes. Because of this reason, a lot of time and water are consumed in these processes along with complexity.

It was noted that cationization and pretreatment of cotton substrate took place in an alkaline environment;[47] however, different alkalis were used in these two processes, that is, $NaHCO_3$ was used for the pretreatment process, and NaOH was employed for the cationization process. Combining the cationization and pretreatment process will save a lot of water and time, and the process will become simpler.[38]

### 6.5.1 Impact of one-bath pretreatment for increased color yield of inkjet prints using reactive inks

They are mainly employed to print cotton substrate and cellulosic fabrics due to the superior fastness characteristics and broad shade scale of reactive dye inks; however, reactive dyes merely have a 50%–80% fixation due to constrained hydrolysis and reactivity during the fixation process. As a result, excessive water is wasted.[48] Furthermore, monochloro-s-triazine dyes with limited reactivity are primarily used in ink formation, creating stability. The outcomes, however, are the low color yield of prints. Thus, one of the reactive dyes' most important inkjet printing problems is unacceptable color yield. Because of this reason, the employment of reactive inkjet printing in generating deep color is constrained.[49]

A study by Ma et al.[38] was conducted to increment the color yield of inkjet printing using reactive ink and one bath pretreatment of the cotton substrate. The pretreatment formation includes glycidyl trimethylammonium chloride, sodium alginate, NaOH, and urea for acknowledging cationization and sizing simultaneously. The pretreatment parameters entail concentrations of glycidyl trimethylammonium chloride and alkali, optimization of time, and baking temperature based on outcomes of color yield on the cationized cotton substrate for magenta ink.

Ma et al.[38] developed one bath pretreatment to merge GTA cationization and sizing pretreatment of the cotton substrate. The sizing means the treatment of the cotton surface with alginate paste. In this methodology, GTA, sodium alginate, NaOH, and urea are evenly mixed in a single bath to pretreatment the cotton substrate before inkjet printing. Their study's main aim was to investigate the single bath pretreatment methodology's effectiveness in improving the cotton substrate's color yield. The effect of pretreatment parameters like concentration of NaOH, GTA, baking time, and temperature on the yield of color was also examined.

### 6.5.2 Impact of concentration of glycidyl trimethylammonium chloride on color yield of magenta ink

The outcomes showed that the concentration of GTA is a critical parameter for the increment of the color yield. There was a substantial increment in values of K/S when the concentration of GTA was more than 2%. Compared to the two-bath methodology, the one-bath methodology can significantly improve color yield.

The primary explanation for the increment in the value of K/S might be because cationic fibers have greater substantivity to dyes and lesser penetration of dyes in the fibers confirmed from previous studies, which might be due to significant ionic interaction between anionic dyes and cationic fibers and other reason was the more excellent fixation of dye. This indicated that a tremendous amount of dyes was covalently bonded to fibers in the company of GTA.

### 6.5.3 Impact of concentration of NaOH

Alkali ionizes cellulosic hydroxyl groups to fix reactive dyes in inkjet printing.[31] Alkaline conditions are essential for the cationization of cotton with GTA. Because of this fact, the cationization process and sizing pretreatment process were blended into a single step. On the other hand, different alkali was used to modify GTA in inkjet printing compared to conventional printing. NaOH is the most efficient alkali for the modification of GTA as it enhances the hydrolysis of reactive dyes because of its intense alkaline nature.

Ma et al.[38] concluded that by adding NaOH in the pretreatment process, cotton fibers got activated, and thus, facilitated interaction between cotton and GTA and acceleration of fixation of dyes. Thus, K/S values are significantly influenced by the concentration of sodium hydroxide.

### 6.5.4 Impact of chitosan and acetic acid on color yield

It was noted by Choi et al.[31] that there was an enhancement in the color yield of printed fabric substrate with the corresponding increment in the quantity of chitosan in pretreatment paste. This indicated that chitosan might enhance color yield in digital inkjet printing. This performance of chitosan might be due to the presence of amino groups, which develop $-NH_3^+$ groups by taking protons from mildly acidic dyebath. The electropositive nature of $-NH_3^+$ group favored the transportation of negative charge dye anion from a bath to a fabric substrate. Because of this reason, color fixation was improved.

Furthermore, it was also observed that by enhancing the quantity of acetic acid in the pretreatment paste, the color of digital printing was strengthened, and maximum color yield was acquired in 40 mL of chitosan employed in the pretreatment printing paste.

### 6.5.5 Impact of sodium bicarbonate on color yield

The use of alkali in the printing process is essential for developing reactive dye colors of all shades.[50] The existence of the alkali group is important to creating ionization of nearby cellulosic. OH, groups can interact with the reactive dyes in the fixation stage. $NaHCO_3$ is less expensive, can impart adequate stability in pretreatment paste, and is the most reactive dye accessible in the marketplace. During steaming, $NaHCO_3$ drops carbon dioxide and increments cellulose's ionization by encouraging the dye's interaction with fiber in the fixation phase. It was noted that by increasing the quantity of $NaHCO_3$ in the pretreatment paste, the color yield of cotton fabric corresponded to an increase. It was also noted that the neutralization impact was negligible in the pretreatment paste when the quantity of $NaHCO_3$ and acetic acid was increased. This might be because the decline in final color yield was not substantial.

### 6.5.6 Impact of urea

The inclusion of urea is significant in pretreatment paste during the steaming process, especially during the use of superheated steam after the substrate is digitally printed. Because of the presence of urea, cotton fibers' swelling takes place, facilitating swift penetration of dye in fibers.[50] Hence, urea performs as a solvent for reactive dyes. Furthermore, it acts as a moisture-absorbing agent in the pretreatment printing paste to improve the moisture regain during the steaming process. Consequently, urea hastens the immigration of dye from the thickener film to the cotton fiber.

Moreover, the urea also decreases the yellowness of cotton substrate in scorching, dry alkaline environments. Choi et al.[31] determined that when the amount of urea was 10 g in pretreatment paste, the maximum color yield was acquired on a cotton fabric substrate. The higher color yield was shown by black color, followed by cyan, magenta, and yellow.

On the other hand, it was noted that as the amount exceeded 10 g in the pretreatment paste, there was a substantial reduction in the color yield of the cotton substrate.[31] This might be because an increase in urea enhances the moisture regain of cotton fibers during the steaming process. A sizable amount of urea in pretreatment paste could increment moisture absorption properties due to the hygroscopic nature of urea and result in the hydrolysis of reactive dyes during the steaming process. Moreover, black, yellow, and magenta inks have a dye configuration of vinyl sulphone dye and tend to get neutralized in the company of urea.

### 6.5.7 Effect of cationization of cotton in digital printing

Much research has been done on the pretreatment of cotton by cationization for the digital printing process.[4,51,52] A study conducted by Yang et al.[53] and Rekaby et al.[54] mentioned improvement in color strength, increment in color yield, improved sharpness in pattern, better colorfastness, and reduction in duration of the steaming procedure in reactive digital inkjet printing on a cationized pretreated cotton substrate.

Hauser and Kanik et al. investigated the digital printing of cationized cotton substrates by using acidic, reactive, direct dyes and pigments.[39,55–57] Wang conducted a study, and Zhang modified cotton with cationic compounds like Cibafix Eco and pigment adjustment to acquire improved color strength and lesser pigment penetration.[58,59] A study by Grancaric et al. presented the cationization of the substrate during the mercerization process.[60]

When the long-chain cationic compound was applied to cellulosic substrates in the pretreatment or posttreatment process by the padding and exhaustion, there was a blockage of active cellulosic groups on the substrate surface. Because of this reason, there was no uniformity of color after the dyeing process. The cationization of the substrate, however, resulted in the formation of different cellulose. Modification in crystal lattice structure occurred, and uniform distribution of cationic compound entrapped between cellulosic chains resulted in evenness of color.[11,13,61,62]

In a study conducted by Glogar et al.,[14] the effect of cationization of the cotton substrate on the print quality was investigated concerning color depth, color yield, and penetration, respectively. The cationization of the cotton substrate was accomplished during the mercerization process with a cationizing agent, that is, Rewin DWR (CHT Bezema). Cationized and noncationized cotton fabric were digitally inkjet printed with and without the extra layer of the binder.

Glogar et al.[14] mentioned that in the preliminary printing stage, high values of K/S were acquired for the cationized substrate. This higher value of color strength might be due to the reason that by incrementing the concentration of cationic reagent, the ionic appeal between anionic pigment and cationic charged fibers is also incremented. The cationization process improved the penetration properties of the cotton fabric substrate because of the interaction between the OH group of cotton and the cationic reagent. In those specimens in which the binder was used, the cationization process improved durability, and bond strength between layers of polymers such as pigment carrier was established with cotton fabric substrate.

### 6.5.8 Effect of hydroxypropyl methylcellulose pretreatment on cotton and polyamide fabric in inkjet printing

A study by Qiao et al.[63] mentioned the utility of hydroxypropyl methylcellulose (HPMC) in the pretreatment process of cotton, polyamide, and interwoven polyamide fabric substrate. The outcomes indicated substantial modification in the surface properties of cotton and polyamide fibers. The hydrophilicity of cotton fiber was lessened owing to modification in functional groups on the surface of fibers. On the other hand, there was an increment in the hydrophilic property of polyamide fiber.

In the case of cotton fibers, there was a substantial decline in surface oxygen content. When the cotton substrate undergoes pretreatment with HPMC, a uniform film is formed on the surface of the cotton fiber. Many OH groups of Cotton fiber were wrapped by HPMC film, and very little oxygen was on the surface from the OH group of HPMC. Simultaneously, there was an increment in carbon content due to the presence of the methyl group in HPMC. Because of these facts, cotton fiber's surface becomes hydrophobic, prohibiting diffusivity along cotton fibers. The hydrophobic nature of polyamide fiber is due to a significant quantity of methylene in a long chain of polyamide macromolecules. The oxygen content on the cotton substrate arrived from amide groups and a small number of terminal carboxyl groups. When polyamide substrate undergoes pretreatment of HPMC, the nature of fiber modifies from hydrophobic to hydrophilic due to the covering of the methylene chain by an extended cellulose chain.

Thus, the properties of the surface of cotton fiber and polyamide fiber in the interwoven substrate of cotton/polyamide are separate. Because of these reasons, there was a modification in the characteristics of the cotton/polyamide substrate surface, which altered the diffusivity and permeability of reactive dye inks on the fabric's surface. These modifications are very fruitful for inkjet printing.[63]

### 6.5.9 The color performance of inkjet printing

The values of K/S for cotton/polyamide fabric pretreated with HMPC were compared with the K/S values of cotton/polyamide fabric pretreated with sodium alginate. It was noted that HPMC pretreated fabric showed increased values of K/S for cyan, magenta, yellow and black colors. The hydrophobicity of HPMC is greater than that of cotton fibers, due to which reactive dyes of HPMC films are transferred to cotton fiber along with the aid of water vapors during the steaming process; however, HPMC is more hydrophilic as compared to polyamide fibers. Thus, many reactive dyes remained in the HPMC film dye throughout the steaming process. Also, HPMC fills upvoid spaces in the yarn to decline ink scattering along the yarn's surface.[63]

### 6.5.10 Use of Plasma pretreatment for depositing printing paste on the cotton substrate for digital inkjet printing

For the last two decades, the use of plasma technology as a pretreatment process for modification in surface characteristics of the textile substrate has been very successful without affecting alterations in bulk properties of the textile substrate.[15,64–66] It was established from prior studies that plasma treatment at a lesser temperature might increment the printing of cotton substrate by using sodium alginate as a printing paste. On the other hand, the plasma-treated substrate became stiff and could not be incorporated into an open system for fabric processing.

Using atmospheric pressure plasma (APP) for cotton substrates currently delivers flexibility and potential opportunity for open and continuous processes.[67,68]

In a study by Kan et al.,[45] APP treatment was employed as a pretreatment procedure to increment the accumulation of printing paste to enhance the color characteristics of digital inkjet printing on a cotton substrate. Three different printing pastes were developed separately, that is, sodium alginate-chitosan mixture, sodium alginate, and chitosan. At first, cotton fabric was pretreated with APP. Afterward, the cotton substrate was padded with printing pastes before enduring digital inkjet printing.

### 6.5.11 Determination of color yield

It was determined by Kan et al.[45] that those pretreated samples with APP technology had greater values of K/S compared to untreated plasma cotton substrate. When a printing paste comprised of sodium alginate and urea was employed, cotton fibers' swelling occurred during the superheated steaming procedure for swift penetration of dyes in cotton fiber. Urea was used as a moisture-absorbing agent to accelerate the transportation of dyes from printing paste to cotton fibers. Because of this reason, an increase in color yield took place.[69] In contrast, plasma employing oxygen gas might increment the number of hydrophilic groups like O-C = O, O-C-O, C = O, and -C = O on the surface of cellulosic fibers.[66,70–72] The number of hydrophilic groups might increment the absorption rate of dyes and color yield.

In the case of printing paste of sodium alginate/chitosan mixture, approximately 98.9% of the final color yield was acquired compared to the control fabric. The mild decline in color yield of SA/Ch specimens might be due to the deactivation effect of chitosan solution with sodium bicarbonate in printing paste.[44] In contrast, the employment of plasma pretreatment might cause the etching of cotton fiber substrate due to the formation of cracks.[70,72] Because of the deposition of printing paste on the cotton substrate, the filling of cracks of cotton fibers took place. Owing to this reason, more interaction of dyes with cotton fiber occurred during dyes' fixation, resulting in an increment of uptake of dyes.

Furthermore, hydrophilic groups formed on the surface of fibers after treatment with oxygen plasma. This might increment the distribution of printing paste, allowing a more significant amount of dye to interact with fibers. As a result, the color yield was enhanced.[45]

## 6.6 Pretreatments of wool fabrics for digital printing

In a study conducted by Yuen et al.,[73] optimum conditions for pretreatment printing paste inkjet printing of wool substrate were investigated. The wool substrate was embedded with a pretreatment printing paste—the ingredients of the pretreatment printing paste comprised sodium alginate, urea, and ammonium tartrate.

### 6.6.1 Impact of sodium alginate on color yield of wool fabric

Sodium Alginate is pertinent to producing printing pastes of acidic dyes because of its limited attraction to them. This might be due to the nonavailability of OH groups in the sodium alginate and the abhorrence of dye anion due to ionized carboxyl groups of wool fiber in an acidic climate. Because of this reason, sodium alginate was used to investigate its impact on the color yield of various colors.[73]

It was noted that there was a minor increase in color yield values with an increment in the amount of sodium alginate in the printing paste. This showed that sodium alginate was not a significant parameter affecting the color yield of inkjet-printed wool substrate using acidic ink. Sodium alginate was used as movement deterrence in printing paste to manage the final shape of the inkjet-printed pattern.[74]

### 6.6.2 Impact of ammonium tartrate on color yield of wool fabric

It was noted by Yuen et al.[73] that there was an improvement in color yield with an increase in the quantity of ammonium tartrate. This might be because ammonium tartrate behaved as an acidifier in printing paste by creating a positive charge in wool fibers.[6] The positive charge behaves as driving power for acid ink to disseminate and transfer into the wool fibers. With the increase in ammonium tartrate, there was a more remarkable development of a positive charge on wool substrates. Because of this reason, improved transmission and diffusivity were acquired.

### 6.6.3 Impact of urea on the color yield of wool fabric

It was observed that the presence of urea deepened the color of the wool substrate, and the highest yield was acquired when the quantity of urea in the pretreatment paste was 20 gm. On the other hand, it was noted that there was a decline in color yield when the amount of urea was increased from 20 to 30 g. This might be because there was an increment in moisture absorption during steaming, which might result in the hydrolysis of the water-soluble acid dye bath.[6,50]

## 6.7 Pretreatment of polyester fabrics for digital printing

### 6.7.1 Impact of pretreatment on color strength

Research conducted by Noppakundilograt et al.[75] used pretreatment solutions of chitosan (CH), N- [(4-dimethyl amino benzyl)imino] chitosan (DBIC), N- [(2-hydroxy-3 trimethylammonium)propyl] chitosan chloride (HTACC), glycine (Gly), and a mixture of CH and Gly for impregnating polyester substrate before encountering digital printing. They mentioned that lower values of K/S were acquired for CH pretreated polyester substrates. K/S values of Gly pretreated fabrics were more prominent than the untreated polyester substrate. On the other hand, when a mixture of CH and Gly was employed, lesser values of K/S were acquired compared to those untreated values. Also, the values of K/S were not greater than that of pure CH or Gly at the same loading levels. Polyester substrate pretreated with DBIC yields lower values of K/S compared to untreated fabrics. On the other hand, polyester substrate pretreated with HTACC delivers excellent and greatest K/S values with a poor concentration of HTACC, that is, 0.1% (w/v).

### 6.7.2 Impact of pretreatment on color saturation

By launching the quaternary ammonium group, the solubility of CH was improved. The conjugation of glycidyl trimethylammonium chloride (GTMAC) to CH generates a CH derivative, HTACC, a polymer soluble in water at all pH values. Thus, HTACC has two nitrogen atoms, one quaternary ammonium group, and one secondary amino group. Because of this reason, pretreatment with HTACC can generate higher values of K/S, gamut volume, and improved color saturation compared to the less effective primary amino group of CH in neutral or dilute acidic solutions.[75] The durable nature of the positive charge might intermingle with the negative charge of polymer-encapsulated pigmented ink through amidation or electrostatic attraction. The greater values and wider color gamut boundaries and volume produced by pretreatment of HTACC might be due to the ease of availability side due to the long side chain. It was mentioned that this long side chain might permit surface adsorption of pigment inks and chemisorption. The protonation of CH and HTACC may deliver swift surface adsorption succeeded by diffusivity and chemisorption of altered pigment molecules in the CH web through electrostatic attractions.[76] In addition, HTACC could produce antibacterial cellulose fibers adding the advantages of the utility of HTACC for the textile substrate.

## 6.8 Pretreatment of silk fabric for digital printing

Pretreatment of silk fabric with sodium alginate and urea is known to control the ink droplet spreading. For digital printing, ink droplet spreading on fabric is the main factor determining the accuracy of the final printing results. Water-absorbing capability, surface energy, and fabric pretreatment are vital in ink droplet spreading. Urea works as a moisture-absorbing substance that increases the movement rate of ink droplets from the pretreatment paste to the fabric. In silk fabric pretreated with sodium alginate, it has been observed that the continuity of capillary completely breaks down, resulting in an enlarged contact angle and reducing the wicking rate of water droplets on the silk fabric. The final droplet deposit image of urea and sodium alginate pretreated fabric produce circular and crisscross, with the roundness advanced to 1 when sodium alginate concentration reaches 4%. Compared to untreated and urea-pretreated silk fabric, sodium alginate can reduce the diffusion rate from 85.81 to 47−58 $mm^2$, ultimately limiting droplet spreading.[4]

For urea pretreatment of silk fabric, adding thickeners like polyacrylamide (PAM) and polyacrylic acid (PAA) can change the ink penetration and spread on the fabric. The main factors influencing ink spreading during inkjet printing were found to be.

- Effect of steaming time
- Amount of alkali
- Amount of urea
- The pH of the pretreatment paste
- Concentration of thickener

The factors that influence ink penetration are, however, similar to ink spreading, except the time of thickener with urea also plays an important role.[77]

### 6.8.1  Effect of steaming time

The digitally printed silk fabric's color strength is influenced by steaming duration; however, the effect of steaming duration is different for each type of thickener used for pretreatment. For example, PAA-based thickener color strength increases with the increase of steaming duration to 12.5 minutes. Nevertheless, a decrease in color strength was observed for the PAM-based thickener, increasing steaming time from 10 to15 minutes. It is due to moisture provided by lengthy steaming duration, which helps hydrolyze the reactive dyes and drop dye fixation and color strength.

### 6.8.2  Amount of alkali

Reactive dyes from covalent bonds only under alkaline conditions.[78] Increasing the alkali concentration from 10 to 17.5 g/L increased color strength and fixation regardless of the type of thickener used. A higher amount of alkali concentration provided more binding sites for dyes onto the fabric.

### 6.8.3  Amount of urea

As urea works as a moisture-absorbing agent, it facilitates the dissolution of dyes and prevents them from aggregation. Increasing urea concentration from 80 to 115 g/L increases dye fixation and color strength; however, a further increase in urea concentration led to a decrease in dye fixation and color strength. It is due to excessive moisture absorbance by urea during the steaming process.

### 6.8.4  The pH of the pretreatment paste

In silk fabric, nucleophilic amino groups like arginine, lysine, and histidine are the active sites for dye fixation at acidic, basic or neutral pH conditions. The phenolic and alcoholic hydroxyl groups, tyrosine and serine, are activated at alkaline pH. Color fixation and strength increase to 7.5 pH and are reduced on further pH increases due to a decrease in amino group protonation and dissociation of reactive dyes at higher pH.

### 6.8.5  Concentration of thickener

A particular amount of thickener needed to be padded onto silk fabric to preserve the outline and sharpness of the edges. Although an increasing amount of thickener increases the color fixation, excessive strength of thickener can act as a barrier for dye and prevent its penetration into the fabric.[79]

## 6.9 Pretreatment of blended fabrics for digital printing

Blending fibers means adding two different components, which significantly alters the final properties of the fabric. Thus, blended fabrics offer a greater challenge in digital printing. For example, cotton fibers' surface properties differ from polyamide fibers. Therefore, the cotton and polyamide (nylon) blend requires special consideration for good print quality. Researchers have tried to overcome these problems for good print quality in the case of cotton and polyamide fabrics by pretreating the fabric with HPMC. The authors have reported that with such pretreatment, the surface of cotton became less hydrophilic, while the surface of polyamide fabric became less hydrophobic. The study reported an increase in K/S value of up to 61.2% using reactive yellow dye compared to sodium alginate on the same fabric. Thus, it was observed that the pretreatment with HPMC offered greater benefits regarding digital printing for cotton/polyamide blended fabrics.[63]

## 6.10 Pretreatment free digital printing

If the fabric pretreatment process could be skipped, it could add significantly towards savings and an environmentally friendly process. Researchers have recently prepared ink for the inkjet printing of polyester fabrics, eliminating the need for pretreatment. They formulated the new ink with different combinations of glycerol, monoethylene glycol, SDS (surfactant), antifoaming agent, triethanolamine, and the respective dispersed dye. By controlling the surface tension, particle size, viscosity, density, and other parameters, the researchers produced a formulation that required no fabric pretreatment and produced good print quality on the polyester fabrics.[80]

## 6.11 Conclusions

This chapter discussed the effect of pretreatment processes on digital textile printing. Different types of pretreatment processes like alkali, plasma, enzyme, and cationization have been used for different fibers. For digital printing, ink droplet spreading on fabric is the main factor determining the accuracy of the final printing results. Water absorbing capability, surface energy, and fabric pretreatment play an important role in ink droplet spreading. Pretreatment with plasma or cationic agent induces a positive charge on the fabric's surface, attracting the negatively charged ink droplets and resulting in better print performance. Different parameters like concentration of thickener, amount of active agent, temperature, and pH greatly influence the printability of fabric using sustainable pretreatment processes like plasma pretreatment is beneficial to the environment and economical in terms of water preservation, low cost, and less energy consumption. Now researchers have developed inks that do not require fabric pretreatment, resulting in cost savings and better yield.

# References

1. Choi, S.; Cho, K. H.; Namgoong, J. W.; Kim, J. Y.; Yoo, E. S.; Lee, W.; Jung, J. W.; Choi, J. The Synthesis and Characterisation of the Perylene Acid Dye Inks for Digital Textile Printing. *Dye. Pigment.* **2019**, *163*, 381–392.
2. Hou, X.; Chen, G.; Xing, T.; Wei, Z. Reactive Ink Formulated with Various Alcohols for Improved Properties and Printing Quality onto Cotton Fabrics. *J. Eng. Fibers Fabr.* **2019**, *14* 1558925019849242.
3. Kim, Y. Effect of Pretreatment on Print Quality and its Measurement. In Digital Printing of *Textiles;* Ujiie, H., Ed.; Woodhead Publishing: Cambridge, England, 2006; p 256.
4. Li, M.; Zhang, L.; An, Y.; Ma, W.; Fu, S. Relationship Between Silk Fabric Pretreatment, Droplet Spreading, and Ink-Jet Printing Accuracy of Reactive Dye Inks. *J. Appl. Polym. Sci.* **2018**, *135*, 46703.
5. Mohsin, M.; Sardar, S.; Shehzad, K.; Anam, W.; Iqbal, M. Performance Enhancement of the Digital Printed Cotton Fabric through Ecofriendly Finishes. *J. Nat. Fibers* **2021**, 1–10.
6. Tyler, D. J. Textile Digital Printing Technologies. *Text. Prog.* **2005**, *37*, 1–65.
7. Jhatial, A. K.; Yesuf, H. M.; Wagaye, B. T. Pretreatment of Cotton. In *Cotton Science and Processing Technology: Gene, Ginning, Garment and Green Recycling;* Wang, H., Memon, H., Eds.; Springer Singapore: Singapore, 2020; pp 333–353.
8. Wang, H.; Farooq, A.; Memon, H. Influence of Cotton Fiber Properties on the Microstructural Characteristics of Mercerized Fibers by Regression Analysis. *Wood Fiber Sci.* **2020**, *52*, 13–27.
9. Zhao, H.; Zhang, K.; Fang, K.; Shi, F.; Pan, Y.; Sun, F.; Wang, D.; Xie, R.; Chen, W. Insights into Coloration Enhancement of Mercerized Cotton Fabric on Reactive Dye Digital Inkjet Printing. *RSC Adv.* **2022**, *12*, 10386–10394.
10. Zhao, H.; Wang, D.; Fang, K.; Zhang, K.; Pan, Y.; Liu, Q.; Shi, F.; Xie, R.; Chen, W. Enhancing Digital Ink-Jet Printing Patterns Quality through Controlling the Crystallinity of Cotton Fibers. **2021**. Available from: https://doi.org/10.21203/rs.3.rs-804119/v1.
11. Sutlović, A.; Glogar, M. I.; Čorak, I.; Tarbuk, A. Trichromatic Vat Dyeing of Cationized Cotton. *Materials* **2021**, *14*, 5731.
12. Dutta, S.; Bansal, P. Cotton Fiber and Yarn Dyeing. In *Cotton Science and Processing Technology: Gene, Ginning, Garment and Green Recycling;* Wang, H., Memon, H., Eds.; Springer Singapore: Singapore, 2020; pp 355–375.
13. Čorak, I.; Brlek, I.; Sutlović, A.; Tarbuk, A. Natural Dyeing of Modified Cotton Fabric with Cochineal Dye. *Molecules* **2022**, *27*, 1100.
14. Glogar, M. I.; Dekanić, T.; Tarbuk, A.; Čorak, I.; Labazan, P. Influence of Cotton Cationization on Pigment Layer Characteristics in Digital Printing. *Molecules* **2022**, *27*, 1418.
15. Zhang, C.; Fang, K. Surface Modification of Polyester Fabrics for Inkjet Printing with Atmospheric-Pressure Air/Ar Plasma. *Surf. Coat. Technol.* **2009**, *203*, 2058–2063.
16. Memon, H.; Kumari, N. Study of Multifunctional Nanocoated Cold Plasma Treated Polyester Cotton Blended Curtains. *Surf. Rev. Lett.* **2016**, *23*, 1650036.
17. Ahmed, H.; Khattab, T. A.; Mashaly, H.; El-Halwagy, A.; Rehan, M. Plasma Activation toward Multi-Stimuli Responsive Cotton Fabric Via in Situ Development of Polyaniline Derivatives and Silver Nanoparticles. *Cellulose* **2020**, *27*, 2913–2926.
18. Rani, K. V.; Sarma, B.; Sarma, A. Plasma Treatment on Cotton Fabrics to Enhance the Adhesion of Reduced Graphene Oxide for Electro-Conductive Properties. *Diam. Relat. Mater.* **2018**, *84*, 77–85.

19. Peran, J.; Ercegović Ražić, S.; Sutlović, A.; Ivanković, T.; Glogar, M. I. Oxygen Plasma Pretreatment Improves Dyeing and Antimicrobial Properties of Wool Fabric Dyed with Natural Extract from Pomegranate Peel. *Color. Technol.* **2020**, *136*, 177−187.
20. Ražić, S. E.; Glogar, M. I.; Peran, J.; Ivanković, T.; Chaussat, C. Plasma Pre-Treatment and Digital Ink Jet Technology: A Tool For Improvement of Antimicrobial Properties and Colour Performance of Cellulose Knitwear. *Mater. Today: Proc.* **2020**, *31*, S247−S257.
21. Hossain, M. A.; Chen, W.; Zheng, J.; Zhang, Y.; Wang, C.; Jin, S.; Wu, H. The Effect of O2 Plasma Treatment and PA 6 Coating on Digital Ink-Jet Printing of PET NonWoven Fabric. *J. Text. Inst.* **2020**, *111*, 1184−1190.
22. Zhang, C.; Wang, L.; Yu, M.; Qu, L.; Men, Y.; Zhang, X. Surface Processing and Ageing Behavior of Silk Fabrics Treated with Atmospheric-Pressure Plasma for Pigment-Based Ink-Jet Printing. *Appl. Surf. Sci.* **2018**, *434*, 198−203.
23. Samant, L.; Jose, S.; Rose, N. M.; Shakyawar, D. Antimicrobial and UV Protection Properties of Cotton Fabric Using Enzymatic Pretreatment and Dyeing with Acacia Catechu. *J. Nat. Fibers* **2022**, *19*, 2243−2253.
24. Olczyk, J.; Sójka-Ledakowicz, J.; Kudzin, M.; Antecka, A. The Eco-Modification of Textiles using Enzymatic Pretreatment and New Organic UV Absorbers. *Autex Res. J.* **2021**, *21*, 242−251.
25. Biswas, T.; Yu, J.; Nierstrasz, V. Effective Pretreatment Routes of Polyethylene Terephthalate Fabric for Digital Inkjet Printing of Enzyme. *Adv. Mater. Interfaces* **2021**, *8*, 2001882.
26. Memon, H.; Wang, H.; Yasin, S.; Halepoto, A. Influence of Incorporating Silver Nanoparticles in Protease Treatment on Fiber Friction, Antistatic, and Antibacterial Properties of Wool Fibers. *J. Chem.* **2018**, 4845687. Available from: https://doi.org/10.1155/2018/4845687, *2018*.
27. Wang, H.; Farha, F. I.; Memon, H. Influence of Ultraviolet Irradiation and Protease on Scale Structure of Alpaca Wool Fibers. *Autex Res. J.* **2020**, *20*, 476−483.
28. An, F.; Fang, K.; Liu, X.; Yang, H.; Qu, G. Protease and Sodium Alginate Combined Treatment of Wool Fabric for Enhancing Inkjet Printing Performance of Reactive Dyes. *Int. J. Biol. Macromol.* **2020**, *146*, 959−964.
29. Vinayagamoorthy, R. Influence of Fibre Pretreatments on Characteristics of Green Fabric Materials. *Polym. Polym. Compos.* **2021**, *29*, 1039−1054.
30. Harane, R. S.; Adivarekar, R. V. Sustainable Processes for Pre-Treatment of Cotton Fabric. *Text. Cloth. Sustainability* **2017**, *2*, 1−9.
31. Choi, P.; Yuen, C.; Ku, S.; Kan, C. W. Digital Ink-Jet Printing for Chitosan-Treated Cotton Fabric. *Fibers Polym.* **2005**, *6*, 229−234.
32. Özgüney, A. T.; Bozacı, E.; Demir, A.; Özdoğan, E. Inkjet Printing of Linen Fabrics Pretreated with Atmospheric Plasma and Various Print Pastes. *AATCC J. Res.* **2017**, *4*, 22−27.
33. Memon, H.; Khoso, N. A.; Memon, S.; Wang, N. N.; Zhu, C. Y. Formulation of Eco-Friendly Inks for Ink-Jet Printing of Polyester and Cotton Blended Fabric. *Key Eng. Mater.* **2016**, *671*, 109−114. Available from: https://doi.org/10.4028/www.scientific.net/KEM.671.109.
34. Ding, Y.; Shamey, R.; Chapman, L. P.; Freeman, H. S. Pretreatment Effects on Pigment-Based Textile Inkjet Printing−Colour Gamut and Crockfastness Properties. *Color. Technol.* **2019**, *135*, 77−86.
35. Kim, H.-J.; Hong, J.-P.; Kim, M.-J.; Kim, S.-Y.; Kim, J.-H.; Kwon, D.-J. Improving the Digital to Garment Inkjet Printing Properties of Cotton by Control the Butyl Acrylate Content of the Surface Treatment Agent. *Appl. Surf. Sci.* **2022**, 152322.

36. Kim, H.-J.; Seo, H.-J.; Kwak, D.-S.; Hong, J.-P.; Yoon, S.-H.; Shin, K. Preparation and Evaluation of Low Viscosity Acrylic Polymer Based Pretreatment Solution for DTP Reactive Ink. *Text. Color. Finish.* **2017**, *29*, 122−130.
37. Ding, Y.; Parrillo-Chapman, L.; Freeman, H. S. A Study of the Effects of Fabric Pretreatment on Color Gamut from Inkjet Printing on Polyester. *J. Text. Inst.* **2018**, *109*, 1143−1151.
38. Ma, W.; Shen, K.; Li, S.; Zhan, M.; Zhang, S. One-Bath Pretreatment for Enhanced Color Yield of Ink-Jet Prints using Reactive Inks. *Molecules* **2017**, *22*, 1959.
39. Kanik, M.; Hauser, P. J. Ink-Jet Printing of Cationised Cotton using Reactive Inks. *Color. Technol.* **2003**, *119*, 230−234.
40. Montazer, M.; Malek, R.; Rahimi, A. Salt Free Reactive Dyeing of Cationized Cotton. *Fibers Polym.* **2007**, *8*, 608−612.
41. Wang, L.; Ma, W.; Zhang, S.; Teng, X.; Yang, J. Preparation of Cationic Cotton with Two-Bath Pad-Bake Process and its Application in Salt-Free Dyeing. *Carbohydr. Polym.* **2009**, *78*, 602−608.
42. Bu, G.; Wang, C.; Fu, S.; Tian, A. Water-Soluble Cationic Chitosan Derivative to Improve Pigment-Based Inkjet Printing and Antibacterial Properties for Cellulose Substrates. *J. Appl. Polym. Sci.* **2012**, *125*, 1674−1680.
43. Yuen, C.; Ku, S.; Kan, C. W.; Choi, P. A Two-Bath Method for Digital Ink-Jet Printing of Cotton Fabric with Chitosan. *Fibers Polym.* **2007**, *8*, 625−628.
44. Yuen, C.; Ku, S.; Kan, C. W.; Choi, P. Enhancing Textile Ink-Jet Printing with Chitosan. *Color. Technol.* **2007**, *123*, 267−270.
45. Kan, C. W.; Yuen, C.; Tsoi, W. Using Atmospheric Pressure Plasma for Enhancing the Deposition of Printing Paste on Cotton Fabric for Digital Ink-Jet Printing. *Cellulose* **2011**, *18*, 827−839.
46. Liu, Z.-T.; Yang, Y.; Zhang, L.; Liu, Z.-W.; Xiong, H. Study on the Cationic Modification and Dyeing of Ramie Fiber. *Cellulose* **2007**, *14*, 337−345.
47. Lugoloobi, I.; Memon, H. Chemical Structure and Modification of Cotton. In *Cotton Science and Processing Technology: Gene, Ginning, Garment and Green Recycling*; Wang, H., Memon, H., Eds.; Springer Singapore: Singapore, 2020; pp 417−432.
48. Khatri, A.; Peerzada, M. H.; Mohsin, M.; White, M. A Review on Developments in Dyeing Cotton Fabrics with Reactive Dyes for Reducing Effluent Pollution. *J. Clean. Prod.* **2015**, *87*, 50−57.
49. Chen, W.; Zhao, S.; Wang, X. Improving the Color Yield of Ink-Jet Printing on Cationized Cotton. *Text. Res. J.* **2004**, *74*, 68−71.
50. Miles, L. Textile Printing. Society of Dyers and Colourists: England, 1994.
51. Kaimouz, A. W.; Wardman, R. H.; Christie, R. M. The Inkjet Printing Process for Lyocell and Cotton Fibres. Part 1: The Significance of Pre-Treatment Chemicals and their Relationship with Colour Strength, Absorbed Dye Fixation and Ink Penetration. *Dye. Pigment.* **2010**, *84*, 79−87.
52. Wang, L.; Hu, C.; Yan, K. A One-Step Inkjet Printing Technology with Reactive Dye Ink and Cationic Compound Ink for Cotton Fabrics. *Carbohydr. Polym.* **2018**, *197*, 490−496.
53. Yang, H.; Fang, K.; Liu, X.; Cai, Y.; An, F. Effect of Cotton Cationization using Copolymer Nanospheres on Ink-Jet Printing of different Fabrics. *Polymers* **2018**, *10*, 1219.
54. Rekaby, M.; Abd-El Thalouth, J.; Abd El-Salam, S. H. Improving Reactive Ink Jet Printing via Cationization of Cellulosic Linen Fabric. *Carbohydr. Polym.* **2013**, *98*, 1371−1376.

55. Hauser, P. J.; Kanik, M. Printing of Cationized Cotton with Acid Dyes. *AATCC Rev.* **2003**, *3*.
56. Kanik, M.; Hauser, P. J. Printing of Cationised Cotton with Reactive Dyes. *Color. Technol.* **2002**, *118*, 300–306.
57. Tabba, A. H.; Hauser, P. Effect of Cationic Pretreatment on Pigment Printing of Cotton Fabric. *Text. Chem. Color. Am. Dyest. Report.* **2000**, *32*.
58. Wang, C.-X.; Zhang, Y.-H. Effect of Cationic Pretreatment on Modified Pigment Printing of Cotton. *Mater. Res. Innov.* **2007**, *11*, 27–30.
59. Wang, C.; Yin, Y.; Wang, X.; Bu, G. Improving the Color Yield of Ultra-fine Pigment Printing on Cotton Fabric. *AATCC Rev.* **2008**, *8*.
60. Tarbuk, A.; Grancaric, A. M.; Leskovac, M. Novel Cotton Cellulose by Cationization During Mercerization—Part 2: The Interface Phenomena. *Cellulose* **2014**, *21*, 2089–2099.
61. Mondal, M. I. H. *Cellulose and Cellulose Derivatives: Synthesis, Modification and Applications;* Nova Publishers, 2015.
62. Grancaric, A. M.; Tarbuk, A.; Pusic, T. Electrokinetic Properties of Textile Fabrics. *Color. Technol.* **2005**, *121*, 221–227.
63. Qiao, X.; Fang, K.; Liu, X.; Gong, J.; Zhang, S.; Wang, J.; Zhang, M. Different Influences of Hydroxypropyl Methyl Cellulose Pretreatment on Surface Properties of Cotton and Polyamide in Inkjet Printing. *Prog. Org. Coat.* **2022**, *165*, 106746.
64. De Geyter, N.; Morent, R.; Leys, C. Pressure Dependence of Helium DBD Plasma Penetration into Textile Layers. *IEEE Trans. Plasma Sci.* **2008**, *36*, 1308–1309.
65. Kan, C.-W.; Yuen, C. Influence of Plasma Gas on the Quality-Related Properties of Wool Fabric. *IEEE Trans. Plasma Sci.* **2009**, *37*, 653–658.
66. Morent, R.; De Geyter, N.; Verschuren, J.; De Clerck, K.; Kiekens, P.; Leys, C. Non-Thermal Plasma Treatment of Textiles. *Surf. Coat. Technol.* **2008**, *202*, 3427–3449.
67. Wang, C.; Qiu, Y. Two Sided Modification of Wool Fabrics by Atmospheric Pressure Plasma Jet: Influence of Processing Parameters on Plasma Penetration. *Surf. Coat. Technol.* **2007**, *201*, 6273–6277.
68. Shenton, M.; Stevens, G. Surface Modification of Polymer Surfaces: Atmospheric Plasma versus Vacuum Plasma Treatments. *J. Phys. D: Appl. Phys.* **2001**, *34*, 2761.
69. Schulz, G. Textile Chemistry of Digital Printing. *Melliand Textilber.* **2002**, *83*, E30–E32.
70. Inbakumar, S.; Morent, R.; De Geyter, N.; Desmet, T.; Anukaliani, A.; Dubruel, P.; Leys, C. Chemical and Physical Analysis of Cotton Fabrics Plasma-Treated with a Low Pressure DC Glow Discharge. *Cellulose* **2010**, *17*, 417–426.
71. Lam, Y.; Kan, C. W.; Yuen, C. Physical and Chemical Analysis of Plasma-Treated Cotton Fabric Subjected to Wrinkle-Resistant Finishing. *Cellulose* **2011**, *18*, 493–503.
72. Wong, K.; Tao, X.; Yuen, C.; Yeung, K. Low Temperature Plasma Treatment of Linen. *Text. Res. J.* **1999**, *69*, 846–855.
73. Yuen, C.; Kan, C. W.; Jiang, S.; Ku, S.; Choi, P.; Wong, K. Optimum Condition of Ink-Jet Printing for Wool Fabric. *Fibers Polym.* **2010**, *11*, 229–233.
74. Achwal, W. Textile Chemical Principles of Digital Textile Printing (DTP). *Colourage* **2002**, *49*, 33–34.
75. Noppakundilograt, S.; Buranagul, P.; Graisuwan, W.; Koopipat, C.; Kiatkamjornwong, S. Modified Chitosan Pretreatment of Polyester Fabric for Printing by Ink Jet Ink. *Carbohydr. Polym.* **2010**, *82*, 1124–1135.
76. Bahmani, S.; East, G.; Holme, I. The Application of Chitosan in Pigment Printing. *Color. Technol.* **2000**, *116*, 94–99.

77. Faisal, S.; Ali, M.; Siddique, S. H., et al. Inkjet Printing of Silk: Factors Influencing Ink Penetration and Ink Spreading. *Pigment. Resin. Technol.* **2021**. Available from: https://doi.org/10.1108/PRT-12-2019-0120. In press.
78. Memon, H.; Khatri, A.; Ali, N., et al. Dyeing Recipe Optimization for Eco-Friendly Dyeing and Mechanical Property Analysis of Eco-Friendly Dyed Cotton Fabric: Better Fixation, Strength, and Color Yield by Biodegradable Salts. *J. Nat. Fibers* **2016**, *13*, 749−758. Available from: https://doi.org/10.1080/15440478.2015.1137527.
79. Faisal, S.; Tronci, A.; Ali, M., et al. Pretreatment of Silk for Digital Printing: Identifying Influential Factors using Fractional Factorial Experiments. *Pigment. Resin. Technol.* **2020**, *49* (2), 145−153. Available from: https://doi.org/10.1108/PRT-07-2019-0065.
80. Gao, C.; Xing, T.; Hou, X.; Zhang, Y.; Chen, G. Clean Production of Polyester Fabric Inkjet Printing Process without Fabric Pretreatment and Soaping. *J. Clean. Prod.* **2021**, *282*, 124315.

# Colorants for digital textile printing and their chemistry

*Abdul Khalique Jhatial[1], Pardeep Kumar Gianchandani[1], Biruk Fentahun Adamu[2,3], Aijaz Ahmed Babar[1] and Hanur Meku Yesuf[2,3]*
[1]Department of Textile Engineering, Mehran University of Engineering and Technology, Jamshoro, Sindh, Pakistan, [2]Key Laboratory of Textile Science and Technology, Ministry of Education, College of Textiles, Donghua University, Shanghai, P.R. China, [3]Ethiopian Institute of Textile and Fashion Technology, Bahir Dar University, Bahir Dar, Ethiopia

## 7.1 Introduction

Introducing the inkjet printing technology has brought advancements in textile printing, offering several advantages such as high efficiency, ease of use, economical and eco-friendly. Over the last two decades, digital fabric printing has rapidly expanded, opening up new prospects for designers, printers, entrepreneurs, ink suppliers, and consumers. The increasing interest in digital printing technologies for short-run productions, particularly dye-based and inkjet print head technology for digital printing of textiles. After the printing machines, the most vital component of this printing technique is the ink. Dyes-based colorants were initially used for digital printing; however, the pretreatment and post-treatment requirements of these colorants hindered their vast range of applications in digital printing.[1] Therefore, scientists looked for a new class of colorants called the inks. These inks have been widely used on textiles.

There is an apparent shift in the textile market's requirements, particularly in the fashion design sector. The US textiles and apparel industry is adopting demand-activated manufacturing architecture to meet customers' needs and survive in the 21st century.[2] In recent years, the rapid production of textile prints *via* digital printing has revolutionized the printing industry. Digital printing was introduced jointly by Canon, Inc., and KANEBO textiles, Ltd in 1996. Digital textile printing is widely performed *via* two methods, the first method uses direct inkjet injection on textiles for printing, and the second method uses the sublimation technique, also referred as the transfer printing technique. The direct printing method has a higher speed as it uses piezoelectric print heads for printing on textiles.

New advancements in the printing sector of textiles, along with design advancements, have completely changed the current market and style. Designers are now bringing changes in the style with the click of their fingers. Digital textile production is becoming the most cost-effective option for firms in the apparel and home markets as a response to the "fast fashion" trend and a switch from mass production to mass customization. This has pushed textile printers to develop new printing techniques with lower cost and process techniques. The inkjet printed textiles and goods found their prominent space considering the emerging need for sustainable marketing,[3,4] with fast fashion and interest in textile waste recycling.[5,6]

Digital printing and colorants have continuously modified to adopt higher speeds and better color performance. Digital printing can print complex design patterns and images directly on the fabric using design software. Scientists at Philadelphia university are working to combine the digital printing process with a loom for fabric printing directly on the loom; this needs the on-loom desizing and coloration facility to prepare the fabric for printing. The reactive dye is mainly used for the coloration of cotton fabric.[7] For different fabrics, different classes of inks are required; for example, cotton, viscose, polyester, and nylon cannot be printed with the same type of ink and printing process. Therefore, a specific class of ink is used for fibers such as reactive, acid, and disperse inks for cotton, nylon, and polyester fibers. In addition, different pretreatment and post-treatment equipment are required for diverse types of inks. Acid and reactive inks require a wet post-treatment, and the printed fabrics need to undergo several steaming, washing, and drying steps. On the other hand, pigment and disperse inks require a dry-post-treatment as they need a heat fixation process after printing the fabrics.

## 7.2 Why digital printing?

Digital printing has brought new textile printing technology and revolutionized it. It has many advantages over conventional printing, such as fast and easy design corrections, eco-friendly and print clarity. With the advancements in computer designing and computer-aided designing in the past three decades, it has become highly creative to design prints with changes and their effect on textiles. Now digital printing has become the tool to get quick samples and productions of the designed patterns. Conventionally it was a hectic process, as for each print design, we had to bring changes to printing machines with the required colors, and the required no of screens, and the printing screens were also changed accordingly. Thanks to digitalization, it has become a matter of clicking on the computer screen to print the required design with the necessary changes. The textile digital printing process, however, has some limitations and are under development, such as production speeds, type of inks for the required fabric to be printed, resolution/drop size and configuration, ink performance, substrate handling, color control, and color matching and washing fastness properties. Second, the traditional printing of fabric requires pretreatment and post-treatment for proper fixation and better rubbing properties. This has to be addressed in digital printing to achieve the required color effects and performance.

## 7.3 Colorants for digital printing

Digital printing is widely used and demanded by consumers due to its better properties and colorful designs. The textile sector has undergone a digital revolution and transformation over the last few decades. Printing textiles with digital printing require a particular type of dye called inks in digital printing. The inkjet inks are manufactured to yield specific properties by optimum droplet formation within the nozzles of the printing head. These properties are tuned with pH, surface tension, viscosity, conductivity, and particle size distribution of the ink. Inkjet inks are broadly classified into two classes, base and colorants. The base category is the classification based on the application medium of ink through which the colorants are dissolved or dispersed in the media, whereas the colorants refer to the type of color used in the media.

Further classifications of base and colorants are given in detail in Fig. 7.1. These inks are categorized as aqueous, nonaqueous phase change, and UV-cured-based ink designed for inkjet printers to print on textiles. Aqueous inks, also referred to as water-based inks, offer environmentally friendly attributes over their nonaqueous-based counterparts due to reduced emissions of volatile organic compounds (VOCs). Nonaqueous, also referred to as solvent-based inks, have high VOC emissions. Some hybrid inks containing water and solvents for the application medium are also used for digital printing;[8] however, considering the environmental concerns and legislations, water-based inks are preferred for digital printing.

Water-based inks require porous or specially treated substrates for their applications; these inks do not adhere to nonporous substrates. Water-based inkjet dyes have a wide range of colors with improved performance. These inks reflect a specific range of wavelengths, various classes of inks have been developed, such as reactive, acid, disperse, direct, basic inks, and pigment inkjet inks are used for a specific class of fibers; the use of traditional dyes fail in the majority of circumstances, due to their physical and chemical characteristics. It is, therefore, specially

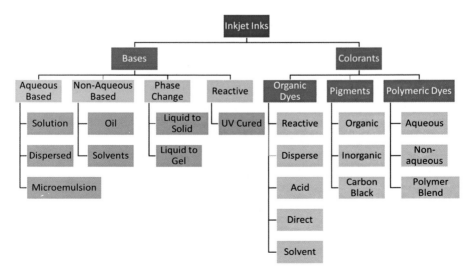

**Figure 7.1** Classification of inkjet inks.[9]

designed inkjet ink is used in digital printing. The physical and chemical properties of ink influence the quality and performance of inkjet printing. Besides dye inks, pigment inks are also widely used; pigment-based inks share more usage in inkjet printing than dyes. More than 50% of digitally printed textiles use pigment inks; however, their stability, reliability, nozzle clogging, performance (washing and rubbing), and comfort hinder their extensive usage. The development of high-speed printing machines and production demands have made colorists and industrialists provide specially designed colorants for digital printing. Initially, digital printing was limited to photography and paper printing. The digital printing of textiles is similar to the inkjet printing of papers using the computer. The printing speed on textiles is, however, slow at 12 m/h. and is continuously improving. Moreover, the printing process has advantages, such as a vast range of designs printed within the same printer and quick pattern changes.

The selection of ink type is based on the type of fiber present in the textile substrate; therefore, several inks have been developed for printing the different substrates, as given in Table 7.1. Inks, however, are under continuous development and modification to formulate a single class of inks that can be printed on all substrates. If this happens, it will significantly contribute to digital printing and benefit such as lower production cost, better productivity, improved color prediction, and reproducibility.

## 7.4 Preparation of substrate for digital printing

Pretreatment of the fabric before printing is carried out to achieve good color strength, better fastness properties, neat, clean print avoiding halloing and overprinting effects, optimum image quality, and better absorption in digital textile printing, especially with reactive dyes, acid dyes, and pigment inks. Fabric pretreatments are vital in the ink-substrate interaction; various chemicals have been used to boost the ink and substrate surface properties for their interaction and fixation.

Table 7.1 Different types of ink colorants and their applications.

| S. no | Ink | Application | Ink-fiber interaction |
|---|---|---|---|
| 01 | Acid ink | Silk, wool, and nylon | Ionic and hydrogen bonding |
| 02 | Disperse ink | Polyesters | Physical interaction, solid-state mechanism |
| 03 | Reactive ink | Cellulose-based fibers, silk, and wool | Covalent bonding |
| 04 | Direct inks | Cellulose-based fibers | Van der Waals forces and hydrogen bonding |
| 05 | Pigments | All types of fibers | Physical interaction, complex surface polymer bonding |

Usually, before coloration, the fabric is pretreated with alkalis, acids, and other auxiliaries.[10] Why we cannot add alkali with the printing ink can now be asked. The simple answer is that reactive dye hydrolysis will take place. Therefore, the fabric is pretreated early with alkali, and then ink is printed. Second, some chemicals may cause the corrosion of nozzles and affect ink's dielectric properties, such as in the case of thickeners. In this case the cellulosic fabric (cotton and viscose) is inkjet printed with the reactive dyes in two phases, initially pretreated with alkali and thickener and later steamed and washed for better brighter color properties. The pretreatment process is different for the different dye classes.

## 7.5 Post-treatment for digital printing

The post-treatment of digitally inkjet printed fabrics depends upon the type of ink used, some fabrics are steamed, washed, and then dried, and others are washed and then dried. Post-treatment processes are dependent on the type of ink and the substrate; however, for textile inkjet printing, the pretreatment and post-treatment processes are mandatory for preparing textiles. Digital printing on textiles is different from paper printing as textiles are subjected to washing in the daily routine; therefore, fixation of printed designs is necessary; otherwise, the designs will fade out after repeated washes. The beauty of digital printing is its clarity in designs that do not fade quickly once fixed with post-treatment processes such as steaming and UV curing. The post-treatments process for different inks are discussed with the inks in detail.

## 7.6 Types of inks for digital printing

The advancements in digital printing for textiles have created new opportunities and options to print textiles freely with any design without any limitations. The colorants used for digital inkjet printing textiles are, however, underdevelopment, especially for textiles. Still, some inks based on textile dyes are widely used and modified to improve their performance. The widely used inks based on dyes and pigments for printing textiles are discussed in detail below. It should be noted that ink might be contained in different forms; it comes in barrels, then it goes into ink buckets installed on the printers, from where it is transferred to cartridges. The different containers of inks are shown in Fig. 7.2. The color of the ink, brightness, and high saturation can also be seen in Fig. 7.2C.

### *7.6.1 Reactive dye inks*

Reactive dye inks belong to the aqueous-based class of inks. These inks are widely used in the digital printing of textiles. Reactive inks are distinguished due to their printing quality, fastness performance, and range of application on various fibers

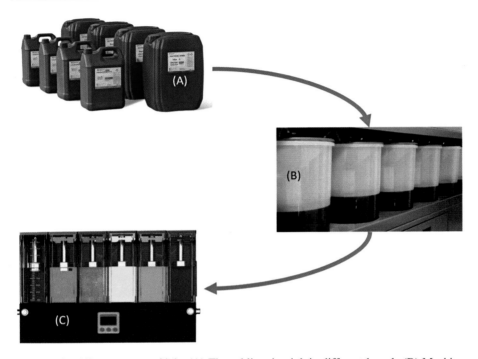

**Figure 7.2** Different stages of inks (A) The sublimation ink in different barrels (B) Machine inking buckets or bottles with ink on the machine (C) The ink in the container (secondary cartridge) This shows the ink's color.

such as cotton, rayon, or silk.[11] The best performance in digital printing is achieved with reactive inks for printing cotton and viscose.

The structure of reactive dyes consists of two parts the chromophore and the reactive system. The chromophore of the dye can be anthraquinone, azo, or phthalocyanine. The reactive dyes are commonly named after their reactive systems, such as vinyl sulfone (VS) and Monochlorotriazine (MCT). Several reactive systems can be attached to the chromophore of the dye to improve its properties and fulfill the needs of various processes. Some dyes are designed for cold application conditions, and others for hot conditions. The application requirements of process and conditions have thrust the dyers to develop several dyes for meeting the process needs. Hence these dyes have continuously been modified with their chemistry to fulfill their purposes. Most chemical structural developments are related to chromophores, reactive groups, and the leaving groups of the dye. Reactive dyes consist of different structural components that perform a specific task. The chromophore of the dye contributes to the substantivity of color for cellulose fibers, whereas the reactive system enables the dye to chemically react with the fiber polymers' hydroxyl (OH) groups.[12] The chromophore and reactive groups are linked via a bridging group. The solubilizing group and their attachment with chromophores make the dye water soluble. The leaving groups are attached to reactive groups, leaving groups to leave the reactive group, and the reactive group covalently reacts with polymers OH groups. The structural components of reactive dyes are highlighted in different colors in Fig. 7.3.

There are two types of reactive dyes, that is, monofunctional and bi-functional dyes; the reactive inks used for digital printing are monofunctional. Monofunctional inks contain one reactive group in their structure. These reactive groups react with the cellulose fiber polymer system and form a covalent bond. Reactive inks containing Monochlorotriazine reactive groups are preferred in textile printing as these dyes are less sensitive to process conditions and have more colors. On the other hand, VS based reactive inks are famous for their dark hues; however, the process sensitivity of these dyes leads to dye hydrolysis, and color variations, especially for new printing supervisors, are challenging to go with these dyes. Second, VS inks need steaming for the fixation; uncontrolled steaming will lead to color variations, such as dark black will become brown or gray; however, controlled conditions and improved dyes would serve the purpose.

Another class of reactive dyes typically used in conventional screen printing is Dichlorotriazine (DCT). These inks are highly reactive, therefore, are not used in digital printing. The high reactivity nature reduces their ink form and printing paste stability.

The MCT inks are low reactive and high substantivity for cellulosic fibers and are primarily used for cotton digital printing. The reactivity of different reactive groups is given in Fig. 7.4. Reactive inks containing MCT reactive groups inks are easier to use than VS but limited to lighter hues; ink manufacturing companies have developed MCT dyes with dark, intense hues but still are not comparable with those shades achieved with VS inks.

The reactive ink application process for digital printing starts loading the inks in the digital inkjet printing machines; the machines are given instructions regarding

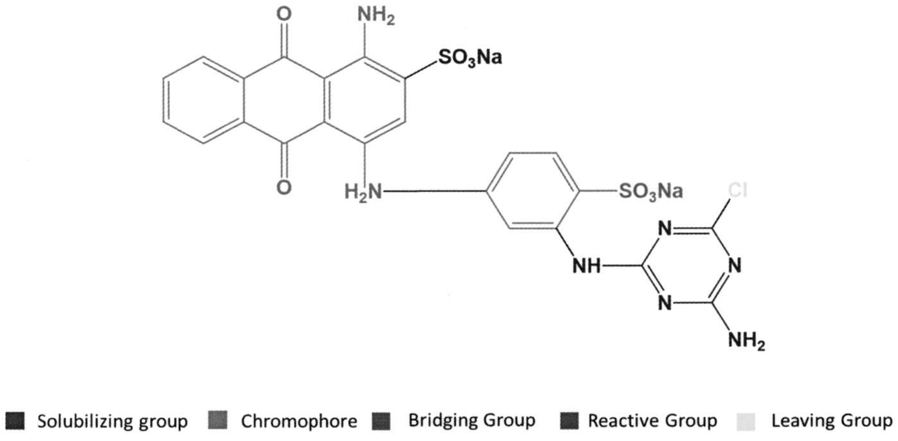

■ Solubilizing group  ■ Chromophore  ■ Bridging Group  ■ Reactive Group  ☐ Leaving Group

**Figure 7.3** Chemical structure of a reactive dye, highlighting its various components.

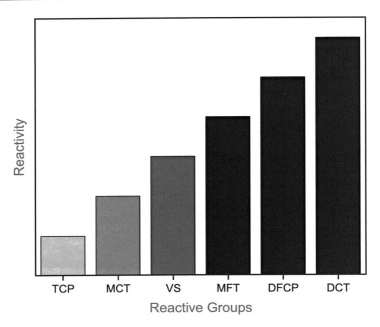

**Figure 7.4** Reactivity of different reactive groups, *DCT*, Dichlorotriazine; *DFCP*, Difloro chloropyrimidine; *MCT*, Monochlorotriazine; *MFT*, Monoflorotriazine; *TCP*, Trichloropyrimidine; *VS*, Vinyl sulfone.

the print designs via their software. The fabric to be printed is pretreated with the electrolyte and alkali; then, the pretreated fabric is printed with the required designs using reactive inks and inkjet heads. After printing, a high-pressure steamer is used to fix the inks covalently with the fiber polymer system. This ensures the deep penetration of inks with higher fastness properties.[13] Finally, the fabric is washed to remove the unfixed dyes. The schematic illustration of the reactive inkjet printing process is given in Fig. 7.5.

## 7.6.2 Acid dye inks

Acid inks are incredibly famous for coloring textiles, especially protein fiber-based materials such as silk, wool, and nylon. The acid inks derived their names as these are applied under acidic conditions. These are the sodium salts of sulfonic acids, highly water soluble and have higher fastness values, suitable for inkjet printing in the fashion and apparel industry. These inks attach to the fibers via an ionic bond. The ionic bond is formed between the Anionic charge ink molecule and cationic groups (Amino groups) present in the fiber polymer.[14]

Moreover, besides ionic bonds, the ink molecules also form hydrogen bonding and Van der Waals forces. The inks also contain hygroscopic agents, surfactants, and rheology modifiers for digital printing recipe formulation. Before printing, the

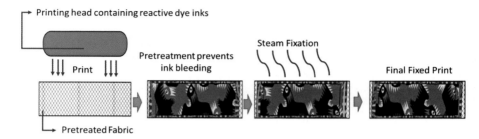

**Figure 7.5** Schematic illustration of process inkjet printing of reactive inks.

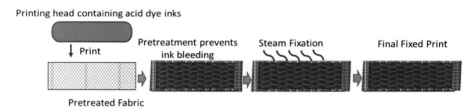

**Figure 7.6** Schematic illustration of process inkjet printing of acid inks.

fabric is pretreated with auxiliaries and inkjet printed with acid inks. The printed fabric is subjected to a steaming process for the fixation of inks with the fiber polymer system; during this fixation process, the steam provides the water molecules and heat energy for the transfer of the dye molecules from the surface of the fibers to the polymer system of fibers. The schematic representation of the application process of acid dye inks is given in Fig. 7.6.

## 7.6.3 Disperse dye inks

Inkjet printing of textiles and fabrics is a rapid process. Disperse inks are widely used for printing synthetic textiles; their chemical structure is shown in Fig. 7.7. Textiles are printed with these inks using a heat transfer method called dye-sublimation printing or can be printed directly on textiles. It is because dispersed inks are insoluble due to the lack of polar groups in their molecular chains. Therefore, these inks are applied as dispersion.[15]

The disperse dyes contain different chromophores such as azo, anthraquinone, etc., These dyes are classified as low, medium, and high energy disperse inks. Low-energy dispersed inks are used for sublimation or transfer printing. Disperse dyes and inks have the smallest molecule size; therefore, these inks can undergo sublimation. The sublimation of disperse inks is their ability to transfer from the liquid phase to the vapor phase without changing their molecular structure. It is solely the property of disperse dye-based inks to sublime at 170°C–250°C.

For transfer printing, the designs are printed on paper and then transferred to textiles using heat energy in the conventional printing process and digital printing or

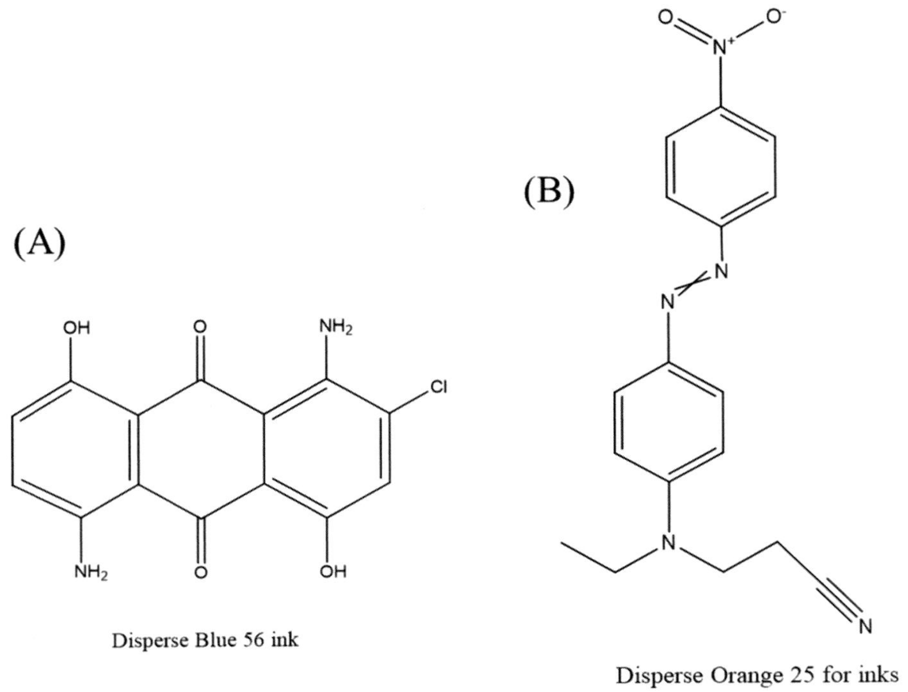

**Figure 7.7** Disperse dye inks (A) Disperse blue 56, (B) Disperse orange 25.

can be printed directly on the fabric. For direct inkjet printing with disperse dyes, fabric pretreatment is necessary to improve the printing and color performance. Polyester fabric is normally pretreated with cationic agents, sodium alginate, and polyvinyl alcohol. The disperse inks are applied through an inkjet nozzle as per the design and color; later in the post-treatment, the fabric is washed, as shown in Fig. 7.8, the schematic illustration of inkjet printing of disperse dye inks.

For direct printing of the fabric, disperse inks are printed and later steamed for the fixation of inks. Directly printing disperse ink via digital inkjet printing is a slow process. Therefore, increasing printing demands of polyester fabric in fashions push the printers to print with transfer printing to meet the production needs. The fabrics printed with disperse inks, however, have good washing fastness properties, fair light fastness properties, and lower fastness to hot pressing.

### 7.6.4 Pigment inks

After dyes, pigments are the colorants widely used for printing textiles. Pigments are insoluble in water and are applied to textiles as dispersion. Digital printing also finds its widespread use even greater than dyes considering their importance in conventional printing. The issue with the dyes lies in their physical properties,

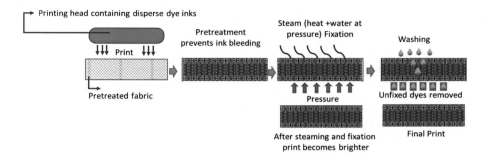

**Figure 7.8** Schematic illustration of process inkjet printing of disperse inks.

especially the particle size and process routes. For example, vat dyes have large particle sizes and lengthy process routes. Pigments have no affinity to textiles; they are physically attached to the substrate using the binder. Binder concentrations may, however, be used carefully as higher values increase the chances of nozzle blockage. Usually, with pigment inks, the dispersions are kinetically stable, and these dispersions are not stable compared to dye-containing inks, which are thermodynamically stable.

The pigment ink system also contains binders, rheology modifiers, surfactants, hygroscopic agents, and defoaming agents in their formulations. After printing, the formulated inks must have good compatibility, printability, and textile properties. The formulation positions the pigment particles in the desired place of the fabric through the print head for printing beautiful images. The binder holds the pigment particles from one side and attaches to the substrate from another. The recipe also helps to improve the performance of pigment inks. The properties of pigment inks are similar to those of conventional printing. The important properties of pigment inks related to digital inkjet printing are pigment dispersion, stability, particle size, and print quality. The particle size and size distribution affect the image quality, especially the color density, as the particle size affects the light reflection and absorption, consequently varying the color strength of the inks. The smaller particle size would yield a glossy surface; smaller particles are more suitable for inkjet printing from a dispersion stability viewpoint or considering the nozzle clogging factors; however, a small particle size (below 0.05 μm) of pigments is also unsuitable for colloidal stability and jetting reliability, as smaller particles have a high surface area-to-volume ratio. Therefore, with high dispersant demand and lower surface area per particle, keeping the charge density the same, smaller particles have less charge and repulsive barrier than the big particles could be less stable. Small particles can also affect the flow behavior and may cause clogging due to particle congestion.

Digital inkjet printing of textiles using pigment inks is carried out without pretreatment of the fabric as typically carried out with other dye-based inks such as disperse, acid, and reactive inks. Pigment inks are used to inkjet print cotton, its synthetic blends, and almost all other fabrics. Pigment printing is economical as it does not require pretreatment and washing. Pigment printing is performed in three stages for example. print-dry-cure process, whereas reactive printing has two

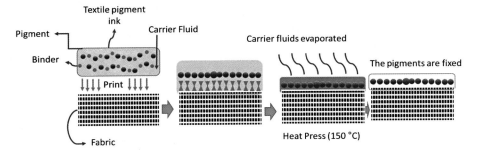

**Figure 7.9** Schematic illustration of process inkjet printing of pigment inks.

additional processes of pretreatment of fabric and washing off process, and fixation is carried out via steaming. Pigments are thermally fixed with fiber for digital inkjet printing using ink systems, as illustrated in Fig. 7.9.

The fastness properties of the fabrics printed with pigment inks have good washing and rubbing fastness; however, these properties depend on the binder system, as the pigment inks are physically attached to the fiber polymer system with the assistance of the binder. In some cases, when the substrate and the pigment have the right pair of chemical groupings, the chemical reaction (cross-linking) also takes place, taking the example of cotton; pigments form the cross-links with OH groups of fibers.[16]

## 7.7 Important properties of inks

### 7.7.1 Viscosity

The viscosity of the printing ink is extremely important. It directly affects the printing quality, high viscosity will cause clogging of nozzles, and lower viscosity will cause inks to flow out from the patterns. The viscosity affects the machine speed and size of droplets. It is used to control the drop formation from the nozzles and the quality of fabric print. Many viscosity modifiers, such as Polyethylene Glycol (PEG), carboxymethyl cellulose, glycerol, and sodium alginate, are generally referred to as thickeners. For PEG, different molecular weights are used. Higher weights may, however, affect the elasticity properties of the printing ink. Thickeners such as PEG with a hydrophobe can act as a surfactant for controlling surface tension, increasing stability, and hygroscopic properties. They act as a humectant and prevent the immediate drying of ink during ejection from the jet nozzles.

### 7.7.2 Surface tension

The surface tension of printing inks is another primary factor besides the viscosity of inks; both factors are interrelated. The printing inks should have a surface tension of 25–60 dynes/cm.[17,18] It can be controlled using an appropriate surfactant and proper solvent composition. For wetting the capillary channels, the surface tension of the inks must be suitable for the smooth flow of inks through the nozzle. The water-based inks comprise 30%–80% of water as a proportion of the total mass of the ink; the remaining portion consists of organic solvents such as monohydric alcohol, the colorant, and the surface-active agent. Surfactants act as dispersants in dispersing ink systems for an aqueous medium. The use of surfactant helps the dye penetration evenly and overcomes the issues of ink stability, especially in disperse inks. The combination of surface tension and viscosity affects the inks flow behavior, referred to as rheology. Rheology determines the printing process, and viscosity and surface tension influence ink flow from the inkjet head.

The inks should not become dry immediately while discharging from the nozzles. Therefore, solvents with high volatile nature and rapid evaporation rate affect the flow behavior of inks. Process conditions such as temperature also affect the viscosity and rate of evaporation. The rheological characteristics of the inks, such as viscosity, thixotropy, and yield value, affect the print quality and performance of printing inks. It is necessary to reach the ductor roller and smoothly transfer to the substrate. The printing inks with lower viscosities and yield values produce pattern irregularities and cause the print quality problems such as mottle. Mottle is the problem of printing associated with uneven distribution of ink on a ductor roller, nonuniform rate of ink drying, and nonuniform absorption of damping solution; however, this printing problem has been eliminated in digital printing due to automation and control systems for viscosity and flow control in inkjet printing machines. The aid of automation reduces cost and time with improved production and quality.

The ink recipe also contains defoaming agents, as, during printing, the formation of bubbles in the inks may cause clogging of nozzles resulting in inappropriate prints.[19] Therefore, defoaming agents prevent bubbles' formation by decreasing the surface tension. It is also considered that defoaming agents shall not affect the stability of ink storage.

### 7.7.3 Particle size

The particle size of ink is also one of the technical parameters that are considered, especially in printing with insoluble dyes, such as with disperse and other functional inks, especially conductive inks,[20] which affects the printing performance significantly. If ink dispersion is not stable and has a precipitation ability, it will aggregate and affect ink's ejection from the printing head, static stability of the ink, and refilling response during operation at higher speeds. The inkjet inks also contain humectants for controlling the evaporation of inks during drying. These hygroscopic agents, such as glycols and alcohols, are used to remove moisture from the air to prevent the print head's clogging during the printer's idle position.[21]

### 7.7.4 pH and electrolytes

The pH is essential for water-based, insoluble, and pigment inks; it significantly affects ink solubility and stability. Inks are a mixture of various chemicals that improve ink performance and colorants. All chemicals and colorants are, however, soluble or dispersed at specific pH levels; changes in this pH value will make dispersions unstable or solubility into precipitation. Therefore, some inks also contain buffers for pH stability. The inks also contain electrolytes in their formulations; the presence of electrolytes also causes severe ink stability problems, especially when stored for prolonged durations. Therefore, very low concentrations of electrolyte can increase the shelf life of inks.

### 7.7.5  *Dielectric properties and conductivity*

Conductive and dielectric properties of inks are also important, especially for the inks used in continuous inkjet printing, as in some printing methods such as continuous, direct (piezoelectric print head), and drop-on-demand printing methods. In these methods, the droplets of ink are deflected due to an electric field, and the charge ability of the inks is controlled with the help of electrolytes and ionic surfactants. For inks, their conductivity must be controlled very precisely as variations in the charge during prolonged storage, salt precipitation, and reaction with the ink container can affect the conductive behavior of inks.[18]

### 7.7.6  *Ink storage and stability*

For the ink's storage, its physical and chemical properties are essential, such as pH, viscosity, surface tension, particle size and dispersion, and dielectric properties.[22] For an ink, its dielectric properties shall remain over the period. This property is fundamental in new nonimpact printers.[23] Its properties shall remain intact for effective ink use over a period. The inks of the soluble type of dyes become unstable when their components react with each other such as polymerization in UV inks, precipitation, phase separation, and reaction with container walls. For the inks based on the dispersion of colored particles, dispersion instability will lead to reduced shelf life. The inks are subjected to a heating test at a temperature of 60°C, referred to as heat aging, to determine the shelf life of inks. One week of heat aging of this test is equivalent to three months of storage at room temperature, and the 1-month (4 weeks) aging by this method is equivalent to one-year storage at room temperature. The shelf life of inks has been continuously improved. Recently, some companies have developed inkjet inks that can be stored for 2 years. The improved shelf life of inks will be economical and eco-friendly without affecting the performance and process.

## 7.8  Challenges in digital printing of textiles

Digital printing technology of textiles has brought the revolution in textile printing. It has addressed most of the challenges of conventional printing primarily related to the design change, the number of colors related issues, and the software integration for designs; however, the digital revolution of printing is still facing challenges, the most of challenges are related to the ink system, printing technology, and production. Dye-based and pigment-based inks have limitations regarding the substrate and printing technology used. Therefore, developments are required in the inks system to introduce inks compatible with most printers for printing textiles. For accomplishments of challenges, interdisciplinary research comprising material scientists, chemists, and mechanical and software engineers is required to bring customer satisfaction through the commercial viability of the developed system. The physical properties of inks, such as their jet ability, performance, size, and image robustness, are continuously modified and have enormous potential to enhance the improvement of the process. The dye-based and pigment-based inks have different fabric pretreatment and post-treatment processes due to their application requirements and are mandatory processes[24]; these two value-addition steps create problems for industrialists having no such facilities. The possibility of combining these several steps in a single machine can serve the issue.[25] This can, however, be improved, and processes can be modified without affecting the functionality and performance of inks. The type of substrate has its viscosity requirements; it also significantly influences the print performance of the same ink on different fibers; therefore, for better image quality (higher resolution) and productivity, it must be increased. Currently both factors are influenced by the viscosity of inks and their factors. Inkjet printings success is based on treating the "inkjet detail" with respect. Although it is elegant in concept, it is exceedingly difficult to implement in practice, especially in very demanding applications such as extremely high throughput systems and printing of functional and unique materials. The future of textile printing will be digital; that day is not far away when conventional processes will be rarely seen in the printing industry.

# References

1. Raymond, M. In *Industrial Production Printers ± DuPont ArtistriTM 2020 Textile Printing System;* Ujiie, H., Ed.; Woodhead Publishing Limited and CRC Press LLC: England, 2006.
2. Fralix, M. Digital Printing and Mass Customization. In *Digital Printing of Textiles;* Ujiie, H., Ed.; Wood hear: England, 2006; pp 293−311.
3. Chen, L.; Qie, K.; Memon, H.; Yesuf, H. M. The Empirical Analysis of Green Innovation for Fashion Brands, Perceived Value and Green Purchase Intention—Mediating and Moderating Effects. *Sustainability* **2021,** *13,* 4238.
4. Memon, H.; Jin, X.; Tian, W.; Zhu, C. Sustainable Textile Marketing—Editorial. *Sustainability* **2022,** *14,* 11860.
5. Memon, H.; Ayele, H. S.; Yesuf, H. M.; Sun, L. Investigation of the Physical Properties of Yarn Produced from Textile Waste by Optimizing their Proportions. *Sustainability* **2022,** *14,* 9453.
6. Wang, H.; Memon, H.; Abro, R.; Shah, A. Sustainable Approach for Mélange Yarn Manufacturers by Recycling Dyed Fibre Waste. *Fibres Text. East. Eur.* **2020,** *28,* 18−22.

7. Dutta, S.; Bansal, P. Cotton Fiber and Yarn Dyeing. In *Cotton Science and Processing Technology: Gene, Ginning, Garment and Green Recycling;* Wang, H., Memon, H., Eds.; Singapore: Springer Singapore, 2020; pp 355–375.
8. Wang, L.; Yan, K.; Hu, C.; Ji, B. Preparation and Investigation of a Stable Hybrid Inkjet Printing Ink of Reactive Dye and CHPTAC. *Dye. Pigment.* **2020**, *181*, 108584. Available from: https://doi.org/10.1016/j.dyepig.2020.108584.
9. Le, H. P. Progress and Trends in Ink-Jet Printing Technology. *J. Imaging. Sci. Technol.* **1998**, *42*, 49–62.
10. Jhatial, A. K.; Yesuf, H. M.; Wagaye, B. T. Pretreatment of Cotton. In *Cotton Science and Processing Technology: Gene, Ginning, Garment and Green Recycling;* Wang, H., Memon, H., Eds.; Springer Singapore: Singapore, 2020; pp 333–353.
11. Hou, X.; Chen, G.; Xing, T.; Wei, Z. Reactive Ink Formulated with Various Alcohols for Improved Properties and Printing Quality onto Cotton Fabrics. *J. Eng. Fibers Fabr.* **2019**, *14*. Available from: https://doi.org/10.1177/1558925019849242 1558925019849242.
12. Memon, H.; Khatri, A.; Ali, N.; Memon, S. Dyeing Recipe Optimization for Eco-Friendly Dyeing and Mechanical Property Analysis of Eco-Friendly Dyed Cotton Fabric: Better Fixation, Strength, and Color Yield by Biodegradable Salts. *J. Nat. Fibers* **2016**, *13*, 749–758. Available from: https://doi.org/10.1080/15440478.2015.1137527.
13. Yasukawa, R.; Higashitani, H.; Yasunaga, H.; Urakawa, H. Dye Fixation Process in Ink-Jet Printing of Cotton Fabric by Reactive Dye. *Sen'i Gakkaishi* **2008**, *64*, 113–117. Available from: https://doi.org/10.2115/fiber.64.113.
14. Choi, S.; Cho, K. H.; Namgoong, J. W.; Kim, J. Y.; Yoo, E. S.; Lee, W.; Jung, J. W.; Choi, J. The Synthesis and Characterisation of the Perylene Acid Dye Inks for Digital Textile Printing. *Dye. Pigment.* **2019**, *163*, 381–392. Available from: https://doi.org/10.1016/j.dyepig.2018.12.002.
15. Kosolia, C.T.; Tsatsaroni, E.G.; Nikolaidis, N.F. Disperse Ink-Jet Inks: Properties and Application to Polyester Fibre. **2011**, *127*, 357–364, https://doi.org/10.1111/j.1478-4408.2011.00334.x.
16. Khan, M. R. Pigment Ink Formulation, Tests and Test Methods for Pigmented Textile Inks. *Chem. Mater. Res.* **2016**, *08*.
17. Akhtar Hayata, Muhammad; Gulzar, Tehsin; Hussain, Tanveer; Kirn, Shumaila; Farooq, Tahir; Ahmed, Akram Eco-Friendly Preparation, Characterization and Application of Nano Tech Pigmented Inkjet Inks and Comparison of Particle Size Effect and Printing Processes. *Am. Sci. Res. J. Eng. Technol. Sci. (ASRJETS)* **2020**, *72*, 197–213.
18. Benjamin, T.; Howard, E. K.; Asinyo, B. K. The Chemistry of Inkjet Inks for Digital Textile Printing—Review. *Int. J. Manag. Inf.* **2016**, *4*, 61–78.
19. Kamyshny, A.; Magdassi, S. J. I.-B. M.; Korvink, J.; Smith, P. J.; Shin, D.-Y., Eds. *Inkjet Ink Formulations;* 2012.
20. Nir, Moira M.; Zamir, D.; Haymov, Ilana; Ben-Asher, Limor; Cohen, Orit; Faulkner, Bill; de la Vega, Fernando In *Electrically Conductive Inks for Inkjet Printing;* Magdassi, S., Ed.; World Scientific Publishing Co. Pte. Ltd., 2010.
21. Cie, C. *Inks for Digital Printing.* Ink Jet Textile Printing Elsevier: Amsterdam, The Netherlands, 201585–97.
22. Memon, H.; Khoso, N. A.; Memon, S.; Wang, N. N.; Zhu, C. Y. Formulation of Eco-Friendly Inks for Ink-Jet Printing of Polyester and Cotton Blended Fabric. *Key Eng. Mater.* **2016**, *671*, 109–114. Available from: https://doi.org/10.4028/www.scientific.net/KEM.671.109.

23. Zhao, H.; Zhang, K.; Fang, K.; Shi, F.; Pan, Y.; Sun, F.; Wang, D.; Xie, R.; Chen, W. Insights into Coloration Enhancement of Mercerized Cotton Fabric on Reactive Dye Digital Inkjet Printing. *RSC Adv.* **2022,** *12,* 10386−10394. Available from: https://doi.org/10.1039/D2RA01053D.
24. Ma, W.; Shen, K.; Li, S.; Zhan, M.; Zhang, S. One-Bath Pretreatment for Enhanced Color Yield of Ink-Jet Prints Using Reactive Inks. **2017,** *22,* 1959.
25. Wang, L.; Hu, C.; Yan, K. A One-Step Inkjet Printing Technology with Reactive Dye Ink and Cationic Compound Ink for Cotton Fabrics. *Carbohydr. Polym.* **2018,** *197,* 490−496. Available from: https://doi.org/10.1016/j.carbpol.2018.05.084.

# Inkjet printing of textiles enhanced by sustainable plasma technology

**8**

*Alka Madhukar Thakker[1], Danmei Sun[1] and Muhammad Owais Raza Siddiqui[1,2]*
[1]School of Textiles and Design, Heriot-Watt University, Galashiels, United Kingdom,
[2]Department of Textile Engineering, NED University of Engineering & Technology, Karachi, Pakistan

## 8.1 Introduction

Typically, a fabric is pretreated before inkjet printing to improve the quality of the print. Traditionally the desizing of sized fabrics is followed by surface modification with cationic agents that are high on water, energy, and chemical demand, jeopardizing the environment and human health. Plasma technology is promulgated as it can be employed to impart functionality to the fabric surface. The process is low on emissions and effluents and is an environmentally friendly approach. After PSM, the fabric is prepared further to be made ready for inkjet printing.[1,2] The field of plasma technology is extensive, involving varied types of plasma, such as thermal plasma and nonthermal plasma (cold or low-temperature plasma).[3] Fig. 8.1 provides a glimpse into varieties of plasma characterized by electron density and temperature.[4] Cold or low-temperature plasma is categorized into atmospheric-pressure plasma, corona discharge, dielectric barrier discharge (DBD), low-pressure glow, radiofrequency and microwave discharges, laser-produced plasmas, and others.[5] Plasma technology has been applied in multiple fields, such as textiles, electronics, automotive, medical, etc.[6] Depending on the type of plasma gas or a mixture of gases used and other treatment parameters, plasma technology can also be used for varies aimed surface effects, such as hydrophilicity, hydrophobicity, and adhesion. Gases such as oxygen, air, nitrogen, and argon can be used to improve the surface free energy; on the other hand, gases such as fluorocarbons are polymerizing plasma gases that can be used to produce hydrophobic surface.[7]

There are several techniques of modifications using plasma, such as plasma polymerization, plasma activation, plasma coating, plasma deposition, plasma implantation, etc.[8] Low-temperature plasma is widely used for the modification of textile-based substrates for enhanced wettability, printability, and adhesion. It enables the acquisition of desired properties without quenching, which is the rapid cooling of materials in water, oil, liquid nitrogen, or air.[9] In addition, the technique can prevent ozone depletion by eliminating volatile organic compounds, controls pollution by being low on emissions and is safe for aquatic flora and fauna as there is no effluent generation.[5]

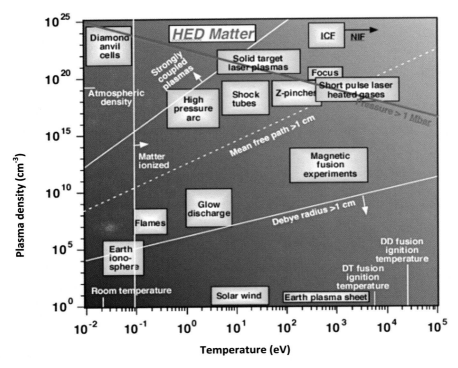

**Figure 8.1** Characterization of plasmas by electron temperature and electron density.[4]

A typical surface activation by plasma technology is shown in Fig. 8.2. The plasma atmosphere consists of free photons, electrons, radicals, ions, atoms, molecules, and a lot of different exciting particles independent of the gas selected. Free electrons gain energy from the source, such as a radio frequency (RF) electric field colliding with neutral gas molecules and thereby dissociating the molecules to form numerous reactive species. The primary process in the generation of activated species, especially ions, involves electron impact and photoionization. Photoionization occurs when a photon is emitted due to the energy falling off an electron from a higher level to a lower one, hitting a gas molecule with high enough energy. The primary ion productions are shown in Eqs. (8.1) and (8.2).[10]

$$e + M \rightarrow M^+ + 2e^- \tag{8.1}$$

$$h\gamma + M \rightarrow M^+ + e^- \tag{8.2}$$

Where "$e$" is an electron, "$M$" is a plasma gas molecule, and "$h\gamma$" is the energy of photons.

Ionization also takes place via collisions of metals species.[13] Free radicals may also be generated by electron impact, thermally, and by photolysis. The impact with a monomer can lead to its excitation and dissociation, generating free radicals. It is

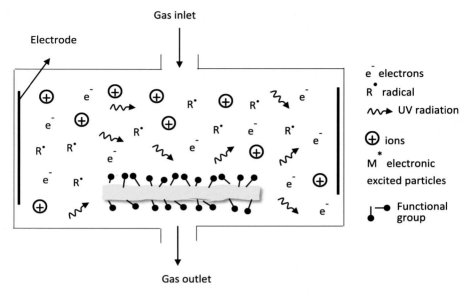

**Figure 8.2** Work principle of plasma surface treatment.[11,12]

the interaction of these exciting species with solid surfaces placed in the plasma reactors that results in the chemical and physical modification of the material surface. All the active species react with the substrate surface. This generates chemical functionality conferring new properties on the substrate surface.[10–12,14,15] The by-products of these reactions are readily removed by the vacuum system. Different reactive species in the plasma chamber interact with the substrate surface; cleaning, modifying, or coating occurs dependent on the selected plasma gases and treatment conditions. The formed reactive particles react directly with the surface without damaging the bulk properties of the treated substrates. In fact, the surface modification is limited to the outermost 10–1000 Å of the substrate materials to be processed with the plasma. The advantage of plasma surface modification is that the treated substrate surface properties can be changed significantly in a short treatment time, and the consumption of chemicals is very low. A disadvantage of plasma treatment is that the properties may alter and deteriorate with time. It was suggested that vacuum packing the treated substrates would be able to prolong the shelf life of functionality induced by plasma.[16]

## 8.2 Plasma treatment of varied fabrics to facilitate inkjet printing

Low-temperature plasma has been employed to improve the surface characteristics of different types of fabrics prior to their inkjet printing. In one of the studies, cotton fabric was pretreated with RF (13.56 MHz) atmospheric pressure plasma (APP) with helium and oxygen gas plasma for 3 seconds. Thereafter, the treated fabric was processed separately with chitosan, sodium alginate, and a mixture of chitosan-sodium alginate paste before inkjet printing. The sodium alginate-chitosan mixture yields better color values, fastness properties, and antibacterial functionality with plasma-treated fabric samples as compared to the untreated samples.[17] In the enhanced fiber surface grooves, shrinkage and cracks were observed on the APP treated cotton fabric samples under the Scanning Electron Microscope (SEM). This enabled the greater deposition of printing paste. Fabric surface elemental analysis using X-ray Photoelectron Spectroscopy (XPS) exhibited an increase in oxygen element content and a decrease in carbon element content. PSM alters the fabric surface characteristics.

Because of APP treatment, $C=C$ and $C=H$ bonds dissociate, and subsequently, elements C and H were replaced by O, forming $-C=O$, $C=O$, $O-C-O$, $O-C=O$ on the treated cotton fabric surface.[10,17] The increase in the polar functional groups is consequently responsible for greater absorption of the resultant fabric. Similarly, the surface contact angle dropped from 98° to 58° and wetting time decreased from 6.3 to 4.5 seconds for untreated in comparison to APP-treated cotton fabric, respectively, indicating plasma treatment has significantly improved the wettability of the fabric. It can be concluded that the APP treated fabric would be more receptive to subsequent inkjet printing. The inkjet-printed fabric ratings for dry and wet rub fastness and wash fastness to color change and staining were noted to increase from grade 4 for the untreated to 4–5 and 5 for APP-treated cotton fabrics correspondingly.[17] Zhang and Zhang treated polyester fabric with plasma under power level of 300 W for 90 seconds. The untreated and treated fabrics were digitally printed with cyan pigment-based ink. The change of surface energy of plasma-treated fabric samples was significantly improved.

The surface energy was increased with an increase in the polar component due to PSM. The mechanical properties were also evaluated. There is no outward change in textile properties. The strength of the fabric in both warp and weft directions has been marginally increased due to a little bit enlarged fiber surface area caused by plasma bombardment on the treated fiber surface. In turn, the increased fiber surface leads to the increased cohesive force among fibers and yarns within the fabric. An excellent sharpness in inkjet printing quality was reported.[18] In another study, a silk fabric was given Dielectric Barrier Discharge Plasma treatment using different gas (es), that is (a) Air, (b) Air/Agron (70/30), and (c) Air/Oxygen (70/30) and treatment time of 30, 60, 90, 120, 150 and 180 seconds correspondingly. The treated silk fabric samples were digitally printed with Cyan, Magenta, Yellow, and Black pigment inks, then baked in the oven at 80°C for 5 minutes. Fabric samples treated with air/oxygen (70/30) gas plasma for 90 seconds gained a maximum K/S value of 6.10, as shown in Fig. 8.3.[19] On the other hand, Fang and his team concluded that silk fabric treated with oxygen plasma gas for 10 minutes, at 80 Watts power level and working pressure of 50 Pa, is optimum in gaining higher color values and excellent design sharpness on digital printing with magenta color ink.[20]

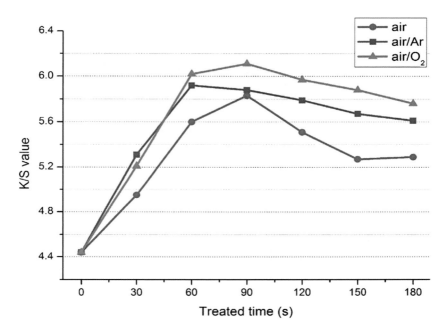

**Figure 8.3** Color strength of magenta color pigment ink obtained on silk fabric samples treated by using different gaseous plasmas against treatment time.[19]

A polyester fabric was given atmospheric air plasma treatment for 30 and 1800 seconds, respectively, and thereafter printed with so-called "particle-free silver ink." The 0.75 mm thickness of PET fabric samples untreated and 27 KJ/m$^2$ plasma-treated were inkjets printed with four-layer silver ink. The surface morphology of the fabric samples was examined using SEM, as shown in Fig. 8.4. It was apparent that the treated fabric (Images C and D) had greater Ag ink deposition and conductivity in contrast to the untreated fabric sample (images A and B).[21] Interesting surfaces of the varied textile materials could be modified with eco-friendly plasma technology.

Materials such as PET have been used as a flexible material base for electronic devices. Molina et al.[22] used plasma to treat polyester fabrics to improve the adhesion of reduced graphene oxide with the fabric surface in order to produce high-quality conductive fabrics. The surface roughness of the treated fabrics was examined by atomic force microscopy (AFM). Negative charges were induced on the surface of fibers in fabric by plasma treatment. In a similar research area, Thomas et al. experimented using PET as a base material and printed their developed conductive inks on the fabric surface. Because of the hydrophobic and low moisture-regaining nature of PET, it has low absorption and adhesion with the conductive inks. They employed air as low-temperature plasma gas and pretreated the PET fabric prior to inkjet printing. It was found that the density and coverage of the printed metal particles, printed ink wettability, and the adhesion with PET fabric samples were significantly improved by a short time atmospheric air plasma surface

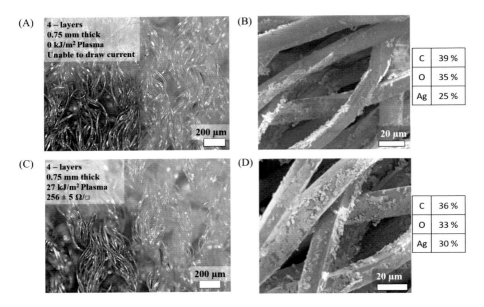

**Figure 8.4** Untreated (A and B) and plasma-treated (C and D) polyester fabrics digitally printed with silver ink.[21]

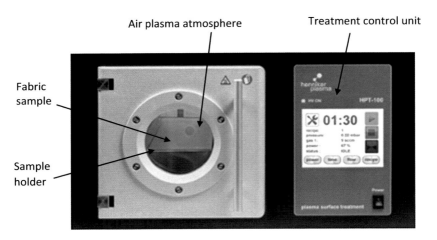

**Figure 8.5** Henniker plasma treatment device illustrating wool fabric surface is activated under plasma atmosphere.[23]

modification.[21] They also studied the influence of plasma treatment parameters on the quality of inkjet printing and identified an ideal treatment power level.

Fig. 8.5 demonstrates the air plasma surface activation of a fabric in progress. During the PSM, the fabric surface is sequentially bombarded with free radicals from air plasma, activated, cleaned, and modified mechanically and chemically. At the set time, the procedure ends and vents out the surplus radicals and indicates the opening of the device door.[11,12,23] The plasma surface treated, and untreated wool and cotton fabrics were further examined by an ATR-FTIR, optical tensiometer for identification of functional groups and using SEM for observation of surface morphology changes of the fabric samples. The polar functional groups and surface morphology of treated fibers in a fabric are the important characteristic parameters directly related to the wettability of the fabric. Both untreated and plasma-treated fabric samples were digitally printed with the formulated plant-based inks that were developed by the researchers. The absorbance spectra of cotton and wool fabric samples plasma-treated at 100 W power level for a period of 3 minutes, obtained by ATR-FTIR, is shown in Fig. 8.6. Comparisons were made to the untreated counterpart samples.

The wool fabric treated by air plasma for 3 minutes exhibited a strong peak of C-O stretch at the wavenumber of 1313.13 $cm^{-1}$, a strong peak of N-O asymmetric stretch at 1514.91 $cm^{-1}$, and a moderate peak of N-H bend 1° amine at a wavenumber of 1608.19 $cm^{-1}$. The characteristic peaks observed on treated wool fabric samples enable the formation of a strong hydrogen bond at the plasma-induced polar functional groups, that is, C-O stretch, as demonstrated in Fig. 8.6A highlighted in a yellow circle. The enlarged image of the peak area is illuminated in Fig. 8.6B. Overall, for the wool fabric sample, wavenumbers between 1000 and 2500 $cm^{-1}$ indicate there is a significant and major shift of functional groups corresponding to polar groups, that is, free hydroxyl, carbonyls, and H-bonded phenols and amines. The functional groups created additional anchoring sites of the fiber surface and are available for bond formation with chromophores when undertaking subsequent inkjet printing. The elevated peak groups were induced by various active particles in air plasma atmosphere. It implies higher color absorbance and fastness properties of the resultant sustainable fabrics. For further quantification of the polar functional groups, specifically, oxygen (O), carbon (C), and nitrogen (N), the XPS analysis have been recommended.[23] It was observed in a study that both the wash fastness ratings to multifiber fabric staining and the color change were improved after plasma treatment. The wash fastness to color change is enhanced from poor (1−2 for untreated) to moderate (3 and 3−4) for the cotton fabric inkjet printed with herbal ink. The observed increase in wash fastness ratings could be attributed to the used plasma surface modification.[23] Correspondingly, the wash fastness to color change was improved from good (3−4 and 4) for untreated samples to very good (4−5 and 5) for plasma-treated wool fabric samples. In another research, a silk fabric was given Atmospheric Air DBD plasma treatment at 300 W for 90 seconds. The color bleeding phenomenon was observed after inkjet printing of the untreated and treated fabric samples, as depicted in Fig. 8.7. It was conspicuous that the treated fabric samples exhibited sharp printed edges as compared to the untreated fabric samples that had high color bleeding. Likewise, the highest color values were gained on plasma-modified silk fabric inkjet-printed, as indicated in Fig. 8.8. These findings were further supported by surface chemistry analysis using XPS, indicating a decrease in hydrophilic groups, such as C−O, C=O, C−N, and N−C=O that

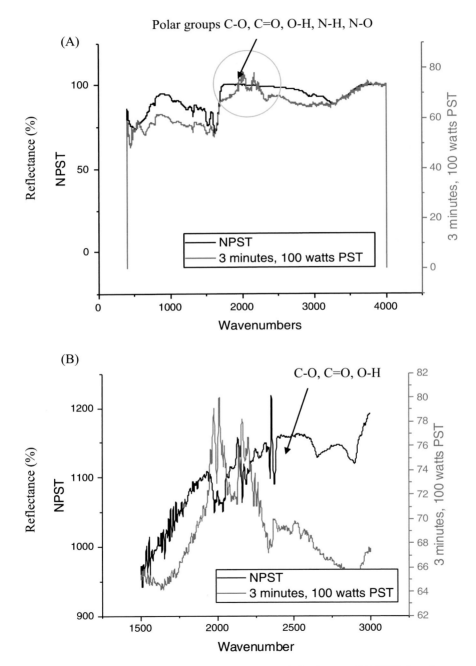

**Figure 8.6** (A) ATR-FTIR analysis—Air plasma-treated (3 minutes, 100 W PST) and untreated (NPST) plain weave wool fabric samples, and (B) detailed view of the picks corresponding to polar functional groups.[23]

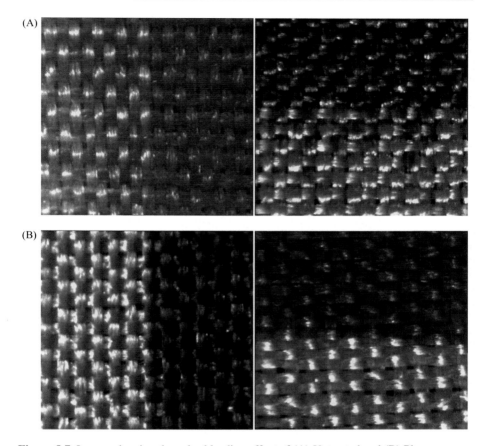

**Figure 8.7** Images showing the color bleeding effect of (A) Untreated and (B) Plasma-treated fabric samples.[24]

were induced by PSM.[24] It is noticeable that material scientists could justify the changes in surface chemistry due to PSM by employing varied efficient technologies that corroborate and validate the material analysis.

Polyester fabric is hydrophobic in nature. Hydrophilic polar groups were reported to be induced by PSM at 300 W for 180 seconds with (a) Pure Air and (b) mixed gases Air/Argon (50/50). The surface morphology examination on AFM revealed an etching effect of the plasma-treated fabric surface in comparison to the smooth surface of the original polyester fabric. The air plasma demonstrated the highest etching effect than the mix of Argon and air gas plasma because noble gas Argon weakened the etching effect, as illustrated in Fig. 8.9.[25]

The DSA-100 drop shape analyzer illustrated that the untreated PET fabric could not absorb water droplets; however, at merely 457 Ms of Air/Argon treatment, the PET fabric completely absorbed the water droplet as indicated in Fig. 8.10. PSM generated an etching effect on the fiber surface, and new polar groups were

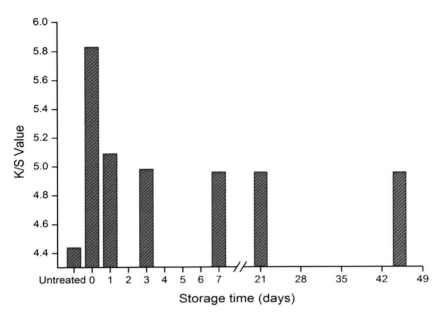

**Figure 8.8** The K/S values obtained on untreated, and plasma-modified silk fabrics stored for 0, 1, 3, 7, 21 and 45 days.[24]

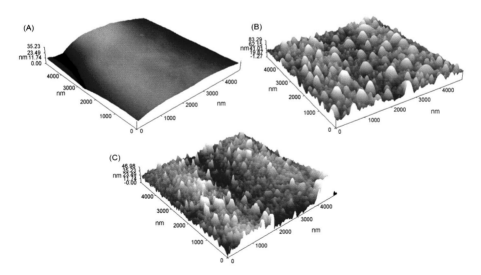

**Figure 8.9** Atomic Force Microscopy of (A) Original PET fabric, (B) Air plasma-treated PET fabric, and (C) Air/Argon plasma-treated PET fabric.[25]

**Figure 8.10** Surface Contact Angle Test: a droplet dropped on Untreated (left), and Air/Argon Plasma-treated (right) polyester fabric samples.[25]

implanted onto the PET fabric surface, thus, enhancing the wettability of the PET fabric. This facilitated the further inkjet printing process. The highest K/S value of 3.82 was acquired on Air/Argon (50/50) PSM-PET inkjet printed fabric, followed by Air plasma-treated fabric exhibited a K/S value of 3.58, and the untreated PET fabric had a much lower K/S value of 2.57. The high K/S value would reinforce the benefits of PSM for improving the hydrophilicity of the PET fabric.[25] Furthermore, the greater K/S value and better antiseepage effect are achieved with the increased plasma power as during which the high-speed electrons in the plasma atmosphere are elevated.[1] On the other hand, the mixed gas Air/Argon (10/90) plasma-treated fabric at 300 W power level for 150 seconds yielded a maximum K/S value of 4.38 as compared to the untreated fabric with a K/S value of 4.20. Zhang, Fang.[26] Furthermore, they exposed four layers of PET fabrics in atmospheric pressure air plasma and studied the depth of penetration of plasma on inkjet printing. The wetting time of each layer in comparison to the untreated fabric has been significantly, and the K/S values are depicted in the Graph presented in Fig. 8.11. It was noted that the plasma could permeate through all the four layers of the PET fabric; however, the effect was reduced from the top down. The same trend was observed for the wetting time, and K/S values attained for the four fabric layers.[27] Equally, Zang et al. experimented with Air and Air /Helium (10/90) plasma treatment on a PET fabric under 300 W power level and treated for 90 seconds. The color values of plasma treated PET fabric on inkjet printing were investigated and comparisons made to the untreated fabric. As anticipated, the K/s value gained with plasma-treated fabric is 23% higher than the untreated fabric.

The XPS spectra shown in Fig. 8.12 and the empirical data given in Table 8.1 indicate that the Air-He plasma surface modified fabric obtained the densest polar functional groups namely, $O=C-O$.[28] The PET fabric being hydrophobic is extensively studied by a team of authors implementing PSM with noble gases such as helium and argon. At the same time, air and oxygen gas plasma were also experimented with and proved to be most viable and efficient. Concurrently, Tsumaki et al. demonstrated a unique approach to plasma-aided inkjet printing wherein the conductive inks were released and implanted on the polyimide (Synthetic resin polymer)[29] substratum by APP in one step rapid process involving no pretreatment

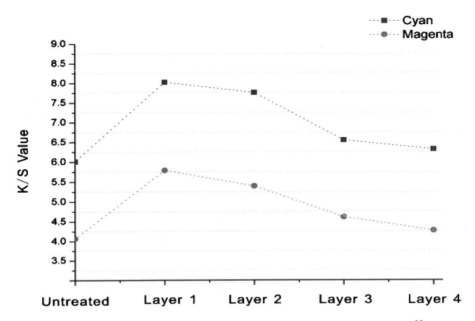

**Figure 8.11** K/S values of the four layers of plasma surface modified PET fabrics.[27]

or posttreatment.[30] A conductive silver line with an electrical resistivity of $2.9 \times 10^{-5}$ $\Omega$/cm was constructed on a polyimide substrate in only 1 second, including drying up, solidifying, and frittage (Sintering) of the ejected silver ink droplets.[30] The digital printing of challenging surfaces such as polyimide films with conductive inks could be accomplished with PSM.

The cotton fabric surface was modified with plasma from oxygen, nitrogen, and sulfur Hexafluoride, to improve the absorption of water-based pigment inks on inkjet printing. A wide color gamut was noted with oxygen plasma-treated cotton fabric, as illustrated in Fig. 8.13. A partial color gamut focusing on the "a*" and "b*" values that were acquired on inkjet printing is demonstrated in the Figure. Also, in one of the experiments, the cotton fabric was given low-temperature plasma treatment, pretreated with sodium alginate paste, and after that, inkjet printed with synthetic inks. It was noted that PSM cotton fabric exhibited uniform uptake of sodium alginate paste as per SEM examination, and the maximum color yield was gained under 2 minutes of plasma treatment.[31]

While plasma activation modifies surface chemistry and increases the surface energy of the treated substrate surface, nanoparticle plasma thin coating could provide hydrophobic functionality to paper. Additionally, improved printability was observed with regard to increased ink-paper adhesion and less ink demand. In this study, ToF SIMS analysis[33] was used. Equally, the static contact angle of PET after Helium and Argon gas Atmospheric Plasma (AP) reduced to 24 degrees from 85 degrees. The K/S values obtained by digitally printing the AP-treated PET and

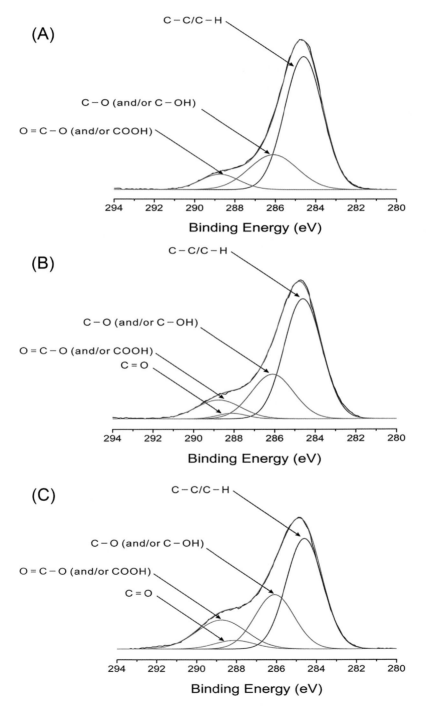

**Figure 8.12** XPS spectra of (A) Untreated, (B) Air PSM, and (C) Air-Helium PSM-PET fabric.[28]

**Table 8.1** Results of deconvolution of C1s peaks for polyester fabrics.[28]

| Binding energy (eV) | Untreated (%) | Air PSM (%) | Air-He PSM (%) | Plausible functional groups |
|---|---|---|---|---|
| 284.6 | 71.0 | 56.2 | 53.8 | C–C |
| 286.1 | 15.6 | 21.3 | 21.7 | C–O |
| 288.1 | 0 | 2.4 | 2.8 | C=O |
| 288.75 | 13.4 | 20.1 | 21.7 | O=C–O |

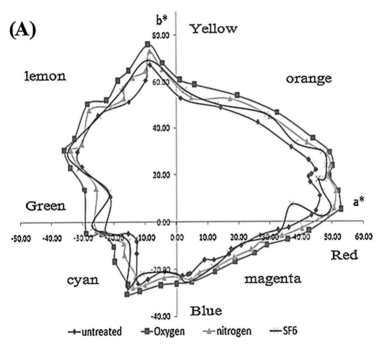

**Figure 8.13** Color gamut obtained on PSM cotton fabric on inkjet printing.[32]

cotton fabrics were higher than the untreated fabrics. For the black color reactive inks, the highest K/S value acquired on the AP-treated cotton fabric was 7.82 in comparison to 1.87 for the untreated cotton fabric.[34] Chitosan is a hydrophilic mediator by its nature. An experiment was conducted in which the hydrophobic PET fabric was PSM, and after that, processed with low molecular weight chitosan. The PET fabric treated at 100 W of plasma power level and 20 Pa pressure for 1.5 minutes was noted to gain the maximum hydrophilicity. PSM etched and roughened the PET fabric surface, consequently adding hydrophilic groups onto the treated fabric surface. This enabled higher adhesion of chitosan to PET fabric on subsequent processing. The PET fabric treated with 1.5% of chitosan concentration and 1.5% of acetic acid concentration at 110°C for 80 seconds was concluded to be the best chitosan treatment conferring hydrophilicity to the plasma surface modified PET fabric.[35] Fig. 8.14 reinforces the difference in the uniformity of the color obtained on inkjet printing of nonwoven PET fabric without and with plasma surface modification. Polyamide (PA) 6 filament from waste was also implemented in the study for coating the plasma surface modified nonwoven PET fabric as it confers hydrophilicity to the hydrophobic PET surface. Together with greater pigment, ink adhesion and color yield were derived.[36] The absorption percentage of 3.5% PA 6 coating solution increased from 11.97% to 19.92% due to PSM. The absorption of 6.8% PA 6 coating solution increased from 5.68% to 10.50%. In this study, the oxygen gas PSM was conducted under vacuum at 300 W plasma power level for 10 minutes.[36]

A low-temperature oxygen gas plasma surface modification was performed on a PET fabric and thereafter treated with N, O-carboxymethyl chitosan (NOCS). Altogether higher absorption, antistatic and antibacterial properties were acquired on the treated PET fabrics as compared to the untreated fabrics. The mechanism of the adhesion of NOCS and the addition of functional polar groups induced by PSM is depicted in Fig. 8.15. PEI and GA are Poly-ethylenimine and Glutaraldehyde, respectively.[37] In this research, the inkjet printing of the prepared fabric was not carried out. It was, however, found that the functional PET obtained was more suitable for digital printing due to the improved antistatic properties gained on the resultant fabric.

Razic et al. investigated the effect of oxygen and argon gas plasma treatment at different power levels of 500 and 900 W on cotton and lyocell knitted fabrics correspondingly. The fabrics were then inkjet-printed with water-based pigment inks. The plasma surface treatment performed at 500 W for 5 minutes was concluded to be the best in obtaining higher K/S value on cotton fabrics indicating the highest of 107.5% of the increase in the color value obtained with cyan color ink, 23.5% for magenta, 50% for yellow and 7.46% for blue color ink. The effects on the untreated and treated lyocell fabric samples were not significant. Besides, it was reported that the antimicrobial efficiency of the fabric was greater for both cotton and lyocell fabrics treated by plasma, inkjet-printed, and treated with 0.5% of the silver-based

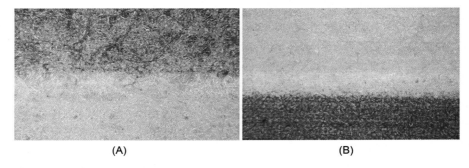

(A)          (B)

**Figure 8.14** Difference in uniformity of color obtained on the nonwoven PET fabric (A) Untreated, and (B) Plasma treated.[36]

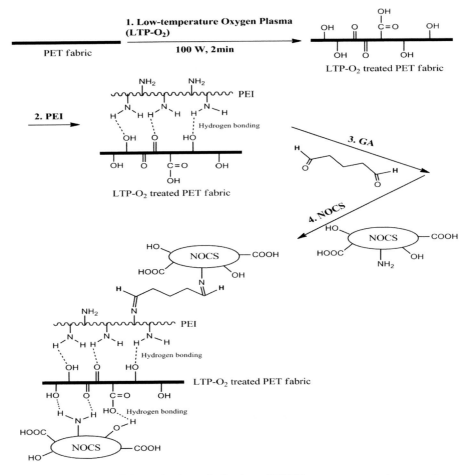

**Figure 8.15** The plausible mechanism of adhesion of NOCS on oxygen gas plasma surface modified PET fabric in the presence of PEI and GA.

antimicrobial solution in contrast to untreated fabric.[38] Polyimide films were treated by oxygen gas plasma and remarked difference was made in the uptake of copper conductive ink as indicated in Fig. 8.16. The fabricated material had an ink resistivity of $62.26 \times 10^{-6}$ $\Omega$/cm at 200°C for 1 hour in an $H_2$ atmosphere.[39]

The PET fabric was modified in three ways by using (1) Sodium hydro oxide alkali, (2) enzyme cutinase, and (3) Atmospheric air PSM at an electrical power of 1 kW, frequency of 26 kHz, and inter-electrode distance of 1.5 mm. Among the three approaches, PSM was concluded to be most efficient in improving the wettability without hampering other fabric properties.[40] To end, plasma surface modification is observed to illustrate the physical and chemical changes of the treated textile materials. The resultant fabrics possess enhanced functionality.

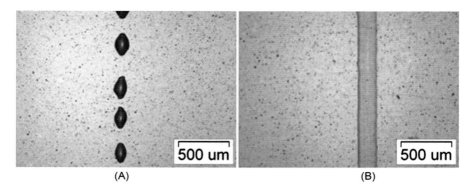

**Figure 8.16** The copper conductive ink on (A) untreated and (B) plasma-treated polyimide film.[39]

## 8.3 Conclusions

Plasma surface modification, as one of the most effective and economical surface treatment techniques, has been widely applied in the textile industry for fabric surface modification and activation. The review demonstrated the increase in wettability and improved quality of inkjet printing of textile substrates undertaken plasma treatment in a short period of time. Varied fabric substrates such as polyester, silk, cotton, wool, paper, lyocell, polyimide, and others demonstrated the same effect of an increase in polar functional groups, namely $C-O$, $C=O$, and $O-C=O$ that can be induced by a short time plasma treatment. Aided by PSM, more anchoring sites were available for chromophores and auxochromes to attach during the subsequent inkjet printing process. Therefore, improved sharpness of printing, color fastness, and functional properties were gained for the treated fabrics.

The review demonstrated wide-ranging techniques can be employed to evaluate the effects of plasma surface modification, such as surface contact angle test, drop-test, scanning electron microscope, AFM, and others. Overall, the plasma surface modification facilitated the inkjet printing process. Both the technologies, namely, the PSM and inkjet printing, are low on air and water pollution and hence are beneficial to both the environment and human health alike.

## References

1. Shenzhen CRF Co., Ltd., 2021. Support Material Testing and Equipment Testing, Inkjet Printing of Low-Temperature Oxygen Plasma Treated Silk Fabric by Plasma Cleaning Machine. [Online] http://en.plasmacrf.com/en/news_of_competitor/609.html [accessed May 19, 2022].
2. Christina, C. *Ink Jet Printing. Elsevier Limited;* Woodhead Publishing: England, 2015.
3. Hrabovsky, M., n.d. Thermal Plasmas: Properties, Generation, Diagnostics and Applications. Praha, Czech Republic: Institute of Plasma Physics.

4. Leer, B.V., 2019. Plasma Dynamics Modeling Laboratory, Aeronautics and Astronautics, Stanford University. [Online] https://pdml.stanford.edu/about [accessed May 22, 2022].
5. D'Angola, A.; Colonna, G.; Kustova, E. Editorial: Thermal and Non-Thermal Plasmas at Atmospheric Pressure. *Front. Phys.* **2022**. Available from: http://doi.org/10.3389/fphy.2022.852905.
6. Coulson, S. Plasma Treatment of Textiles for Water and Oil Repellency. In *Plasma Technologies for Textiles*; Shishoo, R., Ed.; Woodhead Publishing Limited: Cambridge, 2007; pp 183–201.
7. Marcandalli, B.; Riccardi, C. Plasma Treatments of Fibres and Textiles. In *Plasma Technologies for Textiles*; Shishoo, R., Ed.; Woodhead Publishing Limited: Cambridge, 2007; pp 282–300.
8. Inagaki, N. *Plasma Surface Modification and Plasma Polymerization*, I ed.; Technomic Publishing Company, Inc: Basel, Switzerland, 1996.
9. Materials Science & Engineering Student, 2022. What is Quenching?. [Online]: https://msestudent.com/what-is-quenching/ [accessed May 22, 2022].
10. Sun, D.; Stylios, G. Fabric Surface Properties Affected by Low Temperature Plasma Treatment. *J. Mater. Process. Technol.* **2006**, *173* (2), 172–177.
11. Henniker Plasma, 2021. Plasma Surface Activation. [Online]: https://plasmatreatment.co.uk/pt/plasma-technology-overview/plasma-surface-activation [accessed Jun 12, 2021].
12. Henniker Plasma, 2021. Atmospheric Plasma Treatment Explained. [Online]: https://plasmatreatment.co.uk/pt/plasma-technology-overview/atmospheric-plasma-treatment [accessed Jun 12, 2021].
13. Tolmachev, Y. A. *Mechanism of Species Generation in Plasma*; Technomic Publishing: Lancaster, PA, 1986.
14. Chi-wai, K. *A Novel Green Treatment for Textiles, Plasma Treatment as a Sustainble Technology*; CRC Press: Boca Raton, 2015.
15. Mather, R.; Wardman, R. *The Chemistry of Textile Fibres*; The Royal Society of Chemistry: Cambridge, 2011.
16. Plasma Etch, 2020. What is Vacuum Plasma?. [Online]: https://www.plasmaetch.com/vacuum-plasma.php [accessed Jun 12, 2021].
17. Kan, C. W.; Yuen, C. W.; Tsoi, W. Y. Using Atmospheric Pressure Plasma for Enhancing the Deposition of Printing Paste on Cotton Fabric for Digital Ink-Jet Printing. *Cellulose* **2011**, *18*, 827–839.
18. Zhang, C.; Zhang, X. Nano-Modification of Plasma-Treated Inkjet Printing Fabrics. *Int. J. Cloth. Sci. Technol.* **2015**, *27* (1), 159–169.
19. Zhang, C., et al. Study on the Physical-Morphological and Chemical Properties of Silk Fabric Surface Modified with Multiple Ambient Gas Plasma for Inkjet Printing. *Appl. Surf. Sci.* **2019**, *490*, 157–164.
20. Fang, K.; Wang, S.; Wang, C.; Tian, A. Inkjet Printing Effects of Pigment Inks on Silk Fabrics Surface-Modified with O2 Plasma. *J. Appl. Polym. Sci.* **2008**, *107* (5), 2949–2955.
21. Jones, T. D. A., et al. Plasma Enhanced Inkjet Printing of Particle-Free Silver Ink on Polyeser Fabric for Electronic Devices. *MNE, Micro Nano Eng.* **2022**, *14*, 1–7.
22. Molina, J., et al. Plasma Treatment of Polyester Fabrics to Increase the Adhesion of Reduced Graphene Oxide. *Synth. Met* **2015**, *202*, 110–122.
23. Thakker, A. M.; Sun, D.; Bucknall, D. Inkjet Printing of Plasma Surface–Modified Wool and Cotton Fabrics with Plant-Based Inks. *Environ. Sci. Pollut. Res* **2022**, *29*, 68357–68375.
24. Zhang, C., et al. Surface Processing and Ageing Behavior of Silk Fabrics Treated with Atmospheric-Pressure Plasma for Pigment-Based Ink-Jet Printing. *Appl. Surf. Sci.* **2018**, *434*, 198–203.

25. Fang, K.; Zhang, C. Surface Physical−Morphological and Chemical Changes Leading to Performance Enhancement of Atmospheric Pressure Plasma Treated Polyester Fabrics for Inkjet Printing. *Appl. Surf. Sci.* **2009**, *255*, 7561−7567.
26. Zhang, C.; Fang, K. Surface Modification of Polyester Fabrics for Inkjet Printing with Atmospheric-Pressure Air/Ar Plasma. *Surf. Coat. Technol.* **2009**, *203*, 2058−2063.
27. Zhang, C. M.; Fang, K. J. Influence of Penetration Depth of Atmospheric Pressure Plasma Processing into Multiple Layers Of Polyester Fabrics on Inkjet Printing. *Surf. Eng.* **2011**, *27* (2), 139−144.
28. Zhang, C.; Zhao, M.; Wang, L.; Yu, M. Effect of Atmospheric-Pressure Air/He Plasma on the Surface Properties Related to Ink-Jet Printing Polyester Fabric. *Vaccum* **2017**, *137*, 42−48.
29. Oxford University Press, Lexico.com, 2022. Polyimide. [Online]: https://www.lexico.com/definition/polyimide [accessed May 22, 2022].
30. Masanao, T., et al. Development of Plasma-Assisted Inkjet Printing and Demonstration for Direct Printing of Conductive Silver Line. *J. Phys. D: Appl. Phys.* **2018**, *51*, 1−5.
31. Yuen, C. W.; Kan, C. W. A Study of the Properties of Ink-Jet Printed Cotton Fabric Following low-Temperature Plasma Treatment. *Color. Technol.* **2006**, *123*, 96−100.
32. Pransilp, P., et al. Surface Modification of Cotton Fabrics by gas Plasmas for Color Strength and Adhesion by Inkjet Ink Printing. *Appl. Surf. Sci.* **2016**, *364*, 208−220.
33. Pykönen, M. Influence of Plasma Modification on Surface Properties and Printability of Coated Paper, *s.l;* Åbo Akademi University, 2010.
34. Park, Y.; Kang, K. The Eco-Friendly Surface Modification of Textiles for Deep Digital Textile Printing by In-Line Atmospheric Non-Thermal Plasma Treatment. *Fibers Polym.* **2014**, *15* (8), 1701−1707.
35. Wang, Y.; Yin, W. Hydrophilic Finishing for Polyester Fabric Using Plasma and Chitosan. *Mater. Sci. Forum, Trans. Tech. Publ. Ltd, Switz.* **2020**, *980*, 154−161.
36. Hossaina, M. A., et al. The Effect of O2 Plasma Treatment and PA 6 Coating on Digital Ink-Jet Printing of PET Non-Woven Fabric. *J. Text. Inst.* **2020**, *111* (8), 1184−1190.
37. Jingchun, L., et al. Environmentally Friendly Surface Modification of Polyethylene Terephthalate (PET) Fabric by Low-Temperature Oxygen Plasma and Carboxymethyl Chitosan. *J. Clean. Prod.* **2016**, *118*, 187−196.
38. Razic, S. E., et al. Plasma Pre-Treatment and Digital Ink Jet Technology: A Tool for Improvement of Antimicrobial Properties and Colour Performance of Cellulose Knitwear. *Mater. Today: Proc.* **2020**, *31*, S247−S257.
39. Goo, Y.-S., et al. Ink-Jet Printing of Cu Conductive Ink on Flexible Substrate Modified by Oxygen Plasma Treatment. *Surf. Coat. Technol.* **2010**, *205*, S369−S372.
40. Biswas, T.; Yu, J.; Nierstrasz, V. Effective Pretreatment Routes of Polyethylene Terephthalate Fabric for Digital Inkjet Printing of Enzyme. *Adv. Mater. Interfaces* **2021**, *8*, 2001882.

# Technological barriers to digital printing in textiles: a study

9

Md Aktarul Hasan
School of Informatics, Zhejiang Sci-Tech University, Hangzhou, P.R. China

## 9.1 Introduction

The beginning of digitalization Textile printing on an industrial scale relating to technology using inkjets has revolutionized textile production in the previous 10 years. While the Milliken Millitron digital carpet printer was one of the first attempts at digital printing in the 1970s, the first industrial textile inkjet printers were first seen in the 1990s, in defiance of the earliest models, including the Stork TruColor Jet Printer, which appeared in 1991, and the Ichinose Image Proofer, released in 1999. The Ichinose machine, as the name implies, was designed to produce design samples and proofs that would afterward be printed on printers that use rotary screens. It appeared that digital technology would never be suitable for the mass production of printed fabrics for quite some time.

The emergence of several solutions early in the new century points in the future direction, as though the first-time digital machines were appropriate for manufacturing, although on a limited scale and at a disproportionately high price. DuPont/Artistri 2020, Monna Lisa, and Reggiani DReAM systems allowed for excellent-grade new product proofing ideas and the production of low-volume, high-end products for the fashion industry, such as silk scarves. The presence of low-volume, high-value products in the market provided a platform for digital printing technology, allowing manufacturers of inkjet systems and components to justify more investment in next-generation systems while also providing a means to introduce textile mills to the technology's possible benefits and drawbacks. This initiation has resulted in a significant enhancement in the capabilities of inks and printing systems, allowing solutions to be introduced since approximately 2010 that can compete with traditional printing systems in terms of quality, capability, printing costs, and throughput. In the following sections, we will review the factors that will continue to increase performance and save costs in industrial digital textile printing over the next few years.

## 9.2 Pretreatment and posttreatment of the fabric

Most fabrics indisputably need pretreatment before they can be printed using inkjet printers. There are several reasons for pretreating substrates. One reason is that inkjet printing is a noncontact printing process, where the ink drops are jetted out through very fine nozzle heads. The ink is mainly designed to prevent drying and clogging. Some chemicals function as thickeners and binders, which cannot be jetted, and therefore, are added to the fabric rather than directly incorporated into the printing ink. In addition, some inks can only be printed on textiles efficiently under certain conditions. For example, reactive dyes are usually applied under alkali conditions.[1] The fabric is typically pretreated with a thickener for pigment and dye printing and bicarbonate for reactive and acid printing. Direct printing of dispersed dyes may also include an antimigrant and UV inhibitor in the pretreatment.

When using digital textile pigment inks, pretreatments are just as necessary as any other digital textile print process; however, unlike other processes where pretreatment is required for ink fixing, the purpose of pretreatment, considering pigment inks, is confined to enhancing the printed fabric's performance.[2] Fabrics marked "Prepared for Print (PFP)" are usually treated to remove any impurities like various sizing agents, oils, waxes, and other additives used at different stages of the manufacturing process. These chemicals react with one another, preventing the inks from being adequately fixed. These undesired substances must be removed for the inks to work reliably in color and fixation. In addition to the substances used in the process, the "lint" or "protruding fibers" from the fabric's surface must be removed. If this is not done, lint obstructs the placement of ink drops shown in Fig. 9.1, reducing the print quality. The pretreatment begins after the fabric is ready. "Dip and Nip Coating" and "Spray Coating" are the two most commonly used traditional procedures. The Dip and Nip coating techniques are widely employed for padding standard-width fabrics. After that comes the drying process.

The fabric is hydrated with padding liquor and then pressed to remove the surplus liquor. The latest pretreatment setups use the Spray Coating technique to coat the fabric's surface for broader widths. The Dip and Nip process is usually appropriate for low and medium GSM fabrics having concentrated levels of 60%−100% of fabric GSM. The fabrics are simpler to dry due to the reduced use of padding chemicals,

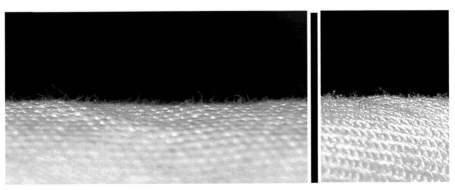

**Figure 9.1** Lint obstructs the placement of ink drops.

but fabric drying necessitates additional energy. The fabric's surface could also be pretreated with function-specific or more intense pretreatments. This makes it easier to work around and overcome difficulties (Fig. 9.2), including coating thickened fabrics; however, a precise method is required to provide an equal coating over the fabric surface. Adhesion boosters, penetration enhancers, and polymerization catalysts make up the majority of pretreatment additives since they collaborate to speed up the curing and adherence of the inks. A further pretreatment option is to improve certain prints' appearance by preventing ink infiltration. Pretreatment also creates a foundation for the ink to settle, which increases print sharpness.

Discovering the correct pretreatment mix ratio for the fabric's characteristics is critical. Bright colors might arise from excessive adhesion promoters, but wash and rub performance will suffer. Too much penetration enhancer usually results in faded or pale colors with a good wash and rub fastness. The fabric's hand feel and texture are also influenced by pretreatment and ink mix. Although all providers give a common formula, it is vital to evaluate the results before moving on to large-scale printing. This assures performance and aids in the reduction of printing costs. Pretreating the fabrics is recommended, but it is not always possible. Some fabrics should not be pretreated because the surface texture is essential, or the coating procedure may change the fabric's feel. Higher print settings must be used to deposit more ink to compensate for the lack of pretreatment. Another important reason to apply pretreatments is that most inks are water-based, so the colorants are attracted to hydrophilic surfaces. Using pretreatments can change the substrate's surface energy, and hydrophilic surfaces can become more hydrophobic. This is particularly imperative when it comes to the hydrophobic polyester substrate. The selection of pretreatment depends a great deal on the printing inks. Several compounds used in pretreatment aim to inhibit dye migration for dye-based inks, while other chemicals are designed to control pH. In pigment printing, however, pretreatment functions as a binder to improve the stability and fastness of colors.[3]

After printing, a fixing phase is performed using either saturated steam (acid and reactive inks) or dry heat (disperse and pigment inks). Furthermore, one needs to wash out the unfixed dye and the primer for any textile produced digitally with reactive, dispersed, or acid ink. This adds to the need for energy and water and cannot be omitted. The leading cause is the unfixed dye's high-water solubility,

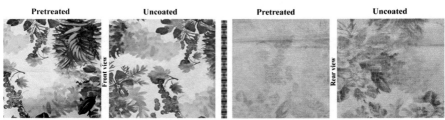

**Figure 9.2** Print output of pretreated and uncoated fabric.

which would result in rapid customer complaints. This cleaning procedure is not required for pigment inks, which is one of their key advantages, aside from their better lightfastness. Additional measures, such as using a softener or sanforizing, may be required before proceeding.

Water is one of the main elements for the pretreatment of the fabric. Fashion is the world's second-largest water-intensive industry after agriculture[4]; estimated consumption is over 79 billion cubic meters per year. This implies that a large volume of water is employed to fuel the fast fashion industry, although billions of people lack access to safe drinking water. Keeping things in context requires total water of 2700 L to produce a standard cotton t-shirt, which is enough freshwater for one individual to consume for around 900 days.[5] For every kilogram of fiber, traditional textile processing necessitates 100–150 L of water.[6] Textile processing uses water and pollutes it.[7]

Consequently, the environment is degraded, and workers are exposed to unhealthy work environments and severe health hazards for individuals who reside close to processing plants. There are approximately 8000 synthetic compounds used in the textile treatment, dyeing, and the transformation of raw materials into fabrics, which accounts for roughly 20% of industrial water pollution on a global scale.[8] Billions of gallons of water are contaminated by these substances used in the textile manufacturing process, which are ultimately released into waterways (Fig. 9.3). Most of these processing facilities are in Pakistan and India,[9,10] with high population densities and poverty rates, and their destruction wreaks havoc on vulnerable populations.

Textile industries must make reasonable efforts to ensure the most efficient use of resources like energy and water. The textile sector is expected to develop environmentally friendly technology and procedures in the future. Training the employees and raising awareness about the need for water and energy-saving is also necessary. There are numerous savings options available. Textile industries can

**Figure 9.3** Pollution and waste caused by the fast fashion sector cause environmental problems.

indeed save money by conserving energy and water, but they can also contribute to mitigating climate change. The clothing industry is conscious of such a diminishing water supply and is introducing new chemical alternatives and new technology, but the biggest problem will be transitioning existing technologies used in textile factories into innovative water-saving technology.

Another problem is persuading current textile manufacturing generations to use synthetic organic alternatives instead of the chemicals used in prior decades. This is going to be a lengthy procedure; however, it will be necessary if the textile industry keeps up with output levels and expands in the coming years.

## 9.3 Printheads

The printhead is the most essential component of any printing system, and even the development of better printheads is a critical component in enabling industrial printing. Maximum jetting frequency, jetting straightness, uniformity, number of nozzles, drop volume, operating window, and cost are the most important criteria in printhead performance. For a long time, piezo printheads using drop-on-demand (DOD) technology have offered the optimum combination of quality, speed, available ink kinds, and robustness, and they are employed in practically every textile printing technique. Continuous inkjet, used in previous years, is another feasible technology and thermal DOD, which has been developed for industrial uses by Hewlett Packard and Memjet and possibly shows some promise in textiles. The evolution of industrial applications of digital textile printing machines seems to have been exclusively reliant on a single printhead, the Kyocera KJ4 (Fig. 9.4).

Successful printers might be constructed around their combination of aqueous compatibility and high speed, many nozzles, and the optimal range of drop sizes and grayscale image capabilities required for textile printing. They range from

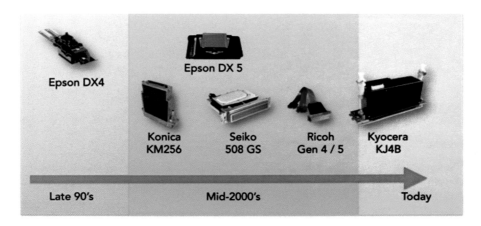

**Figure 9.4** Industrial printhead evolution.

single-pass machines with a few hundred printheads to scanning machines with a single printhead per color. The printhead, however, remains pricey per nozzle component due to its manufacturing (and possibly partly due to its dominant market position).

Alternative printheads, typically based upon microelectromechanical systems (MEMS) made of silicon technology, are increasingly becoming available to system manufacturers and have become a common technique for developing piezo printheads for industrial use. With the development of topologies for single-pass printers, printheads, including a more significant number of nozzles, smaller drop sizes, and increased packing density, has become necessary. The necessity for miniaturization aligns nicely with the photolithographic and micromachining technologies used in MEMS technologies, which allow for exact feature size control. Most ink types used for textile inkjet printing have great chemical compatibility with silicon and silicon oxide. Furthermore, applications that need to be compatible with complicated functional materials, inks that crosslink, and vigorous cleaning fluids necessitate a proper evaluation of upstream building elements and bonding adhesives to explore the possibilities. Eventually, silicon MEMS technology will benefit from the same cost savings as the semiconductor sector. As long as the overall number of units supplied expands, and manufacturers of printers will find a way to take the privilege of this when system manufacturers acquire printheads, the expensive fixed operational expenses of a MEMS fab could be distributed over a larger quantity of the product, cutting the per-component manufacturing costs, and perhaps, per-nozzle printhead costs. It is not yet clear how quickly and to what extent the textile industry will adopt Si-MEMS printheads, and this is a revolutionary technology.

### 9.3.1 Single-nozzle and multinozzle printheads

The application determines whether single-nozzle or multinozzle printheads are used. Single nozzle printheads are the ideal option for material testing and small prototyping. In reality, their functionality is adaptable and straightforward to implement. Because there is just one nozzle to operate, the electronics controller is quite basic. Furthermore, fluid pumped through the capillary will automatically clear any obstruction when a single-nozzle dispenser becomes clogged. As long as the jetting remains stable, the drop velocity does not need to be regulated. Single-nozzle systems have poor throughput as their primary disadvantage. Furthermore, the requirement for scanning a material with a single nozzle causes a significant time lag between each drop being deposited. As a result, distinct solvent evaporation patterns may occur throughout the substrate, compromising the device yield.

On the other hand, multinozzle printheads have a substantially higher throughput since they can print numerous droplets in a line. It is possible to create printheads with plenty of nozzles to print a massive design in one run, which reduces the evaporation profiles over the material's surface. On the other hand, multinozzle printheads are substantially more challenging to operate and maintain than Single nozzle printheads. These print heads are exceptionally hard to clean. Although solvent is injected into the nozzles, the liquid does not drain out of the clogged ones but instead drains out of the open ones, making it incredibly difficult to get rid of particles obstructing the flow of ink.

Furthermore, multinozzle printheads are typically intended to work with a limited number of fluids. It is also critical because all nozzles have to fire droplets in the same direction and speed. If this criterion is not met, the printed pattern may be distorted. Finally, controlling many nozzles involves complex electronic controllers.[11]

## 9.3.2 Drop-on-demand and continuous inkjet printheads

When discussing printhead technology, the speed through which the printhead creates ink droplets (DOD or continuously) and the drop production mechanism are essential considerations (thermal or piezo). As the name suggests, continuous inkjet (CIJ) printheads continuously create and discharge droplets regularly (Fig. 9.5). The drops are given a charge electrically as they escape the printer nozzle. The drops pass through an electrostatic charge that regulates the droplets that are allowed to settle on the substrate's surface. Droplets not needed for the image's production are diverted and redistributed for later use. Continuous inkjet technologies consistently outperform traditional inkjet systems, enabling high levels of ink delivery and being incredibly efficient when operational. These systems have been expensive, and in some cases, challenging to implement. Consequently, CIJ printhead printers have had little success in the textile industry. Stork's Amethyst printer (Hertz printhead technology) and Osiris's Isis printer (Markem-Image printheads) are two continuous inkjet printers.

Textile technology developers generally prefer DOD systems to continuous inkjet printhead systems. As the name implies, DOD printheads produce ink droplets on demand, as requested by design or image. An electrical pulse triggers the emergence of the ink droplet and the technique used to manufacture it. It is possible to classify a drop as either piezoelectric or thermal. The term "thermal inkjet" is a drop-making technique that employs thermal energy (heat) to produce many gas bubbles inside an ink tank. When the gas bubble expands inside the chamber, the ink is pushed, and thus a droplet forms at the nozzle's opening. As the vapor bubble collapses as the ink cools, the droplet separates. As a result, the ink lands on the surface of the substrate.

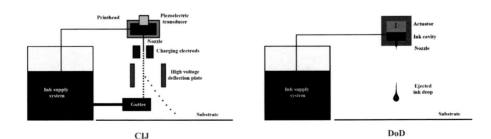

**Figure 9.5** DOD and CIJ technology.

The procedure will then be repeated. The term "bubble jet" is used to describe the thermal inkjet printing technique, which has become widely employed in the desktop printing industry thanks to innovations by companies such as HP and Canon. While printer hardware companies like Canon, Encad, and MacDermid Colorspan investigated thermal technology for inkjet textiles in the early stages of development, the ink's continual heating and cooling made textile chemistry challenging printhead problems became widespread. As a response, much recent research has been focused on piezo inkjet techniques. Inkjet printing systems that print on textiles and signage materials are dominated by DOD piezoelectric technology. This technology works within the inkjet printhead through the electrical stimulation of a piezoelectric substance. The piezo material is physically deformed due to this action, and a corresponding ink volume is forced out of the chamber. Because the formation and ejection of such ink droplets are caused by mechanical rather than thermal action, this device has been discovered to have more possibilities than the heating approach. Advantages include high printhead durability and the flexibility to expand the number of ink types that may be used.

Various manufacturers and suppliers have produced printhead solutions for textile printing applications. For many years, the technology developed by Epson dominated and ruled the market. Today, textile print heads from Kyocera, Konica Minolta, Ricoh, Fujifilm Dimatix (Spectra), Seiko Printek, and others, along with Epson, are used. Two ongoing development aspects in this field are the combined total of nozzles on each printing machine and the spacing of the nozzles. For example, the Kyocera KJ4B printhead features a 600-droplet-per-inch native resolution and 2656 nozzles.[12]

An inkjet printer's printhead is one of the most complicated and technically challenging components and has a crucial role in industrial uses. Here, dependability and longevity are critical. Currently, thermal inkjet printers have a short lifespan and strict ink limits. Piezo printheads are getting popular in practically every commercial use. The number of nozzles and drop volume are the most apparent printhead specs. For example, printheads that move back and forth will be occupied by fixed printheads covering the entire print length to achieve fast speed and efficiency in commercial production. As a result, just the workpiece needs to be moved. A large number of nozzles must be present in such printheads. Many piezo technology manufacturers are working on such print girders or line heads. Such beams can be made up of several single printheads.[13] High-performance printheads for functional materials must meet other parameters. Because the components are frequently dissolved in corrosive solvents, the printhead must be composed of materials resistant to various solvents.[11] Another important characteristic is the highest jetting frequency that a printer can reliably print, which is an important consideration for devices used in a production environment. Degassing equipment and particle filters are frequently included in printheads to prevent particles and air pockets from entering the nozzles and blocking them. The printhead's design development and manufacturing constraints impact the drop positional accuracy. This is a critical factor that influences a printer's drop placement precision. The printhead construction also determines the highest discharge speed that may be achieved. Finally, when using printheads with multiple nozzles, crosstalk must be reduced between nozzles.

## 9.4 Inks

Inks are another area that has advanced rapidly in recent times.[14] More research is needed, but it is expected to provide all of the features that textile printers demand. Fabric printing for sports clothing and spongy signs has been transformed thanks to sublimation inks, while reactive dye inks have improved color performance and reliability. The production of pigmented inks with better color range, performance in terms of feel, and agility at a reasonable price per square meter is an area that requires additional evolution in an attempt to reach the full potential of digital printing in furnishing for the household. Although there is now a significant market void, this is an area where ink companies are inventing technology, and it is believed that over the next few years, further improvements in performance and cost reduction will occur.

Another problem is that the textile industry nowadays uses DOD technology. This technology enables a fall of ink to be delivered through small nozzles along the surface of the printhead. Once the printhead's piezoelectric mechanism delivers them on demand, they appear on the head surface. The printer software calculates the demand for every pixel, and one such command is whether to release a drop or reject it. Because under the gravitational pull, the ink droplet falls on the surface, and the substrate surface, appears as a dot. The ink-dropping pattern becomes disordered and can be noticed on the nozzles' test picture pattern if the drop emerging on the surface is interrupted by internal or external sources. The ink is pushed (purged) during a cleaning operation. Flushing, vacuum purging, and high-pressure purging are some of the several maintenance procedures. Because of these activities, a significant volume of ink is wasted.[15]

Although it is a renewable technology, it produces considerable ink waste due to the cleaning process. According to survey results, every day during operation, an inkjet printhead needs cleaning 2−5 times, leading to a minimum flush of 5 mL of ink, resulting in a 20−25 mL ink waste for every printer. As a result, each year, a significant amount of money is wasted on each machine.[16] Nowadays, businesses can either pour ink into a wastewater treatment system or recycle it. Because recycling is so expensive, the current practice releases the ink and then gathers and disposes of the sludge.[17] The waste treatment procedure takes time and money. As a result, recovering discarded ink at a low cost is economically and environmentally beneficial. Several systems for recycling have been proposed. A few have been tried, but none has proven particularly successful.[18]

## 9.5 Ink supply system

Inks are complex chemical fluids that can contain various ingredients, including particles and binding substances, primarily in instances of colored inks. As a result, finding elements for parts of the ink platform that contact the ink yet do not chemically react is not easy. In an industrial production setting, the ink distribution system that guarantees ink is distributed to the printheads is essential to uphold

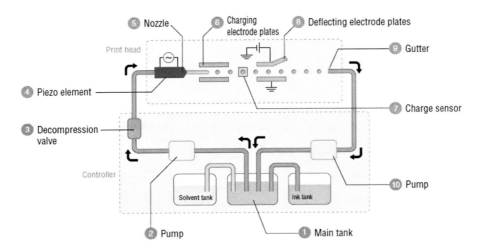

**Figure 9.6** Ink supply system of industrial inkjet printers.

reliability, despite being often treated as a secondary item. Although the ink delivery system appears to be straightforward in theory,

It is frequently a source of complications that are difficult to trace. Under varying external conditions, the ink delivery system (Fig. 9.6) must ensure the constant pressure, temperature, and flow velocity of the ink while also preventing particles and other impurities from accessing the printer and chemical interaction and other difficulties with reliability. Scanning machine tubing must be constructed with care to eliminate variations in pressure that might induce trussing throughout the printed product. It must also be simple to use and refill in a manufacturing environment. As a result of continuing research and development of ink delivery systems and components, inkjet printing methods have only recently gained enough consistency to be considered a viable choice for industrial textile printing. We still need a lot of research and development to ensure the stability of the ink supply system.

## 9.6 Pigments and dyes

The most common colorants often used to add or change the color of anything are dyes and pigments (Fig. 9.7). Textiles, drugs, groceries, skin care, polymers, ink, print, cinematography, and paper are all industries that employ them. Dyes contain colored compounds that are solvent-soluble or dissolve in water. Color is imparted through selective optical emission during the preparation process. Pigments are finely split natural or anthropogenic colored, colorless, or luminous solid particles typically not dissolvable and molecularly unmodified by the carrier or solution they are integrated. Color, on the other hand, is the outcome of the interaction between light and substance and is strongly dependent on the physical and chemical qualities of a substance. The textile sector is increasingly adopting digital textile pigment ink.

**Pigment Paste**          **Dye Powders**

**Figure 9.7** Pigments and dyes.

**Table 9.1** Comparison of pigments versus dyes.

| Pigments | Dyes |
| --- | --- |
| Organic and inorganic, both types | Generally organic materials |
| It is entirely water insoluble | It is water soluble |
| No auxo chrome contains in pigments | It contains Auxochrome in their chemical structure |
| Colorfastness is average to good | Colorfastness is generally average to excellent |
| Need a binder for application | Its application method is straightforward |
| Generally used for printing | Generally used for dyeing |
| Pigments are generally more inexpensive than dyes | Dyes are more expensive than pigments |
| They have no attraction to fibers | Dyes have an attraction to fibers |
| All fibers can be color | It applies selectively to textile materials |

Increased efficiency, faster printing speed, graphic clarity, and flexibility throughout the clothes that may be printed are the key drivers of growth (Table 9.1). In addition, industrial garment pigment colorants are among the less water-intensive fabric printing processes. Because using pigment simply requires the materials to be dried and the binder solution to be crosslinked, there is no requirement to launder afterward. This enables fabric printing using pigment inks on digital textile printers, an environmentally friendly choice; however, there is some complexity to using pigment ink. Nozzles should be changed; if not, they will become blocked to provide for such a seamless flow of pigment particles. If the printing esthetic is compromised, nozzles must be modified. The size of the pigment particle is not as essential as the modification of nozzles. Print color restraint and pattern vividness suffer when pigment

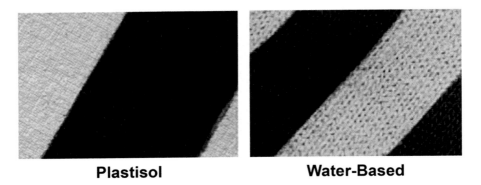

**Figure 9.8** Water-based versus plastisol ink depository.

particles are reduced to less than 1 micron in size. Success can be attained when pigment composition and nozzle construction are considered simultaneously. Secondary solvent chemicals will become increasingly crucial for pigment inks in the near future.

## 9.7 Textile printing and ink depository

Is there a limit on how much ink one may use? Traditional analog printing techniques, including printing with a screen, lay down a surplus of colorant to create color saturation. They can lay down enough binders to ensure satisfactory pigment adherence to fiber. Alternately, inkjet printers lay down tiny layers of ink that become susceptible to scratching and discoloration (Fig. 9.8). The number and kinds of polymers that could influence pigment adhesion to the fabric are limited in such machines since they function at lower viscosity ranges. The trend toward better resolution and smaller droplet diameters that inkjet manufacturers pursue in document printing differs from the need for increased ink volumes in textile printing to achieve optimal dynamic contrast and dye infiltration.

## 9.8 Clogging and monitoring of nozzles

Digital inkjet printing on textiles is an ink sediment process. In both the nonoperational and operational periods, some elements of composite ink evaporate quickly, causing an accumulation of nonvolatile elements near the outlet of the nozzle, resulting in a crust or plug (Fig. 9.9). "Nozzle clogging" is the term for this problem.[19] This word is typically related to printhead malfunctions, which are virtually always irreversible.[20] Considering the selected printhead, inbuilt architecture, and fluid tank, clogging happens on paper, PVC, or fabric surfaces in commercial inkjet printers. The printing area's ambient temperature and relative humidity also affect the ink's drying. In order to analyze an inkjet printability, it is important to know whether the

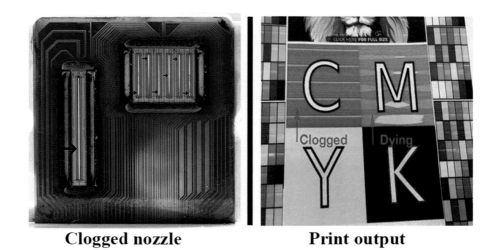

**Clogged nozzle**        **Print output**

**Figure 9.9** The nozzle of the cannon became clogged, resulting in poor print quality.

ink can dry quickly and if it is prone to foam formation. Industrial printers, in particular, have vast volumes of ink in their supply circuit.

Furthermore, quantitative comparisons of different inks, whether they contain additives in drying ink and foam generation, are useful. Fast evaporation of these treated inks around the nozzle results in nozzle blockage, while foam production inside the charge controller frequently causes poor discharge. These are the main roadblocks to massive inkjet-based digital printing implementation.[21] Spraying the nozzle surface using nonwetting laminae is a potential method to prevent inks from completely dehydrating and generating solid formations that lead to blockage.[22]

Another concern that might influence inkjet printing efficiency is the formation of gas bubbles in ink, which can occur when the supply system is flooded or when the nozzle output is used during nonoperational periods. This situation caused foam to form, resulting in poor, heterogeneous, or unsuccessful ink discharge.[23] So, many experimental evaluations are crucial for drying ink and foam formation, such as a quantitative comparison of different inks, ink printability, and evaluating the effectiveness of frequently used compounds.

## 9.9 Automated maintenances

Many textile mill production systems now depend on traditional nozzle cleaning, so developing a dependable and efficient autonomous monitoring and maintenance strategy is critical to inkjet's ongoing acceptance in commercial textile printing. The need for fast, automated maintenance settings grows as the number of nozzles/printhead architectures grows. Furthermore, the need for operational availability limits the amount of time required for nozzle servicing. Manual nozzle maintenance is nearly impossible in many large systems because many of them are unreachable. Satellite ink and mist build upon the printhead faceplate, causing nozzles to become clogged. Other contaminants, such as dust and fibers from the substrate generated in the printing environment, are stuck in nozzles that are not printing; vibration causes ink seepage; ink dries inside the nozzle after air bubbles are sucked into suspended in it. These kinds of factors can put jetting in jeopardy or perhaps lead it to stop.

## 9.10 Handling motion systems and substrates

Motion systems must move the substrate, printheads, or both to run the entire fabric scanning and printing of the finished product. While "sticky rollers" have been used to promote the dimension stability of textile substrates for a long time, the requirements and problems are greater for digital textile printing. Rotary screens keep the fabric in position in traditional printing. With inkjet printing, however, the contactless feature of such printing provides an additional barrier that becomes more problematic at high speeds (Fig. 9.10). Whatever the configuration, smoothness, and constancy of motion are required in an industrial motion system for digital printing, as well as positioning accuracy, substrate handling to maintain dimensional stability during printing, and vibration suppression to avoid obvious print defects. Motion system developers have a number of unique challenges, as distortions in dot size and location are apparent to the naked eye and ultimately unwanted. It is critical to balance optimized structural design with (under certain situations) fault-correcting software for optimal print quality. A significant improvement is necessary for single-pass systems

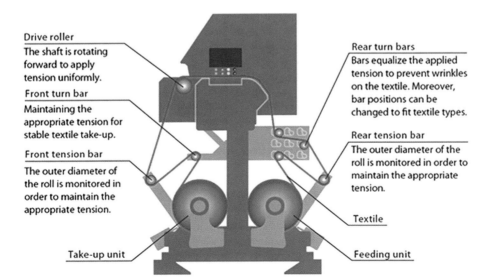

**Figure 9.10** Textile feeding mechanism (motion systems and substrates).

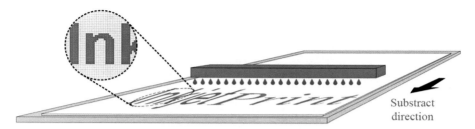

**Figure 9.11** Single-pass printing system.

to provide high textile processing capabilities for manufacturing, and various new difficulties must be overcome.

## 9.11 Single-pass printing

Printing techniques that use a single pass, in which the printheads remain fixed all around the fabric strip at a fixed point with a horizontal path (Fig. 9.11), allow higher productivity from a one-pass printing machine, as the fabric travels underneath it in constant motion. For the first time, single-pass printing systems can match the productivity of rotary screen systems, although at a much higher cost than systems for scanning. One disadvantage of single-pass printers is that the group setting is not possible for multiple printing swaths to help disguise print flaws, as is prevalent in scanning printers. Further considerations include that single-pass printers do not allow nozzle spraying throughout a production run, which is frequent in scanning printers to ensure that all nozzles remain functional. Because of these variables, single-pass print production is more likely to be rejected owing to print defects, and because of the fast-printing speeds, these print flaws can spread across huge areas before being detected. As mentioned above, all single-pass systems place enormous software and digital electronics requirements to accommodate the massive data bandwidth usage expected, and progress has been crucial for the growth of single-pass printing systems in this field.

As a result of these difficulties, single-pass printers are not extensively used for printing in textile industries, and many textile mills prefer to increase output by purchasing more scanning systems instead of following the single-pass strategy. Although some of the largest textile mills have successfully used single-pass systems for a significant period, the emergence of new companies in the region strongly indicates the future potential of single-pass technologies to gain traction in the future.

## 9.12 Software and electronics

Inkjet digital textile printing has many capabilities and advantages for the textile and apparel market and industry, but there are limitations. CAD and CAM

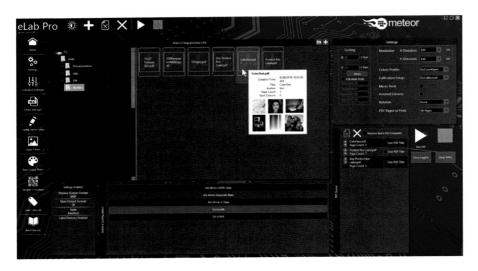

**Figure 9.12** The Meteor eLab Pro user interface. (Courtesy of Meteor Inkjet Ltd.)

technologies have been integrated into the textile and apparel product development lifecycle, but the constant stream of technological advances requires training and expensive updates to technology and software. Inkjet digital textile printing technology has high equipment and material costs and significant initial investments, and printing speeds and capabilities must improve before inkjet printers can be used for production-level textile printing.[24] Some limitations preventing printing inkjet digital textile printing for production include their lower speed and reliability than conventional printing technology.[25] This section will examine the limitations of inkjet digital textile printing technology.

Another crucial area is the printing software (Fig. 9.12), which regulates data flow to the printheads and runs the printing system. The modernization of efficient and flexible operating systems is still a key aspect in accepting commercial textile printers, particularly in a real-world situation such as a clothing factory. A better UI (User Interface) makes it easy to easily find the most critical features, allowing skilled users to make more extensive modifications. Some companies are taking this seriously and improving their software, but there is still much to improve.

For example, garment graphics, separation files, and flat designs with a constant tone require textile-specific software. It is also necessary for CAD modeling or even screen separation systems to accept a diverse range of picture data formats and provide service for process and spot color schemes with increasingly broad process color combinations, as well as the ability to handle quite a picture file types, colorway variants, real-time image repetition, and a slew of additional things. The image pipeline oversees translating a picture file into a data file that controls the firing of each nozzle throughout the printing process. There are several processes for this: color adjustment, for example, to verify that the printed color schemes are accurate; screening (using a grid of dots, the best feasible reproduction of continuous tonal variations); and splitting (depending on the printer configuration, determining the data that is to b delivered to each printhead). Spot color digital printing will necessitate computer-controlled ink distribution and an advanced color database management system, and computerized color analysis. Digital inkjet printer inks are sophisticated mixes requiring meticulous attention to detail, particularly while blended with different colors. A platform like this would need to support 12−16 primaries and exotic color schemes like incandescent or metallic. This type of system must be capable of two quite different pigments and dyes being mixed and distributed. The challenge of pushing magnetic nanoparticles through nozzles would obstruct their usage in inkjet printing.[26] Classical color kitchen technologies do not have the same level of control as inkjet ink technologies. The difficulty of developing a certain technique is a key motivator for keeping process color in textile printing technology. Screen simulation capabilities serve as a bridge between digitally printed fabrics and screen-printed ones. This is particularly helpful while both procedures are still in use. A number of software companies have included capabilities that may be used to simulate and match screen-printed manufacturing textiles, such as mesh size, screen resolution, color trapping, employing gradation contour for tonal cleavage, and color combining and massive printing; however, still, there is plenty of room to improve.[27]

Another sector where software can play a significant role is automatically identifying and adjusting print flaws. Modern imaging and analysis technologies can detect a diverse variety of print flaws. Many printing applications, particularly textiles, still rely on the human eye for much of their work Fig. 9.13. Automation of design flaw identification assures significantly quicker, more accurate, and far more convenient results, as well as allowing for a swift turnaround to the flaw by provoking inspection, reprinting a task, or even nozzle indemnity methods, under which a nearby or replacement nozzle is called into duty to install or isolate a damaged nozzle.

Manual defect detection    Automated defect detection

**Figure 9.13** Inkjet defect detection systems.

## 9.13 Concentrating on printing technology

For more than two decades, digitized textile printing technology has been marketed. It has become an extremely fast-growing sector in the garment machinery industry.[28] Gherzi, an independent, global firm of consultants and technical specialists, undertook a worldwide analysis to determine the primary drivers of digital textile printing investments. One of their survey results is shown in Table 9.2. From the report, it is clear that digital textile printing seems to be more costly than conventional textile printing in terms of pure printing costs due to a lack of efficiency.

In the textile industry, digital print production and workflow can sometimes be difficult for operators. Because most people use a desktop printer, this technology may be overlooked. When visiting textile print companies, it is common to see operators with a lot of textile experience but little knowledge of digital prepress and color management. It appears that debating whether 1200 dpi is required for textile printing is more important than debating how to generate more predictable pale hues through postlinearization or reduce ink usage through profile modifications. This is even more unexpected when considering annual ink expenses; refer to Table 9.3. Technology is the most significant impediment to the mass acceptance of inkjet printing of textiles. Colorant type and color range are still limitations despite recent advances. A processed color scheme is used because inkjet printers are based around a CMYK. Instead of the spot color palette used for traditional textile printing testing, matching colors can be complex.[29] Colorant types must be compatible with current inkjet printhead technology.

Therefore, glitter colorants, flocking, and Devore techniques are unavailable.[30] The availability of pretreated fabric for acid and reactive dye-based colorants is still limited unless the printer operator is willing to pretreat the fabric themselves. Inkjet textile printers currently print at speeds of around 30 $m^2$/h, whereas speeds of over 200 $m^2$/h should be required in order to defeat traditional screen printing in the current market;[31] however, newer digital textile printing models have improved printing speeds to 70 $m^2$/h and even 150–190 $m^2$/h.[32,33] A recently introduced inkjet digital textile printer, LaRio, created by Ms Macchine, can achieve printing speeds of up to 70 m/min or 4200 m/h.[34] The costs associated with the technology need to be lower to increase the competitiveness and acceptance rate in the market for digital inkjet printing on textiles.

One of the limitations of using inkjet textile printing technology for product sampling is how easily and accurately reproducible such samples are. This can depend on various circumstances, such as variations in the substrate and preparation, the ink and printhead quality and consistency, and the performance reliability of the inkjet printheads.[35] While inkjet digital textile printing usually prints without color separation, many CAD/CAM systems can create color separation files from a color-reduced graphic file to move to conventional textile printing.[36] Colorants used for inkjet digital textile printing and conventional printing differ in appearance, with inkjet prints appearing granular and pixel-like, whereas conventional textile printing colorants have uniform coverage. Printed designs, however, are usually visually comparable

**Table 9.2** An Analysis of Printing Expenses for 100% Cotton – 40 × 40 / 132 × 72 ~ 58 finished width, ~ 140 g/m², Reactive printing, each cost mentioned is in Euro per meter.

| Costing parameters | Rotary screen printing | | | | Digital printing | | | |
|---|---|---|---|---|---|---|---|---|
| | China | Italy | Germany | India | China | Italy | Germany | India |
| Dyes & chemicals | 0.1302 | 0.1875 | 0.1875 | 0.1200 | 0.1669 | 0.2000 | 0.2000 | 0.1750 |
| Ink | | | | | 0.8678 | 0.4200 | 0.4200 | 0.3413 |
| Spare parts/maintenance | 0.0134 | 0.0100 | 0.0100 | 0.0136 | 0.1541 | 0.2000 | 0.2000 | 0.2211 |
| Utilities | 0.1328 | 0.1671 | 0.1671 | 0.1596 | 0.1797 | 0.2050 | 0.2050 | 0.1249 |
| * Power | 0.0504 | 0.0504 | 0.0504 | 0.0575 | 0.1175 | 0.1200 | 0.1200 | 0.0368 |
| * Steam | 0.0401 | 0.0600 | 0.0600 | 0.0467 | 0.0113 | 0.0200 | 0.0200 | 0.0353 |
| * Thermopac/natural Gas | 0.0267 | 0.0267 | 0.0267 | 0.0255 | 0.0481 | 0.0500 | 0.5000 | 0.0481 |
| * Raw water | 0.0039 | 0.0100 | 0.0100 | 0.0100 | 0.0007 | 0.0050 | 0.0050 | 0.0016 |
| * Wastewater treatment | 0.0117 | 0.0200 | 0.0200 | 0.0200 | 0.0022 | 0.0100 | 0.0100 | 0.0032 |
| Wages and salaries | 0.0707 | 0.0844 | 0.1144 | 0.0184 | 0.2083 | 0.1350 | 0.1830 | 0.0980 |
| Supervisory and technical staff + administrative | 0.0134 | 0.1000 | 0.1400 | 0.0133 | 0.0134 | 0.1000 | 0.1400 | 0.0133 |
| Packing | 0.0200 | 0.0200 | 0.0200 | 0.0240 | 0.0200 | 0.0200 | 0.0200 | 0.0240 |
| Interest | 0.0027 | 0.0027 | 0.0027 | 0.0667 | 0.0027 | 0.0027 | 0.0027 | 0.0028 |
| Depreciation | 0.0185 | 0.0175 | 0.0175 | 0.0504 | 0.1802 | 0.4167 | 0.4167 | 0.2000 |
| Total cost per meter | 0.4016 | 0.5892 | 0.6592 | 0.4660 | 1.7930 | 1.6993 | 1.7873 | 1.2004 |
| | 100% | 147% | 164% | 116% | 100% | 95% | 100% | 67% |

Remarks:
(1) without screen cost
(2) Chinese digital printing costs are high due to the lack of efficiency of the specific company

*Source*: Gherzi interviews.

**Table 9.3** Digital printing technology barriers.

| Criteria | Technology barrier |
|---|---|
| Print runs | Digital textile printing is not a cost-effective option. The cost of a project rises as the print run lengthens, resulting in a greater click count. |
| Simulate PMS | Unfortunately, PMS colors cannot be printed on digital textile printers. Instead, the Pantone to CMYK values of the specified PMS color will be used to replicate the PMS color. As a result, digital printing's color accuracy is slightly inferior to traditional printing. |
| Belt drive | Belt-driven digital printers are common. The fabric lies flat on a roller driven by a belt to feed it through the machine. This may result in a minor change across prints. |
| Printing speed | Printing speed is to the advantage of a company, and productivity is money. At present, digital inkjet printers are not fast enough compared with traditional printing machines: they are still lagging and need to improve. |
| Ink cost | Although digital inkjet printing does not necessitate the use of a screen, the overall cost of limited production is lower than traditional printing. As a result, after a specific length of printing, the ink price drives the number, and the printing cost increases more than with traditional printing. |
| Merging digital and old technologies | Although digital printing technology is evolving, it still has a tough road to go before it can match the pace of traditional printing machines. As a result, for a considerable time, the textile printing industry's primary pairing method will be using digital inkjet printers for proofing or limited manufacturing, combined with traditional printers for commercial manufacturing. Nevertheless, there are still issues with combining new and old technologies, and more advancements are required. |

between an inkjet digitally printed textile, and a conventionally rotary screen printed textile, especially for darker colors where more ink drops are applied for increased coverage.[36] The physical differences between prints from industrial rotary screen printers and inkjet digital textile printers are the lack of overlap of color areas, usually around 1 mm for conventional printing. This is due to the physical overlap between colors in conventional textile screen printing not being necessary for an inkjet digital textile printer, which prints all colors simultaneously instead of one at a time.[36] Cites coloration and printing techniques as reasons for the color matching and reproduction issues in inkjet digital textile sampling and conventional textile production printing. When using inkjet digital textile printing for sampling, the inkjet printhead resolution and fabric structure should be considered for the sample to be reproducible with conventional printing.[37] While inkjet digital textile prints and conventional rotary screen prints are colorfast to wash, inkjet digital textile prints are less colorfast to artificial light, and wet rubbing or crocking.[36] Inkjet's digital textile printing technology has significantly improved since its introduction to the market, but there are still limitations to its expanded acceptance and ease of use.

One issue people face using inkjet digital textile printers is controlling color and managing color profiles. True "black" in textiles is challenging to achieve across the textile industry, but it is complicated with inkjet digital textile printing. Acid and reactive dye-based colorants have difficulty producing true black, and depending on the absorptive and textural properties of the textile substrate, black colorants will disperse throughout the textile and appear more gray than black.[38] Users must also be familiar and competent with instrumental calibration of printers and monitors, and the company must be able to support the continued printer downtime and textile and colorant used for calibrations. Companies and printer operators must also be made aware of the need for repeated calibrations of the same printers due to changes in textile quality.[38] Companies either currently using or seeking to use inkjet digital textile printing must be willing to search for qualified employees or able to invest in training employees in the multiple aspects of the technology, such as CAD software and computer technology, color, and textile choices, and behaviors, as well as the maintenance and use of the inkjet printers.[38] Knowledge of textile and fiber characteristics is one of the most significant limitations and barriers to implementing mass customization and engineered designs with inkjet digital textile printing. Textiles often stretch during printing and can be distorted by the feeding mechanism of the printer. Wet postprocesses, such as steaming and washing, can further distort printed textiles, causing bowing, skewing, and shrinking.[39] Designers and print operators face the technological learning curve and a learning curve of textile behavior and qualities as they pertain to print quality and the quality of the finished engineered printed product.[33]

### 9.13.1 Learning curve

Inkjet digital textile printing technology users face a learning curve in using it to its full potential. Whether sampling textile products or manufacturing print-on-demand products, users and customers of inkjet digitally printed textiles must re-evaluate the design and manufacturing process. By using digital technology for design and product development requires esthetic knowledge and skill and knowledge of tools such as software and hardware systems.[40] An inkjet digital textile printer's machinery and hardware system are different from conventional textile printers, and the design process and software systems used are either new or use distinct functions. The product development process for inkjet digital textile printing differs from conventional printed textiles. This includes color management and colorant formulation. Although conventional textile printing has simplified technology for product development over many years, the technological demands placed on inkjet digital textile printers are still very high.[40] Those who choose to use inkjet digital printing technology must educate themselves about the process in its entirety. This can often cover not only the inkjet printer's mechanisms but also coloring agents, substrates, material characteristics, and computer application frameworks. By understanding and re-educating themselves about the inkjet digital printing process, they can create an integrated and coherent product development and color management strategy.[40]

LH Nicoll states, "The establishment of new creative and craftsman roles is underway. This new technology transfers an increasing number of distinct professional skills.[30]" As well as requiring adopters of this technology to re-educate and train their employees on how to use and incorporate inkjet digital textile printing into their product design process, new fields and job opportunities are being created. Inkjet digital textile printing requires an integration of many aspects and skills ranging from computer-aided design to digital and physical color management skills and knowledge of dye chemistry and textile properties. Current and future users of this technology will be required to be knowledgeable of printing and software technology and be creative and color savvy.

### 9.13.2 What can go wrong? a real-world example

Some companies have found success with industrial digital printing in textiles; however, many textile mills have lost much money in their digital endeavors. Here is one real-world example: A textile mill in Asia declared at the beginning of 2018 that they would quit digitally printing using reactive inks. The company had previous printing experience with conventional printing and wanted to join the digital revolution, so it purchased a high-grade industrial inkjet printer with reactive dyes. They searched for assistance in early 2017 after failing to satisfy the quality requirements of a British retailer. After analyzing the current print quality and the upstream and downstream processes involved, they determined that changing any in-house processes or chemicals would not allow them to meet the external quality requirements. The mill used PFP fabric for digital printing, which was simply a too-low grade. Even after optimizing the PFDP stage, posttreatment, and washing out unfixed dyes, they noticed considerable mottling. They assumed it was due to a highly cost-effective spinning method with high short-fiber content. In order to enhance quality, the textile mill sought a better, more expensive PFP fabric. It had to ask its digital print customers (very few) to pay more for better fabric quality, which they all declined. They stopped printing digitally a few months later, sold the machine's parts, and returned to traditional screen printing.

### 9.13.3 Digital textile printing-a threat, an opportunity, or a risk?

Risk is defined as "the effect of uncertainty on objectives."[41] An unpredictable event or collection of circumstances that, if they occur, will benefit or hurt an organization is referred to as either an opportunity or a threat;[42] as a result, digital textile printing has a great deal of uncertainty and risk. The crucial question is, what risks in manufacturing and the market pose a threat and represent an opportunity? To answer this question, a textile mill would be wise to spend weeks or months building a sophisticated business plan before investing a large quantity of money in tremendous but challenging technology.

Limitations of inkjet digital textile printing pertain to the technology and its use. The knowledge and skill level of the users can be considered a barrier to the wider use and acceptance of inkjet digital textile printing. While the technological aspects can be improved through research and advancements, users' learning curve requires time and monetary investment in the training and education of users and designers.[43,44]

## 9.14 Awaiting chemistry

Many technical aspects are not presently signed in production textile printing, but they could be in the next few years, allowing digital technology to make even more inroads into the textile (and clothing) industry.

### 9.14.1 Digital finishing

Finishing with digital technology, which involves depositing functional materials using inkjet or other technology as a portion of the manufacturing procedure, is now a niche interest; however, it is projected to become more beneficial eventually. Businesses and organizations demonstrated a wide range of materials featuring conductance, IR and UV shielding, water repellency, etc. Digital deposition's ability to deposit functionalities precisely at the locations where they are required has the potential to produce intricate "smart" textiles and clothing; however, this necessitates a rethinking of the traditional garment manufacturing cycle. Thus, it is now in its initial phases of implementation.

### 9.14.2 Digital dyeing

Digital dyeing (Fig. 9.14), which uses inkjet technology to color fabrics in a single color, is another projected sector to flourish. This may seem contradictory, given that inkjet is usually employed for (and is the finest at) generating intricate designs, yet technology factual is vast. Dyeing, which uses much water per kilo of cloth, is one of the textiles' most environmentally destructive operations. The energy and water reductions are enormous when the inkjet method is employed. Because inkjet

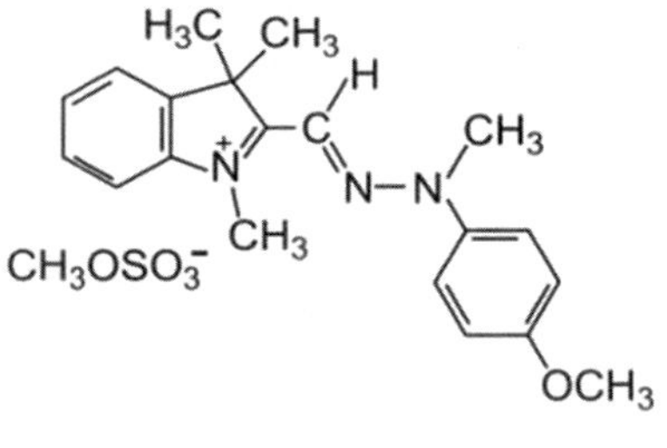

**Figure 9.14** Dye structure of C.I basic yellow 28.

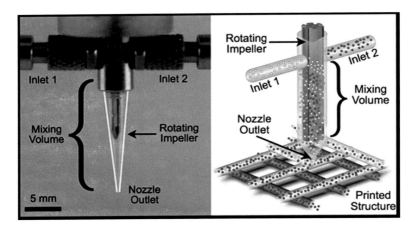

**Figure 9.15** Active mixing and switching printheads.

printing technologies are now geared toward pattern production, they may not be suited for this purpose. Decorating solid sections is a flaw because it reveals print defects that would otherwise be undetected in printed designs. In addition, the overall coverage potential for manufacturing rich neutral tones is constrained. Digital thread dyeing, in which threads are printed on as they move through a stitching or some other machine, has begun to gain traction, enabling colors to be adjusted creatively.

### 9.14.3 Technology advancements in deposition

Developing novel deposition technologies may assist overall digital dyeing and finishing. Like inkjets, they are digitally addressable, but they are built to deposit far more significant amounts of material, providing them with better coverage and even the ability to release substances that an inkjet printer cannot (corrosive materials, high viscosities, large particulates, etc.). Digital deposition technologies have been demonstrated by companies such as The Alchemie Technology, Technology Partnership, and Archipelago Technologies for deposition applications where fine detail resolution is less necessary (Fig. 9.15). They are all based on various electro-mechanical setups that allow for digital control (partial or complete) of the material ejection. In the future, we might see specific capabilities implemented, allowing for even more explorations into the field of textile manufacturing with unlimited possibilities.

## 9.15 Conclusion

Due to the multiple processes and intermediates, a thorough operational impact analysis is generally impossible for any textile print firm wishing to digitalize. Minor adjustments in upstream processes that were insignificant in conventional printing may now be substantial in inkjet printing. Other parameters important for conventional printing might be less critical for digital printing. Along with removing the necessity of repeating patterns, the capabilities of inkjet digital textile printing, including printing in thousands of colors and photo-realistic prints, will impact not only the current users of this technology but the students and future users of this technology who will continue to innovate and stretch the boundaries of textile and apparel design. If inkjet is only seen as a screen-printing replacement, one may find oneself right in the eye of a hurricane. The shortened time to market, the improved supply and demand match, and the enormous creative freedom in design need to be fully utilized. Businesses that take advantage of every update in digital textile printing technology, which is expensive and time-consuming, are the ones to watch. Some of them have already begun to appear on the horizon.

# References

1. Dutta, S.; Bansal, P. Cotton Fiber and Yarn Dyeing. In *Cotton Science and Processing Technology: Gene, Ginning, Garment and Green Recycling;* Wang, H., Memon, H., Eds.; Springer Singapore: Singapore, 2020; pp 355−375.
2. Jhatial, A. K.; Yesuf, H. M.; Wagaye, B. T. Pretreatment of Cotton. In *Cotton Science and Processing Technology: Gene, Ginning, Garment and Green Recycling;* Wang, H., Memon, H., Eds.; Springer Singapore: Singapore, 2020; pp 333−353.
3. Yuen, C.; Kan, C. W.; Jiang, S.; Ku, S.; Choi, P.; Wong, K. Optimum Condition of Inkjet Printing for Wool Fabric. *Fibers Polym.* **2010,** *11*, 229−233.
4. Shaikh, M. A. Water Conservation in Textile Industry. *Pak. Text. J.* **2009,** *58*, 48−51.
5. The Conscious Challenge. Water & Clothing. Available from: https://www.theconscious challenge.org/ecologicalfootprintbibleoverview/water-clothing (accessed Apr 28).
6. Objective, C. The Issues: Water. Available from: https://www.commonobjective.co/article/the-issues-water (accessed Apr 28).
7. Memon, H.; Khatri, A.; Ali, N.; Memon, S. Dyeing Recipe Optimization for Eco-Friendly Dyeing and Mechanical Property Analysis of Eco-Friendly Dyed Cotton Fabric: Better Fixation, Strength, and Color Yield by Biodegradable Salts. *J. Nat. Fibers* **2016,** *13*, 749−758. Available from: https://doi.org/10.1080/15440478.2015.1137527.
8. You, G.O. Fashion: The Thirsty Industry. Available from: http://duxinmiaomu.com/index-292.html (accessed Apr 28).
9. Dutta, S.; Bansal, P. Textile Academics in India—An Overview. In *Textile and Fashion Education Internationalization: A Promising Discipline from South Asia;* Yan, X., Chen, L., Memon, H., Eds.; Springer Nature Singapore: Singapore, 2022; pp 13−34.
10. Ali Hayat, G.; Hussain, M.; Qamar Khan, M.; Javed, Z. Textile Education in Pakistan. In *Textile and Fashion Education Internationalization: A Promising Discipline from South Asia;* Yan, X., Chen, L., Memon, H., Eds.; Springer Nature Singapore: Singapore, 2022; pp 59−82.
11. Singh, T.; Gili, E.; Caironi, M., et al. Picoliter Printing. *Reference Module in Materials Science and Materials Engineering* **2016,** 1−23. Elsevier.
12. King, K. M. Inkjet Printing of Technical Textiles. *Advances in the Dyeing and Finishing Technical Textiles* **2013,** 236−257. Elsevier.

13. Gianchandani, Y. B.; Tabata, O.; Zappe, H. P. *Comprehensive Microsystems*, Vol. 1. Elsevier: Amsterdam, 2008.
14. Memon, H.; Khoso, N. A.; Memon, S.; Wang, N. N.; Zhu, C. Y. Formulation of Eco-Friendly Inks for Ink-Jet Printing of Polyester and Cotton Blended Fabric. *Key Eng. Mater.* **2016**, *671*, 109–114. Available from: https://doi.org/10.4028/http://www.scientific.net/KEM.671.109.
15. Viluksela, P.; Kariniemi, M.; Nors, M. Environmental Performance of Digital Printing. *VTT Research Notes* **2010**, 2538.
16. Kulkarni, A. N.; Dedhia, E. M. Management of Digital Printing Ink Waste with Cotton Fabric using Screen Printing Method. *J. Sci. Res.* **2021**, *65*, 186–189. Available from: https://doi.org/10.37398/JSR.2021.650430.
17. Cayumil, R.; Khanna, R.; Ikram-Ul-Haq, M.; Rajarao, R.; Hill, A.; Sahajwalla, V. Generation of Copper Rich Metallic Phases from Waste Printed Circuit Boards. *Waste Manag.* **2014**, *34*, 1783–1792.
18. Leonard, S.; Michael, L.; Floom, J., et al. Estimating Method and use of Landfill Settlement. *Proceedings of Geo-Denver* **2000**, 1–15. Available from: https://doi.org/10.1061/40519(293)1.
19. Hoath, S. D. *Fundamentals of Inkjet Printing: The Science of Inkjet and Droplets*; John Wiley & Sons, 2016.
20. Sengun, M. Z. Impact of Ink Evaporation on Drop Volume and Velocity. *In International Conference on Digital Printing Technologies* **1997**, 681–684. IS & T Society for Imaging Science and Technology.
21. Santangelo, P. E.; Romagnoli, M.; Puglia, M. An Experimental Approach to Evaluate Drying Kinetics and Foam Formation in Inks for Inkjet Printing of Fuel-Cell Layers. *Exp. Therm. Fluid Sci.* **2022**, *135*, 110631.
22. DeBoer, C. D.; Wen, X. Non-Wetting Protective Layer for Ink Jet Print. Heads. *U.S. Patent 6,345,880* **2002**. (February 12).
23. Magdassi, S. Ink Requirements and Formulations Guidelines. In *The Chemistry of Inkjet Inks*; Hackensack, NJ: World Scientific, **2010**; pp 19–41.
24. Byrne, C. Textile Ink Jet Printing-Market Information, Potential Outlets And Trends. In *Textile Ink Jet Printing – A Review of Ink Jet Printing of Textiles, Including ITMA 2003*; Dawson, T.L., Glover, B., Eds.; Society of Dyers and Colourists Technical Monograph, 2004; pp 30–37.
25. Dehghani, A.; Jahanshah, F.; Borman, D.; Dennis, K.; Wang, J. Design and Engineering Challenges for Digital Inkjet Printing on Textiles. *Int. J. Cloth. Sci. Technol.* **2004**.
26. Jing, S.; Meifei, Z.; Tingfang, M.; Houyong, Y.; Juming, Y. Preparation and Properties of Digital Printing Inks Based on the Cellulose Nanoparticles as Dispersant. *Adv. Text. Technol.* **2019**, *27*, 69–72. Available from: https://doi.org/10.19398/j.att.201803020.
27. Hu, Y. The Influence of Modern Digital Design On Traditional Tie-Dye. *Proc. 6th International Conference on Sensor Network and Computer Engineering. (ICSNCE 2016)* **2016**. Atlantis Press.
28. Khanzada, H.; Khan, M. Q.; Kayani, S. Cotton Based Clothing. In *Cotton Science and Processing Technology: Gene, Ginning, Garment and Green Recycling*; Wang, H., Memon, H., Eds.; Springer Singapore: Singapore, 2020; pp 377–391.
29. Ross, T. A Primer in Digital Textile Printing, **2004**. Available from: https://kipdf.com/a-primer-in-digital-textile-printing_5aac5fe61723dd4d5a80bc7a.html (accessed Apr 28).
30. Nibert, A. T.; Nicoll, L. H.; Smith, S. P. Preface to the Second Printing of this Supplement. *CIN: Comput. Inform. Nurs.* **2006**, *24*, 2S–3S.

31. Polston, K. Capabilities and Limitations of Print-on-Demand Inkjet Digital Textile Printing and the American Craft Niche Market, **2011**. Available from: http://www.lib.ncsu.edu/resolver/1840.16/7036.
32. Caccia, L.; Nespeca, M. Industrial Production Printers–DReAM. *Digital Printing of Textiles* **2006,** 84–97 Elsevier.
33. King, B.; Renn, M. Aerosol Jet Direct write Printing for Mil-Aero Electronic Applications. In *Proceedings of the Lockheed Martin Palo Alto Colloquia*, Palo Alto, CA, 2009.
34. Pikul, J. H.; Graf, P.; Mishra, S.; Barton, K.; Kim, Y.-K.; Rogers, J. A.; Alleyne, A.; Ferreira, P. M.; King, W. P. High Precision Electrohydrodynamic Printing of Polymer onto Microcantilever Sensors. *IEEE Sens. J.* **2011,** *11*, 2246–2253.
35. Dawson, T. Spots before the Eyes: Can Ink Jet Printers Match Expectations? *Color. Technol.* **2001,** *117*, 185–192.
36. Mikuž, M.; Šostar-Turk, S.; Pogačar, V. Transfer of Inkjet Printed Textiles for Home Furnishing into Production with Rotary Screen Printing Method. *Fibres Text. East. Eur.* **2005,** *13*, 54.
37. Dungchai, W.; Chailapakul, O.; Henry, C. S. Electrochemical Detection for Paper-Based Microfluidics. *Anal. Chem.* **2009,** *81*, 5821–5826.
38. Campbell, J. *Controlling Digital Colour Printing on Textiles;* Woodhead Publishing Ltd: Cambridge, 2006.
39. May-Plumlee, T.; Bae, J. Behavior of Prepared-for-Print Fabrics in Digital Printing. *J. Text. Appar. Technol. Manag.* **2005,** *4*, 1–13.
40. Treadaway, C. Developments in Digital Print Technology for Smart Textiles. *Smart Clothes Wearable Technology* **2009,** 300–318. Elsevier.
41. Olechowski, A.; Oehmen, J.; Seering, W.; Ben-Daya, M. The Professionalization of Risk Management: What Role can the ISO 31000 Risk Management Principles Play? *Int. J. Proj. Manag.* **2016,** *34*, 1568–1578.
42. Hillson, D. *Effective Opportunity Management for Projects: Exploiting Positive Risk;* Crc Press, 2003.
43. Abdulla, A.; Schmidt di Friedberg, M. Textiles as Heritage in the Maldives. In *Textile and Fashion Education Internationalization: A Promising Discipline from South Asia;* Yan, X., Chen, L., Memon, H., Eds.; Springer Nature Singapore: Singapore, 2022; pp 145–174.
44. Yan, X.; Chen, L.; Memon, H. Introduction. In *Textile and Fashion Education Internationalization: A Promising Discipline from South Asia;* Yan, X., Chen, L., Memon, H., Eds.; Springer Nature Singapore: Singapore, 2022; pp 1–12.

# Quality of digital textile printing 10

Biruk Fentahun Adamu[1,2], Esubalew Kasaw Gebeyehu[3,4], Bewuket Teshome Wagaye[1,2], Degu Melaku Kumelachew[1,2], Melkie Getnet Tadesse[3] and Abdul Khalique Jhatial[5]

[1]Textile Engineering Department, Ethiopian Institute of Textile and Fashion Technology, Bahir Dar University, Bahir Dar, Ethiopia, [2]Key Laboratory of Textile Science and Technology, Ministry of Education, College of Textiles, Donghua University, Shanghai, P.R. China, [3]Textile Chemical Process Engineering, Ethiopian Institute of Textile and Fashion Technology, Bahir Dar University, Bahir Dar, Ethiopia, [4]Key Lab of Science and Technology of Eco-textile, Ministry of Education, College of Chemistry, Chemical Engineering, and Biotechnology, Donghua University, Shanghai, P.R. China, [5]Department of Textile Engineering, Mehran University of Engineering and Technology, Jamshoro, Sindh, Pakistan

## 10.1 Introduction

Textile printing introduces a color pattern into fabrics that need the knowledge of design ideas, colorants, substrate (fabric), and know-how about bringing together these things using some techniques. Different textile printing techniques include block, stencil, screen, transfer, resist, discharge, and digital textile printing techniques.[1]

Digital textile printing is one textile printing method that started in the late 1990s to substitute traditional analog screen printing. It is the procedure of applying a colored design or pattern or an image to textile fabric with the help of graphic design software. Digital textile printing is classified into direct digital inkjet printing and indirect heat transfer digital textile printing. In digital textile inkjet printing, the colorants, which may be dye inks or pigment inks, are jetted in the matrix of dots to create the appearance of a solid color pattern onto fabrics.[2,3] Digital textile printing has many advantages as compared to other printing types, such as saving processing time, reduced process steps, no restrictions to several colors or pattern repeat for designers (unlimited color and design options), fast and easy design corrections, high creativity in design, eco-friendly, reduce the amount of wastewater, reduce electricity usage, high print clarity, and color, produce quality print products.[2,4–7]

Digital textile printing has been selected as the preferred textile printing technology since 2003 when introducing the first production of digital textile printers at ITMA (International Exhibition of Textile Machinery) in Birmingham.[8] The latest technology that has been given more attention, more straightforward design modification, high production speed, and no need for screen production[9] gives an innovative textile business potential,[6] with a significant advantage of its quality of printing.[4] There is a need to know the choice of materials, printing inks, printing software, designs, pretreatments, and the methods to apply to print to produce high-quality digitally printed fabric. As the new digital printing developments and improvements in the technologies and products are growing, there is also the need for quality assessment of digital textile print. Therefore, this chapter discusses digital textile printing quality issues, the factors that affect it, and the methods used to evaluate its quality.

## 10.2 Digital textiles print quality attributes

Printing issues that are common for all types of printing types or techniques may include appearance-related issues (definition, text quality, resolution, image noise, optical density, tone reproduction), color-related issues (color gamut, color matching, color registration), permanence related issues (water and light colorfastness), and usability related issues (defects). In addition to standard printing, inkjet printing may also introduce issues, such as jaggies (digital artifacts edges), color missing on lines, and satellite drops of ink.[10,11]

When we say digital print quality, it is how the printed dots closely resemble the quality of the images printed on the fabric. The printed image has three basic elements called dots, lines, and solid areas (Table 10.1), each element having its quality attributes. Quality, The attributes of print quality, such as color density, color gamut, line quality, sharpness, graininess, digital artifacts, redeye, color balance, contrast, saturation, lighting quality, line width, edge blurriness, and edge raggedness, are essential parameters, which are used to assess the quality of printed fabric.[12] Pedersen et al.[13] explain the different digital textile printing attributes: overall quality, tone quality, detail highlights, detail shadow, gamut, sharpness, contrast, gloss level, gloss variation, color shift patchiness mottle, and ordered noise.

## 10.3 Factors that affect digital textile print quality

Production of high-quality digital textile printed fabric necessitates understanding printing materials or substrates, dyes and pigments for ink formation, printing software and patterns, pretreated paste formulation, approaches to applying the retreated paste, and fixation.[9] Digital textile print quality depends on many factors or parameters, such as fabric type, fabric structure, yarn type, printing ink type,[14,15] dye structure,[16] and pretreatment.[11,17–20]

### 10.3.1 Effect of fiber type and its properties on digital textile print quality

Print substrate surface characteristics vary, implying that not every printing can be applied to any substrate.[21] The performance of the fabric print feature is influenced

**Table 10.1** Essential print image elements, quality attributes, and their definitions.

| Image element | No. | Print quality attributes | Definitions | Correlation with print image quality |
|---|---|---|---|---|
| Dot | 1 | Dot location | The place of dots in a printed image | |
| | 2 | Dot gain | The tonal color value of a dot in the printed image | + |
| | 3 | Dot shape | The shape of the dot in printing | |
| | 4 | Edge raggedness | The appearance of geometric distortion of printed edge from its ideal position measured optically | |
| | 5 | Satellites | The visibility of printed image dots | + |
| Line | 1 | Line width | The thickness of the printed image | − |
| | 2 | Edge sharpness | The image edges clarity | + |
| | 3 | Edge raggedness | The distribution of edges of printed image line geometrics from their ideal position | |
| | 4 | Optical density | Light absorbency of printed color | + |
| | 5 | Resolution (modulation) | Number of dots in the images or pixels in a unit length or area | + |
| Solid area | 1 | Optical density (tone reproduction) | The color tone value of the printed image compared to the required one | + |
| | 2 | Color (lightness, chroma, hue, gamut) | Image lightness, darkness, chromaticity, spatial hue, and uniformity of color | + |
| | 3 | Noise (grainless, mottled, background, ghosting) | The graininess of the printed image | |

Note: +, means positive correlation; −, means the negative correlation.

by the ink-media interactions.[6] The importance of ink-media interactions in inkjet printing on paper is understood and has been broadly investigated; however, digital textile inkjet printing is different from paper printing. While the effect of textile fibers assembly on print quality is perplexing, the real understanding of ink-fabric interactions and their impact on print quality remains a wide open field for academia and manufacturing investigation.[22]

Some fiber properties, such as wicking, structure, absorbency (the hydrophilicity and hydrophobicity), and fiber length, influence how the colorant is applied to the fabric to be printed. The hydrophilic and hydrophobic nature of the fiber is the primary factor that determines the ink-to-substrate interaction, affecting line quality print, which is tested and characterized by wicking and absorption of fibers,[22] the correlation between average line width gain and water/alcohol wicking ratio is shown on Fig. 10.1. A study on polyester and cotton fabric's wicking effect on print quality.[23] The spreading of ink or wicking mainly occurs through fiber or filament direction, hence, in fabric lengthwise direction in the case of warp yarns and widthwise direction in the case of filling yarns. In this study, the best printing quality is found for cotton fabrics because the cotton fabric is hairy by nature, as the ink spreads inside the fibers before substantial wicking (spreading). Cotton fabrics, however, give irregular, inhomogeneous printing. Fabric made from cellulosic, such as viscose rayon, might produce regular, homogeneous best print quality overcoming the problem in cotton fabrics because of the hydrophilic nature, where ink can diffuse faster before it can spread or wick.[23]

A fabric made from polyester fibers is standard for printing amongst synthetic ones using disperse dyes (such as azo, anthraquinones, coumarin, and quinolone disperse dyes with the presence of dispersing agents) due to their appropriate characteristics, low cost,[11] superior strength, and resilience.[5] The influence of fiber size on inkjet digital printing on polyester fabric print quality is investigated by[24]; the image quality of printed fabric in terms of line width, blurriness, and raggedness of longer polyester fiber was better due to that usually large fibers make more capillary space inside a yarn than small ones.

The fabric to be printed may be composed of different fibers, and the composition, with its chemical and physical properties, will affect the penetration of dye or ink into the fabric structure. A fabric made from natural fibers tends to absorb more dye or ink readily than fabric made from synthetic fibers. Because of that, a fabric made from synthetic fibers has properties such as smoothness, and tight surfaces, offering little texture for ink adhesion. Most printed fabrics are made from cotton fibers, printed using reactive-based inks mostly[11,16] because of the bond (covalent)

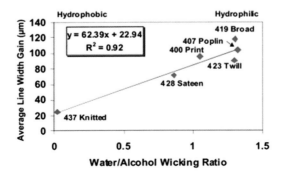

**Figure 10.1** Correlation between average line width gain and water/alcohol wicking ratio[22].

creation between cellulosic fibers and reactive dyes,[25,26] smaller dye molecular size, good solubility, reactive dyes offer full-color gamut, brighter, and diffuse fast.[11] Meseldžija et al.[14] investigated the different fabric composition effects on some print quality parameters. that is, the interaction of fabric made from viscose/cotton (45% viscose/55% cotton), 100% polyamide fiber fabric, and 100% cotton fiber fabrics with printing dye ink and the influence of this interaction on the quality of print parameters (print lines, print surface area, roundness of dot, surface area). This study shows the maximum nonconformity of printed fabric line surface area on viscose/cotton fabric, whereas deviations in the line circumference are visible with all three materials. The maximum dot roundness was perceived on fabric made from 100% cotton due to its highest absorbency. Print quality on fabric made from polyamide gives good print quality, followed by fabric made from viscose/cotton (Table 10.2).

## 10.3.2 Effect of yarn type and its properties on digital textile print quality

The inkjet-based printing technique develops high-grade printed patterns with various substrates; however, not all fabrics are not suitable for this particular printing technique. The digital printing technology must first understand how these fabrics could not be suitable for inkjet digital printing. In order to print digitally, the materials must first be pretreated. A Digital printing machine can read the correct color information from a digital data file and print the colors onto digital printing materials using a printable design from a digital data file. Numerous print heads produce minuscule droplets of ink, which color the fabric. The printing head is placed a few millimeters directly above the substrate. Numerous printing heads in digital textile printing ensure good-quality print designs while allowing for a quick production percentage. Because the printing heads are close to the substrate, some fabrics, however, might not be printed by digital printing.[27] Textile fabrics with many loose threads, for example, will be damaged when fabrics are in contact with heads. A modest static charge develops on the print head due to the architecture of the print heads and the transversal movement overthe width of the fabric. This attracts all loose fibers that cling to the print head. The greater the risk of nozzle blocking, the closer the print head is to the fabric[28]; hence the yarn hairiness in the fabric is a prominent factor in digital printing.

The yarn size is among the different factors that affect the quality of fabric print, with the most significant influence on image graininess and noise[22]; yarn size's effect on the graininess (image noise) of printed fabric study is illustrated in Fig. 10.2. Print color gamut is also affected by yarn size. Fabrics produced from larger yarn sizes showed a better color gamut than fabric made from smaller yarn sizes.[10]

Yarn wicking property is another factor studied as a print quality factor. Too much wicking of yarns showed poor printing quality, whereas lesser wicking may give near-circular drops, which are essential for desired printing effects. Print color

**Table 10.2** Textiles Printing Inks and color fiber-interaction technique.

| Type of dye | (Solubility/Ionic character) | Fiber affinity | Fastness properties | | End-uses | Comments |
|---|---|---|---|---|---|---|
| | | | Light | Wash | | |
| Direct | Water soluble (anionic) | Cellulosic (cotton, viscose) | Poor to good | Poor | Low-quality apparel fabrics/ mattress covers, which are not washed often | After treatment can improve fastness |
| Acid leveling | Water-soluble (anionic) | Protein fibers (wool, silk, nylon) | Good to moderate | Moderate | Carpet yarns, dress material, suiting, overcoats, knitting yarns | |
| Acid (Milling) | Water-soluble (anionic) | Polyamide fibers (nylon), Wool, silk | Good | Good | Carpet yarns, dress material, suiting, overcoats, knitting yarns | |
| Reactive | Water-soluble (anionic) | Cellulosic (also protein and polyamide fibers) | Good to excellent | Excellent | Curtains, furnishings, apparel fabrics, toweling, sewing threads | Excellent shade range. High fastness due to covalent dye/fiber bond |
| Basic | Water-soluble (cationic) | Acrylics (occasionally protein fibers) | Good to moderate | Good | Furnishings, apparel fabrics | Bright shades, excellent tinctorial strength |
| Vat | Insoluble in water (nonionic in insoluble form), anionic on solubilization | Cellulosic (cotton, viscose) | Excellent | | High-quality curtains, furnishing, shirts, towels, sewing threads | Expensive. Bright colors are often difficult to achieve. |
| Disperse | Insoluble in water (nonionic) | All synthetics | Fastness property suitable on polyester, Moderate on nylon | | Apparel fabrics, bed sheets, carpets | Most challenging to print. Good leveling properties |

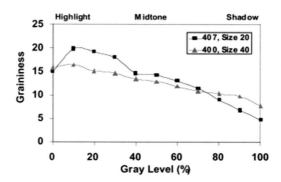

**Figure 10.2** Effect of yarn size on graininess (image noise)[22].

lines investigate the inkjet printing performance of ink or substrate on the substrate.[23] The study revealed that the spreading behavior (wicking) of inks on yarns has a significant effect on printing quality, which, in turn, influenced by the nature of capillaries in the yarn.

### 10.3.3 Effect of fabric type and its properties on digital textile print quality

As a printing substrate, the fabric properties, such as structure, wicking property, and thread density, influence digital textile print quality; different scholars investigated the influence of fabric structure on print quality. Tse et al.[10] explored that the plain weave structure gives good line width gain quality, next to plain, twill, and sateen woven fabrics have good line width gain quality. Knitted structures give the lowest line gain print quality. Park[24] also approved that the plain polyester woven fabric has a better line image than other structures as it has more interlacements, and satin has the worst quality. The researcher also investigated that the fabric made from spun yarn has a better image quality than fabric made from filament yarns. Hajipour and Shams Nateri[5] studied the influence of different weave structures, such as plain and twill, on inkjet printed polyester fabric quality in the print line quality. This investigation showed that the plain structure gives the worst print quality as it gives higher print line width, followed by 4/1 twill, and the best print quality was shown on 2/2 twill fabric.

The wicking property of fabrics, the transfer of liquid through absorbent fabrics measured by capillary forces caused by wetting, affects the print quality of fabrics.[29] Fabric wicking property is different for different fabric structures, long floats in a weave, such as satin, tend to increase wicking property, and plain weave and knit structures decrease their wicking. The bulky fabric structure is more absorptive and requires more significant dye inks. Ding, Y. et al.[15] studied the influence of cotton plain-woven fabric whiteness and wicking properties on pigment-based digital inkjet printing quality. The result revealed that higher cotton fabric whiteness enhanced print color reproducibility, and the fabric wicking affected color performance.

Hajipour and Shams-Nateri[4] investigated the weft density effect on inkjet-printed quality on polyester fabrics printed with water-based disperse dye inks. In the study, the printed line width quality varies with weft density, at which line width increased with increasing fabric weft density resulting in poor line quality.

## 10.3.4 Effect of fabric pretreatment on digital textile print quality

In traditional textile printing, there are some additives, such as alkali (sodium bicarbonate) for fixation and thickening agents (urea and sodium alginate) added within the dye paste (especially for acid and reactive dyes).[11] In digital printing, however, due to the printing head of inkjet printing resistance being poor to these alkalis and additives, some chemicals and auxiliaries, such as thickeners and alkali are applied to the fabric before digital printing as pretreatment that helps for fixation of a dye on the fibers.[30] Therefore, the dye formulation in digital printing is in submicrometer dimension and refined dyes with few compatible additives to the printing heads. As no thickening agents are added to the dye paste in digital printing, the dye ink of inkjet printing's viscosity is low (water-based inks, for example, have a lower viscosity than water). Also, since there is no alkali in the dye paste formulation, the fixation will be very low, resulting in poor color fastness. Printing on nontreated fabric surfaces results in a poor color yield of printed designs, especially in polyester fabrics;[31] as a result, the fabric to be printed needs pretreatment before the digital textile printing process and posttreatment to get a better print quality that allows the ink to fix into the fabric well. The type of pretreatments may depend on the fiber content of the fabrics;[11] however, in the case of printing with pigment dyes, there is no need for pretreatment.

The cotton fabric is made from cotton fibers, and cotton fiber contains cellulose.[32,33] In the textile printing sector, reactive dyes and pigment-based inks, that is, the pretreatments applied on cotton fabric before digital printing, influence print quality.[34,35] Many researchers have investigated the effect of different pretreatments of cotton fabrics on digital textile printing quality.[36] Researchers pretreated the cotton fabric and then applied digital printing to see the pretreatment effect on fabric print quality. Liao et al.[9] explored how the pretreatment of cotton fabric with sodium alginate affects the quality of inkjet printed fabric by reactive dyes (cyan, magenta, yellow and black). The study presented that the high concentration of sodium alginate pretreatment produced a better print quality in pattern sharpness. A high concentration (higher than 20 g/L), however, reduced the color yield because the high concentration of the pretreatment paste reduced the dye spreading through fibers in the fabric. Liu et al.[19] also explored the impact of sodium alginate and sodium alginate plus high fatty acid derivative pretreatment on ink drop spreading on cotton fabrics and print color performance. The study revealed that adding high fatty acid derivatives to the pretreatment enhanced sodium alginate's ink droplet spreading control capability, thus improving color performance. Liang et al.[37] compared the effect of sodium alginate treatment on digital print quality with hydroxyethyl cellulose polymer treatment on the linen fabric via a padding process to create a new fabric surface. After the treatment, linen fabric is printed using reactive dye inks with inkjet printing. It was observed that fabric treated with hydroxyethyl polymer showed more excellent colorfastness performance with sufficient fixation degree, high in-depth, and brighter color than sodium alginate treatment because polymer treatment gives a better fixation rate, more in-depth, and brighter color effect than sodium alginate treatment. Pretreatment of cotton can be replaced by polyamidoamine dendrimer modification of the digitally printed fabric to yield antimicrobial and efficient polymeric materials for inkjet printing.[38] grafted the biocompatible polyamidoamine dendrimer with the reactive dye inkjet printed cotton fabric through anion of cellulose and cyanuric chloride reaction. Modified printed cotton fabric gives improved color strength and colorfastness.

The $CO_2$ laser treatment impact of 100% cotton velvet fabric on digital inkjet-printed by reactive dye ink color yield was investigated by Liang et al.[39] and Jung and Lamar.[40] As the laser treatment by $CO_2$ cotton fabric print color yield intensity increases, the printed fabric's color yield is reduced. This is because the color yield depends on the texture in contact with the dye. The laser treatment can remove pile fibers from the fabric, reducing the surface availability for colorant absorption and light loss.[39,40] The pretreatment of cotton by cationization is used for surface modification.[41] investigated the influence of cotton cationization on digital pigment printed cotton fabric's quality in terms of color yield and depth. It was observed that color strength values are higher for cationized fabrics than for noncationized fabrics. This is because the fiber's anion of anionic pigment attraction will increase when the cationic reagent concentration increases.

Polyester fabric is also used for digital printing, but it has low surface-free energy and poor wettability due to the lack of polar groups in its molecular assembly, leading to poor color yield and bleeding on digital print designs on polyester. As a result, surface modification and functionalization of polyester by pretreatment are required before the digital printing of polyester fabric.[5] Wang and Wang[42] and Park and Koo[43] did a surface modification for polyester fabrics to better digital print quality. Wang and Wang[42] investigated the radio frequency $O_2$ plasma surface-treatment effect on polyester fabric print by magenta pigment ink, resulting in improved print color yield, antibleeding performance, and design sharpness. Park, Y. and K. Koo[43] also applied plasma treatment on polyester fabric before digital printing to increase the colorfastness and yield.

There are studies about silk fabric pretreatment's effect on inkjet-printed silk fabric quality. Faisal et al.[30] pretreated silk fabric with polyacrylic acid and polyacrylamide thickener and investigated the influence of thickener concentration, urea, alkali, pH of pH liquor, and time to steaming on the quality of color strength and fixation percent of reactive dye printed silk fabric. The result showed that alkali concentration, pH, and urea directly relate to digitally printed fabric color strength, while the thickener concentration and steaming time had an indirect relation. Another researcher, Phattanarudee et al.[44] also pretreated silk fabric using amino compounds (serine, glycine, aspartic acid, sericin, chitosan, and Sanfix555) for inkjet printing (using cyan, magenta, yellow and black pigments). The pretreatment on silk fabric provided better print quality in color gamut, wash, and crock fastness than untreated fabrics. Pretreatments by sericin, chitosan, and Sanfix555 gave a wider color gamut than an amino acid. The dry crock fastness of printed silk fabric is better for sericin pretreatments, whereas wet crock fastness was better for serine and glycine. Atmospheric-pressure plasma treatment is a dry pretreatment process used for surface modification of fabrics. Zhang et al.[45] studied the effect of atmospheric air plasma treatment on the quality of water-based pigment inkjet printed silk fabrics' color strength value and print edge. The treatment generated hydrophilic polar groups, such as C-O, C = O, and N-C = O, onto the fiber surface, giving enhanced surface wettability resulting in higher print color strength values (K/S value) and distinct printing edge. Atmospheric-pressure plasma treatment is also used to modify silk fabric surface physical-morphological and chemical properties that impact digital silk print performance.[46] pretreated silk fabric by air, air/Ar, and air/O2 plasma at atmospheric pressure and investigated the effect on inkjet printing. The result revealed that the printing performance of silk was higher in color strength in terms of deeper color, darker shade and larger saturation.

Wool fibers have scales on their structures. These scales on wool fibers affect the quality performance of printed wool fabrics to achieve high-performance images.[20] did a combined pretreatment process on wool fabric using enzymatic (protease enzymes) and sodium alginate and applied inkjet printing by reactive dyes. Printed wool fabrics showed high color performance, that is, deep color, bright and fine color effect, good washing and rubbing fastness, and sufficient strength compared to untreated wool fabrics. In addition,[27] studies the influence of pretreatment solutions rheological of carboxymethyl hydroxypropyl cellulose, sodium carboxymethyl cellulose, and sodium alginate as thickeners inkjet printed quality of reactive dye-based ink printed wool fabrics. Carboxyethyl hydroxypropyl cellulose treatment produced the highest color strength (K/S value) and the sharpest edge than sodium carboxymethyl cellulose and sodium alginate treatments due to the excessive ink droplet spread is controlled and effective higher ink droplet penetration. Sodium alginate treatment, however, showed the poorest fluidity, most viscus, and lowest zero-shear viscosity.

### 10.3.5 Effect of colorants on digital textile print quality

The colorants used for textile printing may be dyes, such as reactive dyes used for cotton, disperse dye used for polyester, acid dye ink used for protein fiber, and pigments used for all types of fibers (water-insoluble colorants).[8,47,48] In textile inkjet printing, the inks are prepared from dyes of reactive, acid, and dispersed and nanoscaled pigments.[49] The natural chemical properties of these colorants will have a different interaction quality with the textile substrate.[50] The types of ink used for digital inkjet printing can be water-soluble and solvent-soluble based on the types of textiles. There are some basic requirements of inks, such as high solubility, thermal and photostability, color purity, high productivity, and good coloring properties, to be used for digital textile printing.[2] Water-soluble inks have a wide-ranging color range, high color strength, and high jetting stability.

Ink formulations for digital inkjet printing of textiles need to fulfill numerous physical and chemical criteria and possess proper rheology and fixation properties based on the different fibers that make the fabric, as stated in Table 10.1.[28] Ink cohesion and adhesion properties are two key qualities for all types of inks. Adhesion refers to the ink's ability to adhere to a different material, such as a substrate, while cohesion refers to its ability to hold together. Adhesion and cohesion of inks remain communal from the standpoint of the pigment particles' physiochemical nature, mainly because their disperse ability is well-defined by the energy required to distribute a single pigment particle in a continuous medium. As a result, each pigment particle is surrounded by the medium and no longer makes contact with other pigment particles.[51]

Reactive dyes have several anionic groups on the dye molecule that will affect the dye absorption into the fabric, and hence, affect color yield. *Soleimani-organic, A. and N. Shakib*[16] investigated the impact of the structure of reactive dye on inkjet printed cotton fabric's color yield. The investigation showed that the anionic groups on dye molecules in the ink formulation played a significant part in printed fabric color yield.

Disperse dyes are used for polyester fabric printing. Disperse inkjet ink properties. For example, the tension surface of inks and viscosity of inks affect the quality of digital print on repeatability and reliability of inkjet printing. The viscosity of the ink is the resistance to flow property associated with ink composition and intermolecular forces (hydrogen bond, van der Waals, and ionic forces) of constituents, which is very important to control the print design sharpness by controlling the spreading of ink.[12] Gao et al.[12] studied the water-based disperse ink viscosity effect on inkjet-printed polyester design sharpness quality. The outcomes displayed an increase in the viscosity of dispersed dye ink and improved the print design. Viscosity is an ink fluid phenomenon inside a print head where an electrical impulse or temperature forces the ink fluid to leave the printing nozzle.

As pigment inks are used in inkjet digital printing, the type of pigments has a different impact on the quality of digital print quality. Textile printing with resin-based pigments has poor crock fastness and poor hand feeling on printed fabrics.[52] *Li, X.*, Powell and Michielsen[1] investigated the effect of six types of pigment inks (Magenta, Black, Yellow, Cyan, Red, and Blue) with the same chemistry formulation on the contrast or clarity of printed polyester woven fabrics. The investigations revealed that black, followed by magenta pigments, showed the greatest print clarity (contrast), while blue showed the poorest clarity on polyester woven fabrics. The six pigment inks print clarity (contrast) order is: Magenta > Red > Cyan > Yellow > Blue > Black. As a cost-effective approach, pigmented polymer lattices with submicron particle sizes are used for textile inkjet printing. Surface-modified and microencapsulated pigments were used for the digital printing of plain silk fabrics to investigate their influence on print quality with improved colorfastness.[53]

Ding et al.[50] analyzed color gamut and color fastness effect by dispersing and pigment inkjet printing on polyester plain woven fabrics. The study reveals that polyester fabric printed by pigment gives good light fastness quality than dispersed dye-printed fabrics because pigments, by nature, have higher lightfastness over-dispersed dye-based inks. Polyester fabric prints by disperse dye ink showed a deeper color shade than pigments due to the higher fixation of disperse dyes; pigment particles could not diffuse into the fibers.

Choi et al.[2] synthesize five perylene acid dyes to formulate water-based inks for digital textile printing applications. Formulated perylene dyes showed outstanding print clarity, greater color strength than commercial azo dyes, and good thermal and photostability.

### 10.3.6 Effect of printing head on digital textile print quality

In digital printing, inkjet prints are categorized into drop-on-demand (DOD) and continuous types depending on the head characteristics of the print. The head types are piezoelectric and thermal inkjet print heads in the DOD system. It has been proven that piezoelectric inkjet print heads can handle an enormous variety of inks than thermal print heads and are mainly used in industrial applications.[54,55]

The print head is the main part of digital printing that movements or motion directions concerning substrate motion. Li et al.[1] showed the impact of print head motion on the print contrast of pigment ink printed polyester woven fabrics', at which the print contrast is good or clear when the print head motion direction is the same as that of the test pattern line and the clarity was decreased when the direction is perpendicular.

## 10.4 Methods of measuring digital textile print quality

Print quality affects the final printed fabric's esthetic appearance and performance. As digital textile printing is a new technology that produces quality print, there is a need to design methods to evaluate the quality of the digital textile print, which helps improve processes, substrates, ink formulations, and printing methods to enhance the digital printing quality.[1]

The printed color is one of the first quality parameters valued by textile consumers and a fundamental element considered in printed fabrics.[56] The digital textile printing quality is also evaluated using different techniques or methods for fabric print line width, gain, blurriness, raggedness, color strength, gloss, and color density. There are two main quality assessment methods for digital textile print: subjective and objective methods (by densitometers, spectrophotometers, gloss meter, and image analysis technique).[13,23,45] Subjective evaluation methods are based on the observation by humans, whereas objective methods do not involve human observers; instead, objective methods are based on measurement devices or algorithms, which are more common in the meantime, save time, are low cost, and need little skill by the user.[13]

## 10.4.1 Subjective evaluation method of digital textile print quality

The subjective print quality evaluation concept is complex since it depends on the objective, measurable variables, such as contrast and graininess, and potential observers' subjective perceptions. It is far from evident that print quality can be characterized in terms of good or bad so that everyone will agree on it, and the question of how to quantify it using objective and subjective criteria remains unanswered.[57]

The subjective evaluation method can predict the quality of the final print image by human eyes within a fixed banding frequency, 1.5 cycles/mm, and finally, the evaluation is to accept or reject the print.[58] Digital textile print quality can be assessed by subjective estimation using a human visual system based on the printed images' professional quality, not the image features.[59] Textile print quality's traditional subjective evaluation technique assesses the printed image's color and tone.[14] Print quality evaluation by humans will depend on many printed image quality attributes, such as lightness, line raggedness, and other effects resulting from the quality of the essential image elements, such as lines and solid areas. In general, subjective quality assessment is not often the choice for researchers as a scientific research analysis due to the fuzziness of the sense of humans.[60] The judgment given by every person is unique, and everyone's perception of color may be influenced by numerous factors, such as age, general health, attitude, surrounding and background color, size of the sample, and other factors. Therefore, a standard evaluation environment is mandatory while visually assessing the print's quality to eliminate any unwanted bias in the assessment.[61]

Colors in the printed fabric are typically evaluated subjectively in clothing manufacturing but give little result due to inconsistency in daylight and individual perceptions. To check the printed fabric against standards, proving cabinets and light booths engaging standard illuminants are used with standard samples[56]

Printed fabrics are frequently subjected to the effects of various conditions throughout their use. Besides the desired appearance, ironing, rubbing, light, and washing are the most critical aspects of treatment, affecting both the inks and printed substrate fibers, emphasizing the need to evaluate digital print quality.[62] The fastness quality of the printed fabric is evaluated both subjectively by trained professionals using the standard greyscale and objectively by calculating color differences using CIE Lab color analysis. Perhaps, the visual quality of the textile prints was also taken into account. Visually, the fineness and sharpness of detail, the fineness of text, and the saturation and purity of the color are all assessed.

The automatic print quality analysis technique was mainly used to measure quality attributes, such as text quality, image noise, line definition, resolution, optical density, tone reproduction, gloss, and color characteristics. These are all defined by color gamut, color matching, color registration, permanence characteristic characterized by lightfastness, wash fastness, application defects, and CIELab color using a print quality analysis system.[22]

## 10.4.2 Objective evaluation method of digital textile print quality

### 10.4.2.1 Measuring digital textile print color values

Measuring the digital printed fabric color patterns is vital from the esthetic point of view and is also used to determine the color change, and it is an indicator for adjustment for better quality control in printed fabrics.[56] Digital textile printed fabric color performances can be expressed in terms of K/S values (color strength) and colorimetric parameters (L*, a*, b*, C*, and h°), which are measured using the spectrophotometer instrument. These instruments for measuring printed color quality consistently use light sources and some form of a photodetector in a defined geometric arrangement (Fig. 10.3).

The spectrophotometer can measure the reflectance values of the material, which is a measure of the light percentage leaving from a material surface divided by the amount of light that strikes the material, and the reflectance for color-printed textiles is inversely proportional to the amount of colorant present on the surface of the textiles.[40] Therefore, the color yield or color strength can be determined using a spectrophotometer. K/S value denotes the color yield of printed fabrics obtained at $\lambda$max according to the Kubelka Munk equation (Eq. 10.1). Printed fabric color yield (K/S value) is calculated at wavelengths of 400–700 nm at 20 nm intervals within a visible spectrum according to equation1.[37]

$$K/S = (1 - R^2)/2R \tag{10.1}$$

where K is the absorption coefficient (depending on colorant concentration), S is the scattering coefficient (caused by dyed substrate), and R is the reflectance of the colored sample at $\lambda_{max}$ obtained from the spectrophotometer. The higher the K/S value, the greater the color yield and uptake.

### 10.4.2.2 Digital print quality evaluation by image analyzer system

In addition to the use of instruments to measure objective print quality, algorithms are used to measure printed image quality metrics, a method due to timesaving is low cost and involves little capability by the user. Quality Attributes are perception,

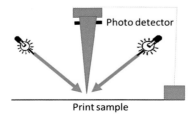

**Figure 10.3** The geometric arrangement of spectrophotometer.

such as lightness, saturation, and details. The image analyzing system determines the quality attributes or parameters, such as print line width, edge raggedness, sharpness, modulation, and ink bleed. The system also measures the optical density, tone reproduction, color gamut, and image noise of the printed fabric, which includes a variety of colors (CMYK and RGB).

### 10.4.2.3 General principles of image analyzing system

Apart from the color and color density qualities of digital print fabric, other print quality attributes, such as dot gin, line and text quality, and nonuniformity of color are critical quality parameters that must be evaluated. The quantitative measurement of these print qualities attributes uses the image analyzing technique. Image analyzers work similarly to spectrophotometers; however, CCD or CMOS sensor replaces the photodetector (Fig. 10.4). The image analyzing technique first requires the images of the printed fabrics either by camera or a scanner (such as CanoscanLiDE110 at 600ppi). Next, the image will be analyzed using different image-analyzing techniques to quantify such print quality attributes. The image analysis technique is supported by standards, such as the ISO-13660 and ISO-19751 appearance-based print quality standards.[4,63,64] After the print image is obtained, the obtained images are analyzed using soft wares, such as MATLAB® program, ImageJ software, and PIAS software[4]; for example, the line width can be obtained in terms of pixels and then converted to millimeters.[4,5] The fundamental step in the print quality evaluation is getting a digital image of the printed fabric surface to be analyzed. The general principle of the image analyzer is illustrated in Fig. 10.4.

### 10.4.2.4 Color image segmentation

The color image segmentation process is a principal task for image analysis and pattern recognition to evaluate the color print image. Color image segmentation divides a digital image into multiple homogeneous area segments, primarily relying on similarity and discontinuity of intensity values, where pixels in a region share similar characteristics, such as color, intensity, or texture.[56] Each pixel has information concerning hue, brightness, and saturation. There are several techniques, such as region-based technique, split-and-merge technique, and thresholding region growing technique.[65]

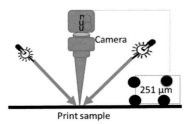

**Figure 10.4** General principle of the image analyzer.

## 10.4.2.5 An automated image analysis system

This method can determine and analyze the printed fabric line width, edge raggedness, sharpness, modulation and ink bleed, optical density, tone reproduction, color gamut, and image noise with various colors (CMYK and RGB).[10] In this method, the fabric was illuminated and analyzed with a CCD camera positioned above, and the structural characteristics will be analyzed from the topical or reflective qualities of the fabrics, see Fig. 10.5.

## 10.4.2.6 Quantitative image sharpness evaluation of digital textile print

The quantitative evaluation method of printed design clarity or contrast based on the estimation of image sharpness was designed by Li et al.[1] First, multiple printed images are being taken by a photo scanner to get an optimum image for evaluating print clarity. Next, data processing of the images is done using different software (Image-Pro and MATLAB) to analyze and compare the image pattern contrast underdifferent conditions. Data is extracted from the images by segmentation systems and exported from the image files to Excel and Text files. After data are extracted from the printed images by Image-Pro, MATLABVR can analyze and compare the image contrast as given by Eq. 10.2. Where: $I_{max}$ and $I_{min}$ are intensities for a corresponding gap bar-gap set.

$$CC_{image} = \frac{I_{max} - I_{min}}{I_{max} + I_{min}} \tag{10.2}$$

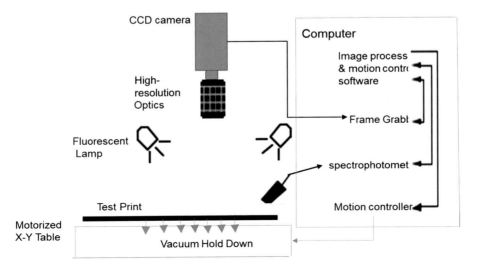

**Figure 10.5** The automatic print quality analysis system.

## 10.5 Digital print quality evaluation standards

People from different countries and organizations may disagree on digital textile print quality analysis methods; for example, textile coloration is a promising discipline for Pakistan,[66] India,[67] Bangladesh,[68] and Nepal.[69] A principle of the image analysis system is shown in Fig. 10.6. For this reason, developing a standard is essential that can bring an agreement with the professionals from any organization or any country about the methods. As a result, the ISO13660 standard on digital print quality attributes was developed by ISO committee members. The ISO/IEC13660 2001(E) standard developed by ISO for digital paper print was also used for textile digital print quality evaluation. This standard can measure the printed fabric image attributes of textile print binary monochrome text and graphic images. The attributes evaluated by this standard are categorized into character and line attributes (blurriness, raggedness, line width, character darkness, contrast, fill, and extraneous marks, and background haze in the character field) and Large-area attributes (darkness and background haze, graininess and mottle, extraneous background marks, and voids). The printed images to be evaluated by ISO-13660 are acquired digitally using a flatbed scanner, a microdensitometer, a CCD camera with a digitizer, or any device capable of digitizing the image at a minimum of 600 dpi (dots per inch).[64]

ISO/IEC 13660 measures characteristics of the print image but has no test charts or reference images except the line quality compliance test method; as a result, a

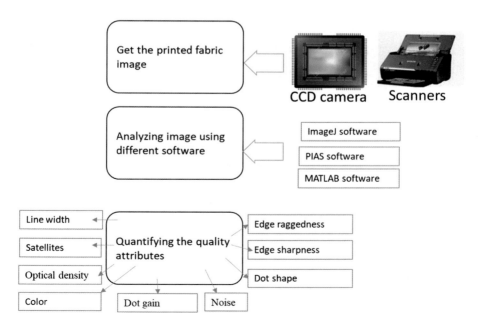

**Figure 10.6** The general principle of the image analysis system.

compliance measurement system test is required. Consequently, JBMIA SC28/WG4 suggested a test chart standard called *ISO 13660 addendum* on the system compliance test chart to enhance compliance test image specifications and goal values for prominent area quality attributes.[70]

Colorfastness to washing, sunlight, and crocking are other central printed fabric quality performances that have their standards. Colorfastness to washing and wet and dry rubbing can be evaluated based on ISO 105-C10:2007 and ISO 105-X12:2001 standards, respectively.[20] There are also international standards, such as ISO5−3 for density measurements, CIE L*a*b* for color measurements, and ISO-2813 standard for measuring digital textile print gloss by gloss meters.

## 10.6 Conclusions

As the new digital printing developments and improvements in the technologies and products are growing, there is also the need for quality assessment of digital textile print. The different factors that could affect digital textile print quality are fiber type, yarn type and its properties, fabric type and properties, fabric pretreatment and after treatments, and colorants. The evaluation methods of digital textile print quality can be subjective or objective. The automatic print quality analysis system, an objective method, is the best method of evaluating the different print quality parameters for digital textile printing. Objective print quality analyzing system quantifies the quality of print based on the basic image elements, namely dots, lines, and solid areas. There are also digital textile print evaluation standards used to measure the quality of digitally printed fabrics.

## References

1. Li, X.; Powell, N. B.; Michielsen, S. Print Clarity on Digitally Printed Textiles—A Quantitative Evaluation. *J. Text. Inst.* **2020**, *111*, 108−121. Available from: https://doi.org/10.1080/00405000.2019.1622273.
2. Choi, S.; Cho, K. H.; Namgoong, J. W.; Kim, J. Y.; Yoo, E. S.; Lee, W.; Jung, J. W.; Choi, J. The Synthesis and Characterisation of the Perylene Acid Dye Inks for Digital Textile Printing. *Dye. Pigment.* **2019**, *163*, 381−392. Available from: https://doi.org/10.1016/j.dyepig.2018.12.002.
3. Ren, J.; Chen, G.; Li, X. A Fine Grained Digital Textile Printing System Based on Image Registration. *Comput. Ind.* **2017**, *92−93*, 152−160. Available from: https://doi.org/10.1016/j.compind.2017.08.003.
4. Hajipour, A.; Shams-Nateri, A. The Effect of Fabric Density on the Quality of Digital Printing on Polyester. *Fibers Polym.* **2017**, *18*, 2462−2468.
5. Hajipour, A.; Shams-Nateri, A. The Effect of Weave Structure on the Quality of Inkjet Polyester Printing. *J. Text. Inst.* **2019**, *110*, 799−806. Available from: https://doi.org/10.1080/00405000.2018.1529722.

6. Gooby, B. The Development of Methodologies for Color Printing in Digital Inkjet Textile Printing and the Application of Color Knowledge in the Ways of Making Project. *J. Text. Des. Res. Pract.* **2020**, *8*, 358−383. Available from: https://doi.org/10.1080/20511787.2020.1827802.
7. Savvidis, G.; Karanikas, E.; Nikolaidis, N.; Eleftheriadis, I.; Tsatsaroni, E. Ink-Jet Printing of Cotton with Natural Dyes. *Color. Technol.* **2014**, *130*, 200−204. Available from: https://doi.org/10.1111/cote.12087.
8. Ujiie, H. Digital Textile Printing: Status Report 2021. *Proc. IS&T Printing for Fabrication: Int'l Conf. on Digital Printing Technologies (NIP37)* **2021**, 47−52. Available from: https://doi.org/10.2352/ISSN.2169-4451.2021.37.47.
9. Liao, S.-k.; Chen, H.-y.; Kan, C.-w. A Study of Quality Factors for Cotton Fabrics in Ink-Jet Printing. *Res. J. Text. Appar.* **2009**, *13*, 33−39. Available from: https://doi.org/10.1108/RJTA-13-03-2009-B004.
10. Tse, M.-K.; Briggs, J. C.; Ma, B., et al. Measuring Print Quality of Digitally Printed Textiles. *In NIP & Digital Fabrication Conference* **1998**, (1), 250−256.
11. Kim, Y. Effect of Pretreatment on Print Quality and its Measurement. In *Digital Printing of Textiles*; Ujiie, H., Ed.; Woodhead Publishing: Cambridge, 2006; p 256.
12. Gao, C.; Xing, T.; Hou, X.; Chen, G. The Influence of Ink Viscosity, Water and Fabric Construction on the Quality of Ink-Jet Printed Polyester. *Color. Technol.* **2020**, *136*, 45−59. Available from: https://doi.org/10.1111/cote.12439.
13. Pedersen, M.; Bonnier, N.; Hardeberg, J. Y., et al. Image Quality Metrics for the Evaluation of Print Quality. *Proc. Image Quality and System Performance VIII* **2011**, *7867*, 11−29.
14. Meseldžija, M.; Vukić, N.; Erceg, T., et al. The Analysis of the Substrate Influence on the Print Quality Parameters of Screen-Printed Textile. *Proceedings of the VIII International Conference on Social and Technological Development, Bosnia and Herzegovina, Balkans* **2019**.
15. Ding, Y.; Zhendong, W.; Chuanxiong, Z.; Ruobai, X.; Wenliang, X. A Study on the Applicability of Pigment Digital Printing on Cotton Fabrics. *Text. Res. J.* **2021**, *91*, 2283−2293. Available from: https://doi.org/10.1177/0040517521997926.
16. Soleimani-gorgani, A.; Shakib, N. The Effect of Reactive Dye Structure on the Ink-Jet Printing of Cotton. *Prog. Color. Color. Coat.* **2014**, *7*, 19−26.
17. Kim, Y.; Lewis, A.; Fan, Q., et al. Effects of Pretreatments on Print Qualities of Digital Textile Printing. *NIP & Digital Fabrication Conference* **2002**. Society for Imaging Science and Technology.
18. Fan, Q.; Kim, Y. K.; Perruzzi, M. K.; Lewis, A. F. Fabric Pretreatment and Digital Textile Print Quality. *J. Imaging Sci. Technol.* **2003**, *47*, 400−407.
19. Liu, Z.; Fang, K.; Gao, H.; Liu, X.; Zhang, J. Effect of Cotton Fabric Pretreatment on Drop Spreading and Colour Performance of Reactive Dye Inks. *Color. Technol.* **2016**, *132*, 407−413. Available from: https://doi.org/10.1111/cote.12232.
20. An, F.; Fang, K.; Liu, X.; Yang, H.; Qu, G. Protease and Sodium Alginate Combined Treatment of Wool Fabric for Enhancing Inkjet Printing Performance of Reactive Dyes. *Int. J. Biol. Macromol.* **2020**, *146*, 959−964. Available from: https://doi.org/10.1016/j.ijbiomac.2019.09.220.
21. Tyler, D. J. Textile Digital Printing Technologies. *Text. Prog.* **2005**, *37*, 1−65.
22. Tse, M.-K.; Briggs, J. C.; Kim, Y. K., et al. Measuring Print Quality of Digitally Printed Textiles. *NIP & Digital Fabrication Conference Society for Imaging Science and Technology* **1998**, 250−256.

23. Mhetre, S.; Wallace, C.; Parachuru, R. On the Relationship Between Ink-Jet Printing Quality of Pigment Ink and the Spreading Behavior of Ink Drops. *Journal of the Textile Institute* **2010**, *101* (5), 423–430.
24. Park, H.-S. Evaluation of Image Quality of Inkjet Printing on the Spun Polyester Fabrics. *Text. Color. Finish.* **2006**, *18*, 61–71.
25. Dutta, S.; Bansal, P. Cotton Fiber and Yarn Dyeing. In *Cotton Science and Processing Technology: Gene, Ginning, Garment and Green Recycling;* Wang, H., Memon, H., Eds.; Springer Singapore: Singapore, 2020; pp 355–375.
26. Memon, H.; Khatri, A.; Ali, N.; Memon, S. Dyeing Recipe Optimization for Eco-Friendly Dyeing and Mechanical Property Analysis of Eco-Friendly Dyed Cotton Fabric: Better Fixation, Strength, and Color Yield by Biodegradable Salts. *J. Nat. Fibers* **2016**, *13*, 749–758. Available from: https://doi.org/10.1080/15440478.2015.1137527.
27. An, F.; Fang, K.; Liu, X.; Li, C.; Liang, Y.; Liu, H. Rheological properties of Carboxymethyl Hydroxypropyl Cellulose and its Application in High Quality Reactive Dye Inkjet Printing on Wool Fabrics. *Int. J. Biol. Macromol.* **2020**, *164*, 4173–4182. Available from: https://doi.org/10.1016/j.ijbiomac.2020.08.216.
28. Yip, J.; Kwok, G. Institute of Textiles and Clothing, The Hong Kong Polytechnic University, Hung Hom, Hong Kong. *Latest Material and Technological Developments for Activewear* **2020**, 193 Woodhead Publishing Limited.
29. Adamu, B. F.; Gao, J. Comfort Related Woven Fabric Transmission Properties Made of Cotton and Nylon. *Fash. Text.* **2022**, *9*, 8. Available from: https://doi.org/10.1186/s40691-021-00285-2.
30. Faisal, S.; Tronci, A.; Ali, M., et al. Pretreatment of Silk for Digital Printing: Identifying Influential Factors using Fractional Factorial Experiments. *Pigment. Resin. Technol.* **2019**. Available from: https://doi.org/10.1108/PRT-07-2019-0065.
31. Memon, H.; Khoso, N. A.; Memon, S.; Wang, N. N.; Zhu, C. Y. Formulation of Eco-Friendly Inks for Ink-Jet Printing of Polyester and Cotton Blended Fabric. *Key Eng. Mater.* **2016**, *671*, 109–114. Available from: https://doi.org/10.4028/http://www.scientific.net/KEM.671.109.
32. Adamu, B. F.; Wagaye, B. T. Cotton Contamination. In *Cotton Science and Processing Technology: Gene, Ginning, Garment and Green Recycling;* Wang, H., Memon, H., Eds.; Springer Singapore: Singapore, 2020; pp 121–141.
33. Siddiqui, M. Q.; Wang, H.; Memon, H. Cotton Fiber Testing. In *Cotton Science and Processing Technology: Gene, Ginning, Garment and Green Recycling;* Wang, H., Memon, H., Eds.; Springer Singapore: Singapore, 2020; pp 99–119.
34. Gebeyehu, E. K.; Sui, X.; Adamu, B. F.; Beyene, K. A.; Tadesse, M. G. Cellulosic-Based Conductive Hydrogels for Electro-Active Tissues: A Review Summary. *Gels* **2022**, *8*, 140.
35. Jhatial, A. K.; Yesuf, H. M.; Wagaye, B. T. Pretreatment of Cotton. In *Cotton Science and Processing Technology: Gene, Ginning, Garment and Green Recycling;* Wang, H., Memon, H., Eds.; Springer Singapore: Singapore, 2020; pp 333–353.
36. Wen, C.; Ranju, M.; Huagen, D.; Chongjia, S. Pretreatment Slurries for Digital Printing of Silk/Cotton Intertexture with Reactive Dyes. *J. Silk* **2021**, *58*, 20–24. Available from: https://doi.org/10.3969/j.issn.1001-7003.2021.06.004.
37. Liang, Y.; Liu, X.; Fang, K.; An, F.; Li, C.; Liu, H.; Qiao, X.; Zhang, S. Construction of New Surface on Linen Fabric by Hydroxyethyl Cellulose for Improving Inkjet Printing Performance of Reactive Dyes. *Prog. Org. Coat.* **2021**, *154*, 106179. Available from: https://doi.org/10.1016/j.porgcoat.2021.106179.

38. Soleimani-Gorgani, A.; Najafi, F.; Karami, Z. Modification of Cotton Fabric with a Dendrimer to Improve Ink-Jet Printing Process. *Carbohydr. Polym.* **2015**, *131*, 168–176. Available from: https://doi.org/10.1016/j.carbpol.2015.04.031.
39. Jung, U.; Lamar, T. The Effects of CO2 Laser Treatment on a Digital Velvet Printing. *J. Text. Inst.* **2021**, 1–11. Available from: https://doi.org/10.1080/00405000.2021.1926119.
40. Jung, U.; Lamar, T. A. M.; Shamey, R. Digital Textile Printing with Laser Engraving: Surface Contour Modification and Color Properties. In International Textile and ApparelAssociation Annual Conference Proceedings; *76*(1); Iowa State University Digital Press, 2019.
41. Glogar, M. I.; Dekanić, T.; Tarbuk, A.; Čorak, I.; Labazan, P. Influence of Cotton Cationization on Pigment Layer Characteristics in Digital Printing. *Molecules* **2022**, *27*, 1418.
42. Wang, C.; Wang, C. Surface Pretreatment of Polyester Fabric for Ink Jet Printing with Radio Frequency O2 Plasma. *Fibers Polym.* **2010**, *11*, 223–228.
43. Park, Y.; Koo, K. The Eco-Friendly Surface Modification of Textiles for Deep Digital Textile Printing by In-Line Atmospheric Non-Thermal Plasma Treatment. *Fibers Polym.* **2014**, *15*, 1701–1707. Available from: https://doi.org/10.1007/s12221-014-1701-y.
44. Phattanarudee, S.; Chakvattanatham, K.; Kiatkamjornwong, S. Pretreatment of Silk Fabric Surface with Amino Compounds for Ink Jet Printing. *Prog. Org. Coat.* **2009**, *64*, 405–418. Available from: https://doi.org/10.1016/j.porgcoat.2008.08.002.
45. Zhang, C.; Wang, L.; Yu, M.; Qu, L.; Men, Y.; Zhang, X. Surface Processing and Ageing Behavior of Silk Fabrics Treated with Atmospheric-Pressure Plasma for Pigment-Based Ink-Jet Printing. *Appl. Surf. Sci.* **2018**, *434*, 198–203. Available from: https://doi.org/10.1016/j.apsusc.2017.10.178.
46. Zhang, C.; Guo, F.; Li, H.; Wang, Y.; Zhang, Z. Study on the Physical-Morphological and Chemical Properties of Silk Fabric Surface Modified with Multiple Ambient Gas Plasma for Inkjet Printing. *Appl. Surf. Sci.* **2019**, *490*, 157–164. Available from: https://doi.org/10.1016/j.apsusc.2019.06.053.
47. Li, X. *New Colorants for Ink-Jet Printing on Textiles;* Georgia Institute of Technology, 2003.
48. Ujiie, H. Digital Textile Printing: Status Report 2021. NIP & Digital Fabrication Conference **2021**, *2021*, 47–52. doi:10.2352/ISSN.2169-4451.2021.37.47.
49. Jing, S.; Meifei, Z.; Tingfang, M.; Houyong, Y.; Juming, Y. Preparation and Properties of Digital Printing Inks Based on the Cellulose Nanoparticles as Dispersant. *Adv. Text. Technol.* **2019**, *27*, 69–72. Available from: https://doi.org/10.19398/j.att.201803020.
50. Ding, Y.; Freeman, H. S.; Parrillo-Chapman, L. Color Gamut Analysis and Color Fastness Evaluation for Textile Inkjet Printing Application on Polyester 1. *J. Imaging Sci. Technol.* **2017**, *61* 50503-50501-50503-50508.
51. Tawiah, B.; Howard, E. K.; Asinyo, B. K. The Chemistry of Inkjet Inks for Digital Textile Printing—Review. *BEST* **2016**, *4*, 61–78.
52. Zhang, J.; Li, X.; Shi, X.; Hua, M.; Zhou, X.; Wang, X. Synthesis of Core−Shell Acrylic−Polyurethane Hybrid Latex as Binder of Aqueous Pigment Inks for Digital Inkjet Printing. *Prog. Nat. Sci.: Mater. Int.* **2012**, *22*, 71–78. Available from: https://doi.org/10.1016/j.pnsc.2011.12.012.
53. Leelajariyakul, S.; Noguchi, H.; Kiatkamjornwong, S. Surface-Modified and Micro-Encapsulated Pigmented Inks for Ink Jet Printing on Textile Fabrics. *Prog. Org. Coat.* **2008**, *62*, 145–161. Available from: https://doi.org/10.1016/j.porgcoat.2007.10.005.
54. Zapka, W. *Handbook of Industrial Inkjet Printing: A Full System Approach;* John Wiley & Sons, 2017.

55. Hoath, S. D. *Fundamentals of Inkjet Printing: The Science of Inkjet and Droplets;* John Wiley & Sons, 2016.
56. Kumah, C.; Zhang, N.; Raji, K.; Pan, R. Color Measurement of Segmented Printed Fabric Patterns in Lab Color Space from RGB Digital Images. *J. Text. Sci. Technol.* **2019,** *05,* 1–18. Available from: https://doi.org/10.4236/jtst.2019.51001.
57. Mangin, P. J.; Dubé, M. Fundamental Questions on Print Quality. *Proc. Image Quality and System Performance III* **2006,** 605901.
58. Dehghani, A.; Jahanshah, F.; Borman, D.; Dennis, K.; Wang, J. Design and Engineering Challenges for Digital Ink-Jet Printing on Textiles. *Int. J. Cloth. Sci. Technol.* **2004,** *16,* 262–273. Available from: https://doi.org/10.1108/09556220410520531.
59. Leisti, T.; Radun, J.; Virtanen, T.; Halonen, R.; Nyman, G. *Subjective Experience of Image Quality: Attributes, Definitions, and Decision Making of Subjective Image Quality;* SPIE, 2009Vol. 7242.
60. Tanaka, K.; Sugeno, M. A Study on Subjective Evaluations of Printed Color Images. *Int. J. Approx. Reason.* **1991,** *5,* 213–222. Available from: https://doi.org/10.1016/0888-613X(91)90009-B.
61. Sharma, P. *A Study on the Effect of Fabric Structure and Finishing on Perceived Image Quality;* Rochester Institute of Technology, 2019.
62. Kašiković, N.; Vladić, G.; Milić, N.; Novaković, D.; Milošević, R.; Dedijer, S. Colour Fastness to Washing of Multi-Layered Digital Prints on Textile Materials. *J. Natl Sci. Found. Sri Lanka* **2018,** *46,* 381. Available from: https://doi.org/10.4038/jnsfsr.v46i3.8489.
63. Briggs, J. C.; Tse, M.-K. Objective Print Quality Analysis and The Portable Personal IAS® Image Analysis System. *NIHON GAZO GAKKAISHI (J. Imaging Soc. Jpn.)* **2005,** *44,* 505–513.
64. Briggs, J. C.; Klein, A. H.; Tse, M.-K. Applications of ISO-13660, a new international standard for objective print quality evaluation. *Japan Hardcopy* **1999,** *99,* 21–23.
65. Zanaty, E. A.; El-Zoghdy, S. F. A Novel Approach for Color Image Segmentation Based on Region Growing. *Int. J. Comput. Appl.* **2017,** *39,* 123–139.
66. Ali Hayat, G.; Hussain, M.; Qamar Khan, M.; Javed, Z. Textile Education in Pakistan. In *Textile and Fashion Education Internationalization: A Promising Discipline from South Asia;* Yan, X., Chen, L., Memon, H., Eds.; Springer Nature Singapore: Singapore, 2022; pp 59–82.
67. Dutta, S.; Bansal, P. Textile Academics in India—An Overview. In *Textile and Fashion Education Internationalization: A Promising Discipline from South Asia;* Yan, X., Chen, L., Memon, H., Eds.; Springer Nature Singapore: Singapore, 2022; pp 13–34.
68. Uddin, M. F. Brief Analysis on the Past, Present, and Future of Textile Education in Bangladesh. In *Textile and Fashion Education Internationalization: A Promising Discipline from South Asia;* Yan, X., Chen, L., Memon, H., Eds.; Springer Nature Singapore: Singapore, 2022; pp 35–57.
69. Singh, R.; Shrestha, A. Namuna College of Fashion Technology: Pioneering in Fashion and Textile Education in Nepal. In *Textile and Fashion Education Internationalization: A Promising Discipline from South Asia;* Yan, X., Chen, L., Memon, H., Eds.; Springer Nature Singapore: Singapore, 2022; pp 103–118.
70. Inagaki, Toshihiko, et al. *Extensive Works of ISO/IEC 13660 and the Current Status (ISO/IEC JTC1/SC28 and JBMIA SC28/WG4);* IS AND TS PICS CONFERENCE. SOCIETY FOR IMAGING SCIENCE & TECHNOLOGY, 2003.

# Western markets for digitally printed textiles

*Muhammad Ayyoob[1] and Muhammad Khan[2]*
[1]Department of Polymer Engineering, National Textile University, Karachi Campus, Karachi, Pakistan, [2]Nanotechnology Research Lab, Department of Textile & Clothing, National Textile University, Karachi Campus, Karachi, Pakistan

## 11.1 Background

The invention of digital textile printing modernizes the process of designing and printing textile products by allowing the opportunity to design and produce products for specialty markets. Digitally printed textiles have been gaining the market rapidly since the commercialization of digital printing technology.[1] Lower lead time, eco-friendly, and design flexibility are the major factors driving the growth of digital textile printing globally. Digital printing technology for textiles was commercialized in the early 2000s. Previously, the rotary printing of textiles was a modern technique to print textiles. With the development of digital printing technology for textiles, the textile printing industry has been revolutionized. Digital textiles printing (DTP) has modernized the printing process of textiles in recent history (Fig. 11.1).

DTP can be defined as printing any design onto fabrics through inkjet printing with the help of digital images of designs. DTP technology has gained tremendous popularity since its introduction in the early 2000s. Textile machinery manufacturers and printers have been working closely to modernize printing technology through modification and market gains in printed textiles. The flexibility and easiness in design change, compatibility with fast-changing fashion trends, and short lead times make DTP more convenient for textile printers. The attributes of digital textile printing have allowed it to supplement, and to some extent, replace traditional textile printing techniques.[3]

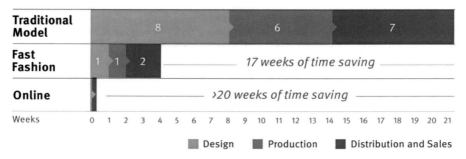

**Figure 11.1** Traditional versus online customized fashion.[2]

DTP is growing very fast. Currently, the digital textile printing market is valued at $2.00 Billion in 2020 and is projected to be 14.5% CAGR to 8 billion USD by 2030;[4] however, DTP has been impacted negatively by the COVID-19 pandemic in recent times. Initially, the DTP was used for soft signages, gaining popularity among textile printers with time. With the advancement of modern digital textile printing technology and the development of different types of inks for textiles and clothing, the DTP has been diversified from casual wear to home textiles and other applications, such as industrial textiles, outdoor tents, automotive textiles, textile products for boating and many more (Fig. 11.2).

Conventional textile printing techniques depend on using separate rollers or screens for each color used in the design. This limits the number of colors used in one design, as the higher the number of colors higher the cost of the development will be. Further, the design is limited by the size of the roller or screen used; the quality and the variety of the products produced using this method are also hindered by the fact that the colors used are mixed individually to meet the design specifications.

**Figure 11.2** Affordable and customized fashion is the future of digitally printed wearables.
*Source*: Courtesy of Miss XinRui Hui @ College of Textile Science and Engineering, Zhejiang Sci-Tech University, China.

A typical IJDTP or inkjet digital textile printer uses a four-color system known as CMYK (cyan, magenta, yellow, black). It could be further extended to 7−12 colors.[5,6] IJDTP, when used in combination with CAD, can produce products with hundreds of different colors, large sizes, and photo-realistic quality, which will be either impossible to produce using conventional printing methods or prohibitively expensive.

## 11.2 Digital textiles printing and sustainability

In this era, when very high volumes of resources are produced and consumed, the textile industry is one of the largest environmental polluters, and within the textile industry dying process has great importance for its impact on the environment and sustainability.[7,8] The process of printing not only uses many chemicals, including dyes, surfactants, solvent binders, and more, which are then released into the environment without processing and are very dangerous for the environment. Moreover, printing requires a lot of energy and water.[9] So, conventional printing technologies release many effluents containing dangerous chemicals into the environment. Contrary to that, DTP is considered environmentally friendly, consuming much less water and energy when compared to screen printing.[9] Eco-friendly pretreatment processes are also designed for fabric preparation before digital printing.[10] Although digital printing has many advantages, it allows designers to work with thousands of different colors and complex designs; however, eco-friendliness is one of the most critical factors driving the adaptation of digital printing technology.[10] DTP is considered a greener and environment-friendly method of textile printing, which has allowed high precision, complex, and low-cost processing compared to conventional printing techniques focused on high quality and high volume production.[5,7,9,11]

In inkjet printing, ink droplets are deposited on the material, allowing tolerance for different types of surfaces. A digital textile printer uses a minimal size of ink as compared to the drop dripping from a faucet. There are two main categories of inkjet printers: Drop-on-demand Inkjet printers and continuous inkjet printers.[11,12] Following is an overview of the mechanics used in ink jet[13] (Fig. 11.3).

About 11%−13% of all textile products produced worldwide are printed products.[14] Comparing the textile market between 2008 and 2019 shows that while digital printing is gaining popularity, it only accounts for about 2% of worldwide textile products. Rotary screen leads the market with about 65% of the textile products produced using this method[15] (Fig. 11.4).

Different processes in the textile industry, like Growing and manufacturing of fiber, weaving, wet processing of fiber and cloth, dying, and spinning, all have some environmental effects. The primary material for the textile industry is fiber, which is a defining factor for selecting sustainability, life cycle, use, and processing technology. Textile products and the textile sector are both fundamentals of human society. Moreover, they have existed throughout the history of human civilization in one form or another. They have applications in everyday life, medical field and

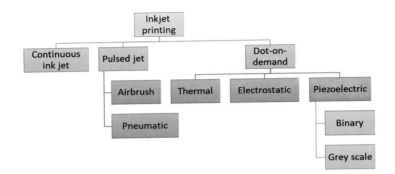

**Figure 11.3** Different mechanisms used by inkjet printers.

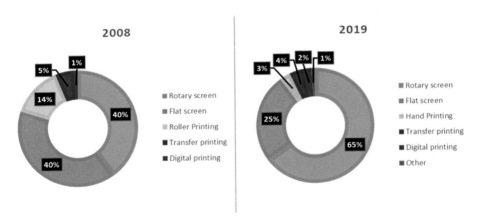

**Figure 11.4** Printing methods in 2008 and 2019.

industry, and high-tech like the space industry and nanofabrication. With the increase in the consumption of textile products, there is an increasing demand for the fiber used for producing those products; the demand for it reached 95.6 million tons in 2015.

The most significant segment of the fibers used in the textile industry is petroleum-based fibers accounting for around 62.1% of the total fiber used, followed by 25.2% cotton fiber, the portion of wood-based regenerated fibers is 6.4%, while 5.1% of fiber comes from protein and cellulose-based fibers and wool-based fibers account for only 1.2% of the total fiber used (Fig. 11.5).

The technology used for manufacturing or growth is the critical factor in understanding the effect of any type of fiber on the environment and its sustainability. Different types of fiber groups have different types of advantages and disadvantages.

When comparing two of the most widely used fibers, cotton-based and polyester-based fibers, one could be mistaken to think that cotton is natural, so it

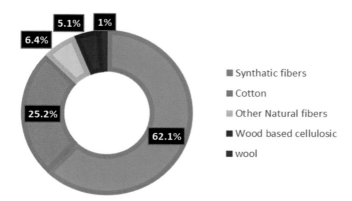

**Figure 11.5** Fiber types and share in world consumption.

will be more sustainable and have a lesser environmental effect. Different chemicals, including fertilizers and pesticides widely used for growing cotton, affect the environment considerably. Another environmental sustainability concern is the enormous amount of water required to grow cotton.[16,17] Even with growing awareness about the environment, the production and use of organic cotton are increasing, which is used without using such chemicals; cotton requires vast amounts of water for irrigation purposes. On the other hand, while polyester fiber production requires less water, this process is very energy-intensive, requiring oil and releasing a tremendous amount of greenhouse gases into the environment and causing global warming.[18]

The second step in manufacturing conventional textile products is yarn manufacturing, and the principles used for this process have remained unchanged for the last two centuries. Nevertheless, much progress has been made toward a high production rate, higher efficiency, less manual labor, and less waste production. The technology used in ring-spinning machinery has improved since its inception, but it has primarily focused on quality and quantity of production while remaining almost unchanged in principle. Along with ring-spinning technology, other yarn-spinning technologies have been introduced; air-jet spinning, rotor spinning, and air vortex spinning are modern technologies. These technologies not only focus on a faster production rate but also require less preand postprocession and are more sustainable.

The textile surface procession is the next step, done by weaving, braiding, or knitting methods. These processes have practically remained the same for the last few centuries and were even used before the textile industry's proper inception. The weaving machinery used has its environmental effects along with the postweaving processes, which are not only very energy expensive but also have a high environmental impact and water consumption.

Wet processing of textile products, including printing, dyeing, and finishing, is very energy consuming and has a very high impact on the environment due to wastewater discharge loaded with harmful chemicals. The production process of textile products, especially wet processing, not only requires a lot of water and energy, but the waste material produced during these processes is loaded with harmful chemicals and has a profound environmental effect.

The conventional dying process requires much water, and it not only consumes the freshwater but also discharges the used water loaded with chemicals into the environment. Processing one kilogram of textile product requires between 100 and 150 L of water; yearly, more than 28 billion kilos are processed.

One of the revolutionary technologies is digital textile printing, which is being adopted rapidly by the industry. It has many advantages over the more conventional dyeing processes like screen printing. It requires as little as one liter of water per line meter printed compared to the 50–100 L required in conventional printing techniques. It also requires less than 50% energy and emits only 10% CO2 compared to screen printing or other conventional dying techniques.[18]

## 11.3 Market overview

The market share of DTP is growing fast, it is being used in many innovative ways, and the main drivers behind its growth are the demand for personalized products, sample manufacturing, shorter print runs, and low cost.[19] The ability to produce personalized and customized products with shorter print runs has allowed the introduction of POD. The paper printing industry is credited with developing POD, but the textile printing industry is now adopting it. It allows the customers to order the product with a unique design, with the order as small as a single piece.[20] POD has enabled a culture that focuses not on consumption but on creation and production. Companies offering customized textile products do not start printing until the order is placed by the customer.[21] The introduction of DTP has opened a new dimension for the textile industry, and the designers' imaginations only limit the designs and their applications. Following are some areas that have seen the most extensive use of DTP.

### 11.3.1 Home furnishing and interior decoration

Almost 40% of printed textile worldwide is based on products related to printed home textiles. This includes but is not limited to curtains, carpets, blankets, rugs, and bed linen. In the traditional process, the factories are forced to carry a large inventory of different designs for several years to fulfill the demands of the customers, while in the DTP factories only maintain the inventory of the white good, and on customer orders, as per their requirements, they print cut and dispatch the order. This also allows the customer to order the same design whenever needed, as the design will never become obsolete. Despite its benefits, digital textile printing only accounts for a fraction of the overall home textile segment of the world's market. The main obstacles in adapting DTP in this segment are the preand postprocessing steps and the availability of a small number of large digital printers.[22]

### 11.3.2 Apparel and fashion

The apparel industry is the leading segment of the textile industry, where DTP offers the most commercial and creative opportunities. The apparel industry has transformed and seen changes in how products are forecasted, designed, produced, and distributed. Now producers are introducing new designs more frequently as the fashion cycle time is reduced drastically. This requires that companies design new products and ship them every 6–8 weeks, which requires drastic cuts in lead time. This model has allowed the DTP to thrive, allowing faster manufacturing and implementation of just-in-time production.

### 11.3.3 Direct-to-garment printing

Direct-to-garment printing is the process where printing is done on the assembled product, and the printer that can perform that task is known as direct to garment printer or DTG printer. This type of system has seen considerable growth in recent years as it allows companies to do short runs and offer customized designs.

### 11.3.4 Soft signage

This nontraditional textile application has seen the fastest adaptation of digital textile printing. Two methods are used for printing substrate based on polyester for soft signage applications, dye-sub, and direct-to-fabric, while the direct-to-fabric method produces designs with lower print definition. It is gaining momentum due to the overall cost and higher productivity benefits.

Among DTP markets, North America continues to be the main driver of the digital printing markets, followed by Europe. North America is said to be the one-third revenue generator throw digital printing mainly due to the prominent digital printing players. North America is the primary consumer market and a significant player in digital printing. India and China, however, are the central hub of digital printed textile manufacturers and exporters, followed by Pakistan and Bangladesh. A large volume of digitally printed textiles is being exported from these countries.

The digitally printed textile market drives the growth of the innovation, R&D, and production of new technologically advanced digital printers. The subtle integration of technologically advanced and old printers in the digital printing industry creates rivalry among printer manufacturers, which will gain much traction in the future. The increasing demand for custom digital printed textiles. Most printer manufacturers are targeting expanding the supply chain and auxiliaries (printing inks and other supplies) to cater to these increased demands. A strong supply chain is significant for the healthy growth of any increasing demand-based market. Japan and China have become the centers for digital printer manufacturing and manufacturing services due to their ability for equipment manufacturing. Both countries have exported a vast chunk of digital printers to the printing industry worldwide.[23]

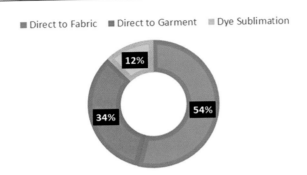

**Figure 11.6** Major three segments of digitally printed textiles.

Fig. 11.6 shows the three major segments of digitally printed textiles. Direct-to-fabric printing accounts for almost 54% of the total digital printing market in the global digital textile printing market. The direct-to-fabric printing is ideal for cotton, silk, nylon, sisal, hemp, and banana-based fabrics. Because of fast-changing fashion trends and emerging demands for custom-based garments, direct-to-garment is the ultimate choice. The segment contributes almost 33.9% throughout the 2022–2029 forecast. Where polyester is involved, however, the last segment, dye sublimation printing, constitutes 12.15% of the digital printing market. Dye-sublimation printing produces high-quality photographic results.[23]

## 11.4 COVID-19 impact on digital textiles printing market

COVID-19 emerged in late 2019 and reached a global pandemic in early 2020. COVID-19 had a wide range of implications affecting and will affect the printer suppliers, printing industries, and export of printed goods. The lockdown in different countries is a single point regarding global market supply chains. The pandemic disrupted the global supply chain. Most countries suffer economic depression, but emerging and developing economies suffer the most.[24]

Other associated issues related to the global pandemic are travel restrictions and short and long-distance movement limitations, which have caused an unmanageable impact on this globally integrated and labor-intensive business segment. The world health catastrophe and response to the pandemic resulted in production losses and significant disturbance in the supply chain of export-related goods, including digitally printed textiles.

Moreover, China and Japan are the prominent manufacturers and suppliers of digital textile printers and other supplies in the global textile production industries. The COVID-19 pandemic caused delays and disruption in the global supply chain of printing equipment and inks. As a result, this also impacted the increasing trends in custom fashion and brandings. The supplies were reduced significantly to the western markets, including North America and Europe.

China holds a significant western market share in digital textile printed products, the lockdown in China, decreased demand for digital products, and delays in freights also significantly affected the market. One example is enough to understand that most orders from top global brands were canceled during the pandemic. In the late 2021 seconds and early 2022 seconds, however, the market again starts growing and gaining momentum right now.

On the other hand, the COVID-19 period was the best time to set back and rethink the feasible solution to the global issues and reduction in market demands. Post-19 is the time to transform challenges into opportunities. Hopefully, the market will fully recover very soon.

The devastating effect of the pandemic on the digital textile industry has been evident due to retail shutdowns, global lockdowns, forcibly closing the production factories, disruption of the global supply chains, reduction of demand in specific segments, and increasing demand in others (face masks). The businesses and customers have been reprioritized, and garment manufacturers, retail partners, and brands have shifted to online businesses and E-commerce.[25] Those who have redesigned their production and supply chain during COVID-19 and now dare to transform the idea into reality will lead the market.

## 11.5 Market driving forces

With the increasing access to the internet and integration of E-commerce in digital marketing supported with artificial intelligence, the approaches toward conventional markets are diversified significantly. With the advent of the internet and the integration of digital technology into the marketing process, technology has paved the way for new-age marketing with the online, offline, and conventional market mix in almost every field, including commodity markets. Consequently, the engagement of the online availability of design, arts, and ideas has expanded the market mix to a new extent.[26] Textile industry 4.0 is another driving force for the new future of the textile market. The companies once were at the driving end in innovation in textile fibers, dyes, and machinery are now nowhere in the lead in digital printing textiles. New players are jumped in with digital inks and printing equipment.[27] The leading brands are another segment of the market trendsetters. Sustainability in fashion can be attained by encouraging brands to shift to green purchase intention.[28] Rapidly changing fashion is another trendsetter in digital printing textiles.[29]

The shift toward digital textile printing is, however, mainly driven by innovation and technology development in the segment. Digital printing technology speeds upthe production process, is compatible with rapidly changing fashion, and a significant reduction in colorant costs and handling factors are responsible for the steady growth in the digitally printed textile market segment.[29] The technological progress and innovations associated with digital printing inks and other consumables, digital printers, printing heads, and specially developed design software and printing machinery are vital to the market growth.

Specific other driving forces and trends set the market demand and supply chain progress goals. Several trends and market drivers have been mentioned in the report published by, Research and Market: Global Textile Printing Market 2021−2024: COVID-19 Unwinds and Accelerates Disruptive Changes within Textile Printing Industry (http://www.researchandmarkets.com). Specific drivers and trends will decide the fate of digitally printed textiles.[30] The same has been reported by some other sources.[1] The main driving forces for digital printing textiles are,

- Rising middle-class consumer segment
- Rapid urbanization
- Eco-friendly inks and sustainable print technology
- Customized design accessibility and the possibility of rapid design change as compared to conventional printing technology.
- Green and sustainable printing practices
- A range of acceptable fabric printing possibilities
- Possibilities of home textiles and décor fabrics
- Artificial intelligence in printing technology.
- Soft signages
- Rapidly changing fashion industry
- Room for constant innovation and change in technology.
- Globalization, E-commerce, and online business trends
- Online accessibility to the direct consumer market
- On-demand inkjet digital printed textiles

## 11.6 Share of digital printed textiles in printed textiles

In the 1990s, digital printing was introduced in the textile industry; initially, the success of digital printing was doubted due to the extreme demands of the textile industry. Some of the significant challenges in textile printing are the following.

There are about a dozen different types of natural and synthetic fibers, each having different characteristics related to ink compatibility.

- The surface needed to be printed could be textured, porous, and stretchable.
- The print should be able to handle extreme use cases, including but not limited to heat, light, wear, cleaning, and contact with abrasive materials.
- Along with the visuals, the texture or feel of the fabric should also be as per requirements.

Even almost 25 years after its introduction, against such odds, digital textile is gaining shear. Because of the slow speed of the technology adaptation in the industry, high initial cost and supply chain has kept the DTP from becoming the mainstream, and its shear in the printed textile is less than the more conventional methods like screen printing. The industry is making progress, and the share of digital textile printing is increasing by leaps and bounds; due to its vast potential, many dye manufacturers and technology providers have invested significant amounts in its development.

The textile fashion market is dominated by digitally printed apparel, but still, it is far less than conventional printing products. The cost is still a significant factor. Because of large production volumes, conventional printing costs are far less than digital printing.[31] Digital printing, however, has very different superiority points than conventional points, as discussed earlier, such as design flexibility, readily available for markets, and new business models, including on-demand printing. Easy color registration, excellent dimensional stability of the design, more accessible and cheaper design development and revisions, shorter sampling time, printing fine details of design, esthetic and luxuries design patterned for apparel and home textile, home textile designs are the other superior areas over the conventional printing of textiles.[31]

The global production of printed textiles is projected at 36.8 billion square meters by 2024, whereas the share of digitally printed textiles will be only 5%–10% of the total printed textiles.[32] Less share means less business saturation and more opportunities. According to a digital textile tracking company, "World Textile Information Network, we still have a long way to go in the industry's analog-to-digital transformation, with just 6% of the world's printed textiles produced digitally." According to the data in 2018, only a small share of 36 billion square meters of fabric was printed digitally, which amounts to roughly 6% (2.57 billion square meters) of the total production. The growth, however, was incredible compared to 3.3% in 2015. The installed printers increased from 27000 in 2015 to 50000 in 2018.[33]

Most digital printers were installed in Europe and Asia, with only 10% in North America. With advancements in technology, today's printers have the capability of more than 8000 square meters per hour as EFI Reggiani BOLT. Reggiani sold four new printers, three of which were ordered from Pakistan and one from Italy.[33]

## 11.7 Western markets

Western markets are very diversified. Europe and North America are technologically advanced regions, and with advancements in digital technology, they are already producing innovative textiles. We can see a considerable increase in European textile exports, upto 10.6% in 2021, and imports will drop by 7%.[34] As a result, Europe's trade deficit decreased, but still, it is at $50 billion. Europe is now the world's second-largest exporter of textiles after China. This boost is due to the strong growth of China, the United States, and the Swiss markets. Contrary to that, sales dropped by 23% in the United Kingdom as Asian suppliers dominated the market after Brexit. On the other hand, European textile imports also dropped by 28% from China and 48% from the United Kingdom due to several issues, including problems created by the global pandemic.

Western wear market size is increasing considerably, with a 4.8% CAGR (compound average growth rate) from 2017 to 2023.[35] It is expected to reach 99.423 billion USD compared to 71.123 billion USD in 2017. With a significant share of digital printing in wear textiles, digital textile growth at least at a similar rate, is also expected in the western markets.

## 11.7.1 Europe

Before the COVID-19 pandemic, Europe was the biggest producer of digitally printed textiles; however, the pandemic impacted the European market very severely. In 2021 European output of digitally printed textiles was trailed by Asia by an astonishing 94% of total production. It does not mean the market was slow overall. That is not the case; output increased by 28% compared to the previous year. Asian production of digital textiles, however, increased before the pandemic, whereas Europe lagged by 10%.[36] Apparel is the largest segment of digital textiles; therefore, in 2021, apparel and home textiles were 88% of the total digital printing outputs. The soft signage market also, however, increased. Direct-to-fabric and sublimation almost share the same output as direct-to-fabric more than sublimation.[36]

Europe is a robust market for digital textiles; however, severely affected by the COVID-19 pandemic. Lockdown, retail market restrictions, and increased inflation are the leading causes of the market crumple. On the other hand, pandemics also influence new business models, such as online marketing and E-commerce. The B2B and B2C businesses started with great zeal during the pandemic. Being adventitious of innovation, technology, and easy internet access, Europe became the hub of online trading. Overall, Europe is a more developed market. After the pandemic energy, however, the crisis hit the production industry, and consumers' buying behaviors were also very severely affected. The impact worsened in early 2022; if the Ukraine-Russia conflict continued, it would hit hard on European markets. The impact will be seen shortly.

Contrary to the abovementioned details, the European market has several advantages as producers and consumers share borders. Currently, world trade plunged due to a considerable lag in freight and freight costs. Shipping agencies give dates of several weeks to several months; however, this is not the case with European markets. There is an excellent opportunity for retailers and brands. They have the advantage of a considerable reduction in transportation costs. Therefore, they can profit more significantly and sustain during challenging times, such as pandemics and energy crises. Textile is an extensive labor industry, and labor costs constitute a significant portion of total production costs. Thanks to technology, the advancements in textiles, such as artificial intelligence, automation, and digitalization, reduced the labor cost many folds. Local availability of materials, digitalization, and nearshoring can cut production costs, which as a result, is making the digital printing of textiles more viable in Europe.

Italy and Turkey are the two most prominent players. Italy has an unyielding passion for fashion and has a very lucrative fashion history. With a 241 million square meters production in 2021, Italy produced 26% more inkjet printed textiles than the previous year; 62% was printed on high-volume printers, and 32% was printed on 650 square meter/hour machines. Ninety-three of the total digital textiles were apparel and home textile due to the rich traditional trends in fashion and décor. Digital inkjet printers' installation was also increased by 12%.

Similarly, the installed capacity of Turkey was also enhanced by 17% compared to 2020. Turkey is determined to provide digital textiles to Western Europe. For that reason, Turkey is already improving its production capacity. Over 200 million square meters were produced, which is about 25% higher than the previous year's production.

**Figure 11.7** Some random clicks from the European Print Exhibition and Eurologo Exhibition by FESPA from May 31 to June 3, 2022, held at the Exhibition Center in Berlin, Germany.

Portugal and Spain from the Mediterranean are also important in digitally printed textiles, as both increased installed production capacity. Less growth, however, was observed in Central Europe and Northern Europe. Interestingly, Germany is also taking an interest in digital inkjet printed textiles, this year there was a very successful European Print Exhibition and Eurologo Exhibition held by FESPA from May 31 to June 3, 2022, at the Exhibition Center in Berlin, Germany, some random clicks of the exhibition are shown in Fig. 11.7.

The market share of digital printing is significantly less in these two areas of Europe. In 2021 a considerable increase, however, was observed in these regions, which is mainly due to nearshoring and sustainability. The effects of war in Europe will undoubtedly disrupt the said growth. Currently, energy crises are the single most challenging for Europe.

### 11.7.2 North America

The United States is one of the largest fabric producers.[37] Digital textile growth is also, however, increasing more recently. For example, in 2021, the installed capacity was increased by 12.5, and output was increased by 28%. At the same time, the growth was almost 5% downfrom that in 2019. The United States will continue to grow its digital market compared to the other regional players. Mexico and other Central American nations will follow.

The United States is the dominant player in North America, and due to the reshoring movement, it is argued long-term that the country will continue to grow its market share compared to Mexico and Central American nations. The United States produced 198.7 million sqm of digitally printed fabric, up by 28% since 2020. Its installed base also grew by 12.2% to 4820. Mexico is also determined to increase its installed capacity for digital printing. Production has grown to 21.4% and 2% since 2020, and installed capacity increased by 18%.

Consumer trends are changing as that fashion. The demands for sustainability and on-demand printing are the main driving forces in North America. The rest of the demands are being fulfilled by Asian nations, mainly China, India, and Pakistan.[38,39] The future of the inkjet printing market in the region is bright.

### 11.7.3 South America and the caribbean

Southern America is crumbling due to economic and political instabilities. This caused uncertainty in the digital textile market. Buying powers and consumer demands are the main driving forces of any market, although demands for new fashion trends and digitally printed apparel are increasing with global digitalization. At the same time, however, it is incredibly challenging to meet such demands from digital equipment suppliers and finished goods suppliers. Outsourcing materials and buying printing machinery are problematic due to economic pressure and political issues. Therefore, the demand for finished goods and online deliveries to the consumer's doorstep is increasing; however, due to the increasing demands for digital printed textiles and apparel, the service providers are also considering the region, looking forward to understanding the consumer behavior and supply chain of the finished good. Companies are considering installing local production units. The future of the production of digitally printed textiles looks stronger in Latin America.

On the contrary, the demand for digital textiles rose in the region, including the Caribbean. Therefore, in 2021, the production surpassed by 2% compared to preCOVID and reached 38% compared to 2020. The total production was 165 million square meters. The installation capacity was also increased by 16%.

On the other hand, transfer sublimation was the primary driving force for Brazilian markets, as demand for sportswear was high. The share of sublimation was 62% of the total digital printed textiles. During the production year 2021, the overall production remained 35% higher than in 2020.

## 11.8 Summary

In the recent past, technology changed at a fast rate. Especially during the COVID-19 pandemic, new manufacturing, marketing, and shopping trends emerged. The world is connected digitally like never before. Shopping trends changed from traditional shopping malls to online shopping platforms. It is convenient, easy, time-saving, and secure. Moreover, adding customized fashion trends and a short lead time for digitally printed textiles change shopping and marketing forever.

Previously, the large-scale production of printed textiles was shifted to the eastern countries like other textile sectors. Increasing demands in fast-changing fashion trends, shipping and freight-related problems and border closures, and the introduction of sustainability in digitally printed textiles, however, forced textile manufacturers to produce and market with the benefit of nearshoring. Many players from western countries also install larger set-ups for digitally printed fashion textiles. It also provides the benefits of technology, as modern textile technology mainly originates from western countries, such as Italy, Germany, Spain, and the United States; however, Japan is also well-known for its advanced technology, and China is emerging too.

The future is digitally printed textiles for fashion and home applications due to easy installation, fewer capital requirements, less workforce, time-saving, and customized designs; however, the current share of digitally printed textiles is not much in overall printed textiles. It will take time to shift a more significant portion of printed textiles to digital platforms. It will require more affordable, high throughput, and modernized printing solutions with a special emphasis on sustainability. Aggressive marketing and promotion will also help adapt digitally printed textiles.

# References

1. Memon, H.; Khoso, N. A.; Memon, S.; Wang, N. N.; Zhu, C. Y. Formulation of Eco-Friendly Inks for Ink-Jet Printing of Polyester and Cotton Blended Fabric. *Key Eng. Mater.* **2016**, *671*, 109–114. Available from: https://doi.org/10.4028/www.scientific.net/KEM.671.109.
2. Micro-run: The Evolution of Digital Textile Production, Kornit Digit. (n.d.). https://www.kornit.com/micro-run-evolution-digital-textile-production/ (accessed June 14, 2022).
3. Dehghani, A.; Jahanshah, F.; Borman, D.; Dennis, K.; Wang, J. Design and Engineering Challenges for Digital Ink-Jet Printing on Textiles. *Int. J. Cloth. Sci. Technol.* **2004**, *16*, 262–273. Available from: https://doi.org/10.1108/09556220410520531.
4. Digital Textile Printing Market Size | Forecast Report, 2030, Allied Mark. Res. (n.d.). https://www.alliedmarketresearch.com/digital-textile-printing-market (accessed May 6, 2022).
5. Colorgate, ColorGATE News, (n.d.). https://www.colorgate.com/about-colorgate/news/color-management-and-rip-software-in-digital-textile-printing-5/ (accessed July 6, 2022).
6. http://www.fibre2fashion.com, Color Management and RIP Software for Digital Textile Printing Managing Color for Optimal Results, (n.d.). http://www.fibre2fashion.com/industry-article/1099/color-management-and-rip-software-for-digital-textile-printing (accessed July 6, 2022).
7. Memon, H.; Jin, X.; Tian, W.; Zhu, C. Sustainable Textile Marketing—Editorial. *Sustainability* **2022**, *14* (19), 11860 [Online]. Available from: https://www.mdpi.com/2071-1050/14/19/11860.
8. Memon, H.; Khatri, A.; Ali, N.; Memon, S. Dyeing Recipe Optimization for Eco-Friendly Dyeing and Mechanical Property Analysis of Eco-Friendly Dyed Cotton Fabric: Better Fixation, Strength, and Color Yield by Biodegradable Salts. *J. Nat. Fibers* **2016**, *13*, 749–758. Available from: https://doi.org/10.1080/15440478.2015.1137527.

9. Chen, L.; Ding, X.; Wu, X. Water Management Tool of Industrial Products: A Case Study of Screen Printing Fabric and Digital Printing Fabric. *Ecol. Indic.* **2015**, *58*, 86–94. Available from: https://doi.org/10.1016/j.ecolind.2015.05.045.
10. Shi, F.; Liu, Q.; Zhao, H.; Fang, K.; Xie, R.; Song, L.; Wang, M.; Chen, W. Eco-Friendly Pretreatment to the Coloration Enhancement of Reactive Dye Digital Inkjet Printing on Wool Fabrics. *ACS Sustain. Chem. Eng.* **2021**, *9*, 10361–10369. Available from: https://doi.org/10.1021/acssuschemeng.1c03486.
11. Polston, K.; Parrillo-Chapman, L.; Moore, M. Print-on-Demand Ink-Jet Digital Textile Printing Technology: An Initial Understanding of user Types and Skill Levels. *Int. J. Fash. Des. Technol. Educ.* **2015**, *8*, 87–96. Available from: https://doi.org/10.1080/17543266.2014.992050.
12. Ujiie, H. 19 - Design and Workflow in Digital Ink-Jet Printing. In *Digital Printing of Textiles*; Ujiie, H., Ed.; Woodhead Publishing, 2006; pp 337–354. Available from: https://doi.org/10.1533/9781845691585.4.337.
13. Hayat, M. A.; Gulzar, D. T.; Hussain, D. T.; Kirn, D. S.; Farooq, D. T.; Ahmed, D. A. Eco-Friendly Preparation, Characterization and Application of Nano Tech Pigmented Inkjet Inks and Comparison of Particle Size Effect and Printing Processes. *Am. Acad. Sci. Res. J. Eng. Technol. Sci.* **2020**, *72*, 197–213.
14. Kan, C. W.; Yuen, C. W. M. Digital Ink-Jet Printing on Textiles. *Res. J. Text. Appar.* **2012**, *16*, 1–24. Available from: https://doi.org/10.1108/RJTA-16-02-2012-B001.
15. Digital Textile Printing Market Continues to Grow, Text. News Appar. News RMG News Fashion Trends. (2019). https://www.textiletoday.com.bd/digital-textile-printing-market-continues-to-grow/ (accessed July 6, 2022).
16. Yesuf, H. M.; Xiaohong, Q.; Jhatial, A. K. Advancements in Cotton Cultivation. In *Cotton Science and Processing Technology: Gene, Ginning, Garment and Green Recycling*; Wang, H., Memon, H., Eds.; Springer Singapore: Singapore, 2020; pp 39–59.
17. Wang, H.; Memon, H. Introduction. In *Cotton Science and Processing Technology: Gene, Ginning, Garment and Green Recycling*; Wang, H., Memon, H., Eds.; Springer Singapore: Singapore, 2020; pp 1–13.
18. Wang, H.; Siddiqui, M. Q.; Memon, H. Physical Structure, Properties and Quality of Cotton. In *Cotton Science and Processing Technology: Gene, Ginning, Garment and Green Recycling*; Wang, H., Memon, H., Eds.; Springer Singapore: Singapore, 2020; pp 79–97.
19. P. Campbell, Taking Advantage of the Design Potential of Digital Printing Technology for Apparel, 4 (2005) 10.
20. B. Hunting, R. Puffer, S. Derby, L. Loomie, Issues Impacting the Design and Development of an Ink Jet Printer for Textiles, (n.d.) 4.
21. Craft Retailers' Criteria for Success and Associated Business Strategies - Paige – 2002– J. Small Bus. Manag. - Wiley Online Library, (n.d.). https://onlinelibrary.wiley.com/doi/abs/10.1111/1540-627X.00060 (accessed July 6, 2022).
22. Market Trends - the vibrant future for Digital Textile Pinting - FESPA | Screen, Digital, Textile Printing Exhibitions, Events and Associations, (n.d.). https://www.fespa.com/en/news-media/features/market-trends-the-vibrant-future-for-digital-textile-pinting (accessed July 6, 2022).
23. Digital Textile Printing Market, (n.d.). https://www.futuremarketinsights.com/reports/digital-textile-printing-market (accessed May 28, 2022).
24. Khurana, K. The Indian Fashion and Textile Sector in and Post COVID-19 Times. *Fash. Text.* **2022**, *9*, 15. Available from: https://doi.org/10.1186/s40691-021-00267-4.
25. D. Lee, Has Covid-19 Increased The Shift To Digital In The Textile Printing Industry?, (n.d.). http://blog.focuslabel.com/has-covid-19-increased-the-shift-to-digital-in-the-textile-printing-industry (accessed May 28, 2022).

26. Andrew-Essien, E. Art as a Dependable Driving Force in New Age Marketing. *PINISI Disc. Rev.* **2021**, *5*, 9–20. Available from: https://doi.org/10.26858/pdr.v5i1.22014.
27. Aneja, A.; Kupka, K.; Militky, J.; Kadi, N. 1B3_0287_ Textile Industry 4.0 – Preparing For Digital Future. *Proceeding of the 19th World Textile Conference— Autex 2019* **2019**. Available from: https://openjournals.ugent.be/autex/article/id/63822/ (accessed May 29, 2022).
28. Chen, L.; Qie, K.; Memon, H.; Yesuf, H. M. The Empirical Analysis of Green Innovation for Fashion Brands, Perceived Value and Green Purchase Intention—Mediating and Moderating Effects. *Sustainability* **2021**, *13*, 4238.
29. Memon, D. N. A. Rapidly Changing Fashion Trends Growth in the Textile Printing Market. *Pak. Text. J* **2018**, *67* 66–66.
30. R. and M. ltd, Textile Printing—Global Market Trajectory & Analytics, (n.d.). https://www.researchandmarkets.com/reports/1244798/textile_printing_global_market_trajectory_and (accessed May 29, 2022).
31. Koseoglu, A. U. Innovations and Analysis of Textile Digital Printing Technology. *Int. J. Sci. Technol. Soc.* **2019**, *7*, 38. Available from: https://doi.org/10.11648/j.ijsts.20190702.12.
32. What is Digital Textile Printing and What are the Advantages for Your Company? https://blog.spgprints.com/what-is-digital-textile-printing (accessed May 29, 2022).
33. Digital Printing of Textiles: A Growth Opportunity. Print. News. https://www.printingnews.com/digital-textile/article/21082893/digital-printing-of-textiles-a-growth-opportunity (accessed May 29, 2022).
34. EU textile and Clothing Exports on the Up, WTIN. https://www.wtin.com/article/2022/may/16-05-22/eu-textile-and-clothing-exports-on-the-up/ (accessed May 29, 2022).
35. Report says, "Global Western Wear Market Size is Expected to Reach $ 99,423 Million by 2023," Text. News Appar. News RMG News Artic. 2022. https://textilefocus.com/report-says-global-western-wear-market-size-expected-reach-99423-million-2023/ (accessed May 29, 2022).
36. 2021 Digital Textile Industry Review, WTIN. https://www.wtin.com/article/2022/april/11-04-22/2021-digital-textile-industry-review/ (accessed May 29, 2022).
37. America's ink market diversifies, WTIN. (n.d.). https://www.wtin.com/article/2022/april/04-04-22/america-s-ink-market-diversifies/?channelid = 17675 (accessed May 29, 2022).
38. Ali Hayat, G.; Hussain, M.; Qamar Khan, M.; Javed, Z. Textile Education in Pakistan. In *Textile and Fashion Education Internationalization: A Promising Discipline from South Asia;* Yan, X., Chen, L., Memon, H., Eds.; Springer Nature Singapore: Singapore, 2022; pp 59–82.
39. Dutta, S.; Bansal, P. Textile Academics in India—An Overview. In *Textile and Fashion Education Internationalization: A Promising Discipline from South Asia;* Yan, X., Chen, L., Memon, H., Eds.; Springer Nature Singapore: Singapore, 2022; pp 13–34.

# Emerging market trends: the cultural designs printed with digital printing technology: an overview of Ajrak design

**12**

*Sippi Pirah Simair[1], Nuzhat Baladi[2], Hanur Meku Yesuf[3,4] and Altaf Ahmed Simair[5]*

[1]Key Lab of Science and Technology of Eco-Textile, Ministry of Education, College of Chemistry, Chemical Engineering and Biotechnology, Donghua University, Shanghai, P.R. China, [2]Department of English, Government College University, Hyderabad, Sindh, Pakistan, [3]Ethiopian Institute of Textile and Fashion Technology, Bahir Dar University, Bahir Dar, Ethiopia, [4]Key Laboratory of Textile Science and Technology, Ministry of Education, College of Textiles, Donghua University, Shanghai, P.R. China, [5]Department of Botany, Government College University, Hyderabad, Sindh, Pakistan

## 12.1 Introduction

The textile business is diverse, and many countries allow to choose a path and direction, whether in medical fabrics or high fashion. However, several trends are emerging in the textile industry's new direction. Countries are refurbishing factories and facilities, hiring new employees, and using cutting-edge 3D-printed clothing, local clothing items, and smart and traditional fabrics to meet the local and international market's growing local demands. Countries like China and the United States focus on domestic and worldwide sales. The development and marketing of national products to locals opens up a larger market for countries traditionally focused on the global market. Because of the heavy use of water, air pollution, and waste, the environment has become an essential issue in the textile industry. Researchers have begun to pay much attention to eco-friendly auxiliaries,[1] textile recycling,[2,3] green synthesis using leaves[4] biobased materials,[5,6] etc. Clothes made from leftover coffee grounds, seaweed, or curds are among the latest developments. These new materials are unique in their class and are gaining ground in the market.

Over the last decade, the textile market has been on a roller coaster ride. Various issues have hampered the growth of the textile industry, including specific country recessions, crop damage, and a lack of products. The textile industry in different countries has grown significantly, but most countries use old methods to make traditional clothes. Although old ways are environmentally friendly, they are costly and cannot fulfill the population's needs.

The rapid development of digital printing technology is opening up new opportunities in many markets. One is the digital inkjet printed fabric market, where companies and customers benefit from new printing methods. Since its introduction a few decades ago, digital textile printing technologies have advanced. Initially motivated by the need to optimize and enhance the sampling procedure, it has resulted in new technology with several application fields in mass customization and lavish markets. Technology-wise, it is still a complicated system that necessarily involves knowledge of many disciplines that contribute to digital textile printing. Print head design, ink chemistry, pretreatment and posttreatment of fabric, design, and prefligting are all critical factors in achieving high-quality digital textile printing. Nonetheless, digital textile printing is a newer technology.

Business owners have used a portion of the vast consumer sector. According to the market, only a few Internet-based companies provide end-users with digital textile printing services. Personal computers have rapidly progressed in the last half-century. The change has gradually affected all aspects of people's lives. Digitalizing multiple tasks, especially in industry, has introduced new capabilities and opportunities. Textile printing has its roots in ancient crafts and has continued to evolve. It has, however, remained a highly technologically complex process. Only specialized industry manufacturers have been able to produce appropriately printable fabrics using techniques like rotary screen printing thus far. Digital textile printing technology has changed all of that in the last decade. It is now possible to reduce sampling time to a few hours by implementing digital printing on paper technologies. These technological developments may pave the way for new textile printing applications such as sampling, strike-offs, and bulk customization.

## 12.2 History of textile in Pakistan

Cotton's first historical traces were discovered near Quetta, Pakistan, making Pakistan among the first cotton-growing countries.[7] Cotton threads on a metal bead were discovered at a Neolithic burial site (6000 BC).[7] The mineralized lines were recognized as belonging to the cotton genus (*Gossypium*) through metallurgical examination using a reflected-light microscope and a scanning electron microscope.[7,8] Cotton farming developed during the Indus Valley Civilization, covering parts of modern-day eastern Pakistan and Northwestern India.[9] Archaeobotanical evidence of seeds dating back to 5000 BC have been discovered in Mehrgarh, although it is uncertain whether they originated in a wild or cultivated variety. Cotton fabric was originally used around 2500 BC in the Indus Valley cities of Mohenjo-Daro and Harappa. Cotton pollen is found in Balakot.[10] Cotton threads were discovered knotted to the handle of a mirror, possibly from a female gravesite, and around a copper razor at Harappa (Mature Harappan period 2500–2000 BC).[11,12] There is also scientific proof of cotton in other forms, such as Malvaceae (flowering plant) pollen type, similar to Gossypium in Balakot (Mature Harappan period, 2500–2000 BC), seeds in Banawali (Mature Harappan, 2200–1900 BC),

Sanghol (Late Harappan, 1900−1400 BC), Kanmer, Kacchh (Late Harappan, 2000−1700).[12]

Cotton is one of the country's four principal crops and is known by popular nicknames such as "King Cotton" and "White Gold." It is the primary raw material used in Pakistan's textile industry.[13] Cotton is crucial to the economic growth of Pakistan. According to a USDA Foreign Agricultural Service report from 2015, it is grown as an industrial crop on 15% of the country's agricultural acreage. It is harvested during the Kharif season, which lasts from May through August. Between February and April, it is also grown on a limited scale.[14,15]

The country's cotton is grown in the Punjab and Sindh provinces, making up 79% and 20% of the nation's cotton-growing land. It is also cultivated in the Khyber Pakhtoon Khawah and Balochistan provinces. During the 2014−15 growing season, cotton was seeded around 2,950,000 hectares. (7,300,000 acres). Small farmers with land of fewer than 5 hectares (12 acres) account for the bulk of growers,[14] and farmers with less than 2 hectares (4.9 acres) account for half of all farms. Cotton-growing landholdings of 25 hectares (62 acres) or less form less than 2% of farms.[16] According to a 2013 estimate, there were 1.6 million farmers (out of 5 million). Cotton farming covers about 3 million hectares.[14,16] Pakistan is the fourth largest in the world in cotton production, trailing China, India, and the United States in that order.[17] Pakistan's raw cotton exports rank third in the world, according to 2012−13 figures.[7] In terms of consumption, it ranks fourth (about 30% and 40% of its production). It is the world's largest cotton yarn exporter.[16] The staple length of cotton farming in the country is medium. As a result, imported long-staple cotton is used to make high-quality export fabrics.[14] Medium staple cotton, also known as normal medium-staple cotton, has a staple length of 1.3−3.3 cm (0.51−1.30 in) in the American Upland type. Long-staple cotton is more expensive, has a longer fiber, and is generally used to make fine textiles, yarns, and hosiery.[18] The cotton industry and its linked textile sector are critical to its economic development, giving cotton a prominent position in the country.[17] Cotton seeds are used to make oil and textiles such as cotton lint, yarn, thread, cloth, and so on.[11]

Although there is a large diversity in cotton genetics and phylogeny of the cotton crop,[19] cotton (*Gossypium hirsutum* L.) is the world's prominent fiber and natural crop, providing one of the world's largest textile industries and generating at least $600 billion in global economic effect each year.[20] Pakistan is the world's fourth-largest yarn producer and second-largest exporter. (USA, ICAC), and the world's seventh-largest cloth producer globally. Cotton goods account for roughly 60% of Pakistan's international earnings.[21] Fig. 12.1 depicts this; however, cotton and its derivatives account for at least 2% of Pakistan's GDP and around 10% of its agricultural value-added.[22,23]

China is the world's largest cotton-producing country, with approximately one million farmers growing the crop.[24] It has at least 7400 textile enterprises that generate cotton cloth worth $75 billion annually. Cotton farming requires a modest amount of rainfall to thrive. Many fertilizers, pesticides, and fertilizers protect the cotton crop from hazards such as chewing and various borers. See the result below in (Table 12.1).

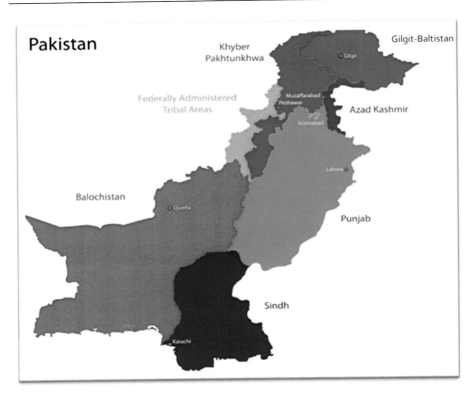

**Figure 12.1** Pakistan cotton production.[13]

In the crop year 2017/18, this statistic shows the world's leading cotton-producing countries. According to Table 12.2, the top cotton-producing countries are China, India, the United States, and Pakistan.

The cotton belt in Pakistan stretches for 1200 kilometers along the Indus River. The soil changes from clay loam to sandy, with clay dominating the south.[25] Cotton cultivation covers approximately 2.79 million/ha in total. Upland cotton is primarily grown in two Pakistani provinces: Sindh and Punjab. Punjab has the most specific cotton cultivation sector, and Sindh is the other; Sindh is also well known for cotton cultivation, as seen in (Fig. 12.1).

## 12.3 The history of traditional clothing in Sindh

Overall South Asia is very rich in culutral heritage; there are different textiles that have been transmitted from generation to generation; such as some weaves in Sri Lanka and Bhutan that have been transformed for generations.[26, 27] In Pakistan, the traditional handicraft business holds a significant place compared to other industries since

**Table 12.1** According to Fan et al.[13] the world's top cotton-producing countries.

| Million 480 lb. bales | 2013/14 | 2014/15 | 2015/16 | 2016/17 | 2017/18 | 2017/18 |
|---|---|---|---|---|---|---|
| China | 31.0 | 29.5 | 25.9 | 27.0 | 28.5 | 28.5 |
| India | 32.8 | 30.0 | 22.0 | 22.8 | 27.5 | 27.5 |
| United States | 12.9 | 16.3 | 12.9 | 17.2 | 21.3 | 21.0 |
| Pakistan | 9.5 | 10.6 | 7.0 | 7.7 | 8.2 | 8.2 |
| Brazil | 8.0 | 7.0 | 5.9 | 7.0 | 8.0 | 8.0 |
| Australia | 4.1 | 2.3 | 2.9 | 4.1 | 4.4 | 4.7 |
| Turkey | 2.3 | 3.2 | 2.7 | 3.2 | 4.0 | 4.0 |
| Uzbekistan | 4.1 | 3.9 | 3.8 | 3.7 | 3.7 | 3.6 |
| Mexico | 0.9 | 1.3 | 0.9 | 0.8 | 1.5 | 1.5 |
| Turkmenistan | 1.6 | 1.5 | 1.5 | 1.3 | 1.4 | 1.4 |
| Mali | 0.9 | 1.0 | 1.0 | 1.2 | 1.4 | 1.4 |
| Burkina | 1.3 | 1.4 | 1.1 | 1.3 | 1.3 | 1.3 |
| Greece | 1.4 | 1.3 | 1.0 | 1.0 | 1.2 | 1.2 |
| Rest of world | 9.8 | 9.8 | 7.7 | 8.5 | 9.0 | 9.6 |
| African Franc Zone | 4.1 | 4.8 | 4.0 | 4.8 | 4.8 | 4.8 |
| EU-27 | 1.6 | 1.7 | 1.3 | 1.3 | 1.5 | 1.5 |
| World | 120.4 | 119.1 | 96.2 | 106.8 | 121.4 | 121.9 |

it embodies rich culture, history, and customs, as well as traditional and contemporary designs.[28–30] Pakistani crafts are diverse and showcase the country's rich cultural heritage. This sector is critical to Pakistan's economy since it gives many employment opportunities to local people, following the agriculture sector.[30] According to recent data, the handicraft industry accounts for 14.60% of total national employment.[31] Because handicrafts involve cultural history and traditions passed down through generations, it justifies the necessity to safeguard traditional knowledge and skills to preserve the identity of local communities. Pakistan is a culturally rich country, so it must preserve its traditions and culture by establishing ambitious handicraft communities.

Industrialized items have captured markets in most developing nations, including Pakistan, making it hard for traditional handicraft communities to make a decent living.[30,32] The expanding global demand for traditional handicrafts signals growth prospects for local and foreign handicraft producers, assuming they are competent enough to seize those opportunities. Many intermediate agencies (wholesalers or importers) request that local craftspeople create alien product designs depending on purchasers' specifications.

Ajrak is an essential part of Sindhi culture. It typically measures 2.5−3 meters long. It features block-printed motifs and patterns in vibrant colors predominantly rich crimson and deep indigo, with some white and black used to achieve geometrical symmetry in the artwork. It is given as a symbol of respect to guests in Sindh; men wear it as a turban or hang it around their shoulders, while women wear it as a dupatta or drape it as a shawl in winter. It is also placed on a coffin as a symbol of

Table 12.2 World cotton exports according to Fan et al.[13]

| Million 480 lb. bales | 2013/14 | 2014/15 | 2015/16 | 2016/17 | 2017/18 | 2017/18 |
|---|---|---|---|---|---|---|
| United States | 10.5 | 11.2 | 9.2 | 14.9 | 14.5 | 14.8 |
| Australia | 4.9 | 2.4 | 2.8 | 3.7 | 4.4 | 4.4 |
| Brazil | 2.2 | 3.9 | 4.3 | 2.8 | 4.2 | 4.2 |
| India | 9.3 | 4.2 | 5.8 | 4.6 | 4.2 | 4.2 |
| Burkina | 1.3 | 1.1 | 1.3 | 1.1 | 1.1 | 1.1 |
| Greece | 1.3 | 1.2 | 1.0 | 1.0 | 1.1 | 1.1 |
| Mali | 0.9 | 0.9 | 1.0 | 1.1 | 1.1 | 1.1 |
| Uzbekistan | 2.6 | 2.3 | 2.3 | 1.3 | 1.2 | 1.1 |
| Turkmenistan | 1.6 | 1.5 | 1.3 | 0.9 | 0.7 | 0.7 |
| Benin | 0.5 | 0.5 | 0.7 | 0.8 | 0.7 | 0.7 |
| Cote d'Ivoire | 0.8 | 0.9 | 0.8 | 0.6 | 0.6 | 0.6 |
| Tajikistan | 0.4 | 0.5 | 0.5 | 0.3 | 0.5 | 0.5 |
| Sudan | 0.2 | 0.1 | 0.1 | 0.3 | 0.1 | 0.5 |
| Rest of world | 4.7 | 4.6 | 4.1 | 4.0 | 3.9 | 3.9 |
| African Franc zone | 4.1 | 4.0 | 4.6 | 4.3 | 3.9 | 3.9 |
| EU-27 | 1.6 | 1.6 | 1.3 | 1.3 | 1.4 | 1.4 |
| World | 41.1 | 35.1 | 35.1 | 37.3 | 38.2 | 38.8 |

Figure 12.2 Varieties of Ajrak shape dupattas.

respect. The Ajrak crafters believe their profession dates back to the early Medieval Era. The debris of printed pieces was meant to originate from Western India. The Khatri community practiced the Ajrak craft, living on the river Sindh, Pakistan (Figs. 12.2 and 12.3).

**Figure 12.3** A variety of Ajrak design scarfs and stallers.

The crafts produced by specific Khatri families and communities were mainly decided by their specific clientele and technical skills developed, with some Khatris crafting block-prints and several other tie-dyed fabrics. Sixteen Each family prefers to cooperate with a specific caste or caste.

Ajrak is believed to represent the entire universe because of the color scheme used. The earth is symbolized by a red, darkness by black, clouds by white, and the universe by blue. Think moonless, midnight, and darkness, a stark blue-black background with a star-strewn sky. Nature plays a vital role in the crafting of Ajrak. The art pieces perfectly harmonize with their surroundings, including the sun, streams, wildlife, forests, and clay in their creation (Fig. 12.4).

Ajrak is a fabric from the Indian Subcontinent's desert regions. This textile produced explicitly in Pakistan's Sindh province and Kutch in India, has been used for thousands of years to create various textile goods. The first instance of Ajrak cloth in the area was discovered at Mohenjo-Daro as part of a Harappan sculpture of a Priest-King wearing a shawl with trefoil motifs similar to those found in contemporary Ajrak prints. The cloth is thought to have been used by the Indus River Valley Civilizations that inhabited the area.[33]

The blend of handloom textiles and vegetable dyes creates beauty. Natural colors declined after the introduction of chemical dyes around the 19th century. Ajrakh printing, which uses natural dyes, is one of India's oldest resist printing processes and one of the most intricate and sophisticated printing techniques, it is explicitly explained by Dutta and Bansal in their work[34] (Fig. 12.5).

**Figure 12.4** Sindhi Ajrak.

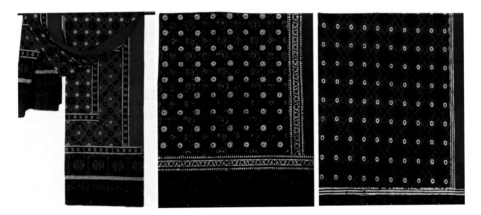

**Figure 12.5** Variety of Ajrak designs: rotary prints, inspiration work, Block printing.

Sindhi and Rajasthani women traditionally wear Ajrak patterned cotton. Aside from lungis and pagris, ladies wear printed dresses and use Ajrak fabric to wrap cradles for babies. Every color has a tale to tell, and the design reflects the status. The Khatris have learned the value of the market, and Ajrak yardage, kurta sets, decor, and scarves can now be purchased.

Artisans have adapted it in various ways, and the traditional Ajrak pattern on Kurtis and bed linen is now being sold worldwide. Its use can be traced back to the Indus Valley culture (3500−1500 BC). The king-priest figurine discovered at Mohenjo-Daro depicts him covered in an Ajrak shawl.[35]

The making of an Ajrak is a laborious 21-stage procedure. Ajrak was printed by traditional artists using indigenous, locally manufactured materials. Although the technique is the same, natural dyes have been replaced to some extent.[36]

Ajrak printing is notable for combining resist printing with other printing and dyeing techniques on a single fabric while keeping the same design. The procedure, which calls for unparalleled expertise, is carried out on both sides of the material with flawless cohesion. Ajrak uses mud-resist in the different stages, but another unique aspect is that the dyeing and printing are repeated twice on the fabric to ensure the vibrancy of coloring. The repetitions are superimposed so precisely that the clarity is enhanced. Block printing is a time-consuming and precise process. The fabric is spread and pinned onto a table first. The printers cover wooden pieces with resistance and hover them above the fabric to achieve a balanced application. Once positioned, the block is hammered on the fabric with a mighty whack. The block outline in each of the three resist bases is repeatedly applied to the fabric until it is completely encased. The fabric is dyed in a base color, dried in the sun, rinsed, and then dyed again until it is converted into a dense mosaic of colors and designs.

The printing blocks need to be finely carved by experts in the industry. A group of three blocks creates a dovetailing effect that results in the artwork. They are made from Acacia Arabica trees native to Sindh. A grid system determines the repeating patterns, which gives the design its identity. The pattern is first imprinted to the block, and then the block-maker, who employs simple equipment, carves it with exquisite precision. The blocks are divided into pairs so the opposite can register an exact mirrored image. There is currently only one surviving member of a family of block-makers whose forbears were experts in this technique.[37]

Old procedures, structural designs, and patterns are increasingly fading or losing integrity when they are transferred into new regional or cultural contexts. Ajrak, with its rich visual expression, is changing as traditional processes, raw materials, patterns, products, and markets evolve due to globalization's global phenomena. Face masks made in China are very common during the Covid-19 pandemic situation (Figs. 12.6 and 12.7).

## 12.4 Digital printing: a fast-growing vibrant sector in Pakistan

Digital printing is fast making its mark in Pakistan's textile industry. In just a few years, the market for digital printing has shown remarkable growth. At best, the global share of digital printing is only 5%; however, this share was only 2% just a few years ago. The growth in this sector seems to be exponential due to several reasons. Digital printing is a process in which prints are directly applied to fabrics with the printer. Similar to the inkjet printers used widely in offices and homes, the technology prints on different fabrics according to demand. The outstanding features of digital printing can be seen in terms of operational and environmental

**Figure 12.6** Variety of Ajrak style female appraisers.

**Figure 12.7** Ajrak designed face masks.

advantages. Digital textile printing processing uses a remarkably reduced amount of water, and the energy cost is a fraction of conventional rotary printing. The most significant advantage from the perspective of a final consumer of products, however, is the ability for customization at a nominal cost.

Digital printers can deliver smaller lots quickly to maintain exclusivity for highly valued garments according to fast-changing designs and sustainable fashion.[38,39] Due to Pakistan's uncompetitive cost of production, which is caused by high energy prices, low liquidity as a result of government export refunds that have been stuck in the pipeline for years totaling about 200 billion Rupees, and an overvalued Pakistani Rupee, the country's textile industry has been experiencing hardship over the past 10 years. The exports have been stagnant and significant and declined in yarn and fabrics, commodities with meagre margins. The progressive and forward-looking textile manufacturers considered this dire export scenario. They started investing in the retail market of Pakistan, with over 200 plus million population and a sizeable middle class of about 80 million people.

With the fast opening of retail outlets and shopping malls, women's apparel demand has grown tremendously over the last 20 years. Traditional apparel printing has been done on flatbed printers and rotary printers. The investment in rotary printing is very high and requires not only space but plentiful water and large print orders to be feasible. Digital printing meets the demands of today's market perfectly. The cost of digital printing production is decreasing daily with the declining cost and availability of digital ink. It is important to note that the high cost of inks has been the main deterrent to the sector's development. Millions of fabrics are being printed daily to meet the seemingly insatiable demand of discerning consumers belonging to all segments of our society. A perfect match for digital printing is the growth of online shopping made possible with high-speed internet and 4 G technologies. The retailers sometimes sell their first lot on the internet before the season starts.

Digital textile printing has been expanding steadily for a variety of reasons. In addition to being a more creative textile printing method than screen printing, digital printing offers more design flexibility. For small prints, designers choose digital printing since it is more economical. One of the significant benefits of digital printing is that it is environmentally beneficial because it uses less physical inventory, which leaves fewer carbon footprints.

These variables contribute to the worldwide textile industry's increasing use of digital printing. Even yet, the deeper integration of digital printing into the creation of commercial prints is crucial for the future expansion of digital textile printing. The machine makers must also create technologically sophisticated production printers to compete with more traditional textile printing machinery.

A technological issue that was resolved was the print speed of digital printing, which conflicted with more conventional printing techniques. Modern digital textile printing machines are sophisticated and offer excellent printing speed, whether by adding more heads in a row to improve the firing frequency or increasing the number of nozzles.

### 12.4.1 Technology takes digital printing to a new level

The price of digital textile inks has decreased along with updated technology and technological advancements in digital printing. Effective sourcing and improved production techniques are also contributing to the overall cost-effectiveness of digital printing. Additionally, acquisitions and investments in the digital textile industry can help the sector as a whole. It will result in the introduction of novel goods and services.

Innovations in digital printing include thermal inkjet, continuous inkjet, thermal transfer, piezoelectric inkjet, electrostatic printing, and electrophotography. Each technique has unique qualities and benefits of its own. Additionally, practically any type of cloth can now be digitally printed using inkjet technology because of advancements in ink chemistry.

### 12.4.2 Starting to make wise decisions

The market for premium garments has also seen success with digital printing. Luxury clothing retailers may affordably make short runs of high-end, limited-edition pieces of digitally printed clothing. Manufacturers also use digital textile printing to offer clients customized apparel items, allowing them to select a design and digitally print it on various textile products.

The use of digital textile printing in home textiles will increase as more people replace their bedding and furniture regularly. Currently, the proportion of digitally printed home furnishings in all global production is 7%–10%. The absence of bigger-width digital printers is the cause of this market's sluggish growth.

Customization can help digital printing flourish. Businesses can let customers pick their designs from a list of products and have them digitally printed on fabric or textile products like t-shirts, stoles, etc., at a retail location. Digital textile printing has many prospects in nonconsumer applications, such as soft displays. Advertising and marketing signs printed on fabric rather than paper or vinyl are known as soft signage. Customers typically purchase soft signage in small quantities, and short print runs can be quickly produced with digital textile printing at a reasonable cost. The development of digital textile printing in developing nations like Pakistan and India has also made a substantial contribution to world growth. The quick fashion trend of today has reduced production and shipping times. In the past, new collections were released twice or three times a year, but nowadays, fashion trends change often. Just-in-time manufacturing and quick changes in fashion trends are essential to this printing. As a result, nations like Brazil, Turkey, and China are focusing on implementing digital printing. The quick fashion notion has become the norm in most European nations; thus, Europe embraces the trend. The adoption of digital technology by the primary supply chain in the textile industry will determine how quickly digital printing develops. Digital printing is being used to its full potential in nations like China and India, which bodes well for the future. Many more countries are anticipated to be intrigued by this trend.

## 12.5 Culture is not a trend, but it has become a significant source of profit worldwide

In Sindhi culture, the Ajrak is revered and customarily given to visitors as a sign of gratitude. It is intolerable to see a brand with no connection to Sindhi culture portray it as standard cloth without regard for its intended role. We are sick of seeing traditional garments from a person's culture and traditions appropriated and repurposed as "festive" attire or rebranded for consumerism. Because it is only a fun fashion trend, cultural appropriation erases the group from which it has taken and is unaware of the background behind the thing they have taken. These trends hurt the feelings of people around the world.

### 12.5.1 Ezri collection, 2016

The "Ezri" collection from the New York-based Sea brand includes a variety of outfits with Ajrak patterns. The clothing, made in China and costing about US$ 200, was sold there. There were no Sindhi artisans employed in the creation of these imitations. Furthermore, rather than the genuine Ajrak origins of Sindh, it was stated that the French directly influenced these costumes.

### 12.5.2 Urban outfitters, 2016

A series of swimsuits with Ajrak patterns were made available by Urban Outfitters and marketed as "Moroccan swimsuits." Because Ajrak is so important to the Sindhi people, this replication was regarded mainly as insulting in addition to giving them bad credit.

### 12.5.3 Forever 21, 2016

Ajrak-printed outfits from Forever 21 have been labeled "Baroque Print." Social media users were outraged by these blunders. Similar to the Urban Outfitters bikinis, it was believed that such rhetoric was used to increase sales for the company.

## 12.6 Conclusion

There are two objectives to writing this review; one is digital trends, and the second is the conservation of cultural legacy. New opportunities are emerging in numerous markets due to the rapid development of digital printing technology. The printed cloth market is one example where innovative printing techniques benefit businesses and consumers. This evaluation used digital textile printing technology for on-demand fabric printing services. The developments in digital textile manufacturing technologies, their uses, requirements for digital textile printing, and potential economic effects. Digital textile printing technologies have advanced since they were first developed a few decades ago. Personal protective equipment produced via 3D printing has assisted in resolving supply chain bottlenecks. Products made for particular consumer castes, most notably farmers and herders, are the focus of the traditional textile market. It looks into postcolonial trends like industrializing wholesale manufacturing and changing factory farming practices that have changed this traditional market. It reinforces the emergence of new, internationally interconnected markets and how regional manufacturers adapt by creating new commodities. Any region has a cultural legacy passed down from generation to generation. It enables the communities to situate themselves inside a timeless quality that bestows specific qualities on a given location and serves as the cornerstone of creating a shared cultural identity. It is advisable to include the essential cultural components, both tangible and intangible; those are associated with a place and are thought to have special spiritual and symbolic significance by the locals.

# References

1. Memon, H.; Khatri, A.; Ali, N.; Memon, S. Dyeing Recipe Optimization for Eco-Friendly Dyeing and Mechanical Property Analysis of Eco-Friendly Dyed Cotton Fabric: Better Fixation, Strength, and Color Yield by Biodegradable Salts. *J. Nat. Fibers* **2016**, *13*, 749−758. Available from: https://doi.org/10.1080/15440478.2015.1137527.
2. Memon, H.; Ayele, H. S.; Yesuf, H. M.; Sun, L. Investigation of the Physical Properties of Yarn Produced from Textile Waste by Optimizing Their Proportions. *Sustainability* **2022**, *14*, 9453.
3. Wang, H.; Memon, H.; Abro, R.; Shah, A. Sustainable Approach for Mélange Yarn Manufacturers by Recycling Dyed Fibre Waste. *Fibres Text. East. Eur.* **2020**, *28*, 18−22.
4. Yu, L.; Memon, Hu. R.; Bhavsar, P. S.; Yasin, S. Fabrication of Alginate Fibers Loaded with Silver Nanoparticles Biosynthesized via Dolcetto Grape Leaves (Vitis vinifera cv.): Morphological, Antimicrobial Characterization and In Vitro Release Studies. *Mater. Focus.* **2016**, *5*, 216−221.
5. Memon, H.; Wei, Y.; Zhu, C. Correlating the Thermomechanical Properties of a Novel Bio-Based Epoxy Vitrimer with its Crosslink Density. *Mater. Today Commun.* **2021**, *29*, 102814. Available from: https://doi.org/10.1016/j.mtcomm.2021.102814.
6. Memon, H.; Liu, H.; Rashid, M. A.; Chen, L.; Jiang, Q.; Zhang, L.; Wei, Y.; Liu, W.; Qiu, Y. Vanillin-Based Epoxy Vitrimer with High Performance and Closed-Loop Recyclability. *Macromolecules* **2020**, *53*, 621−630. Available from: https://doi.org/10.1021/acs.macromol.9b02006.
7. Mensah, R. Travel: International Cotton Advisory Committee (ICAC) 77th Plenary Meeting, Present; New South Wales Department of Primary Industries, **2019**.
8. Moulherat, C.; Tengberg, M.; Haquet, J.-F.; Mille, B. *First Evidence of Cotton at Neolithic Mehrgarh, Pakistan: Analysis of Mineralized Fibres from a Copper Bead. J. Archaeol. Sci.* **2002**, *29*, 1393−1401.
9. Stein, B. *A History of India*: Blackwell Publishing, **1998**; p. 47. ISBN 0-631-20546-2.
10. Ahmed, M. *Ancient Pakistan-an Archaeological History*: Amazon, **2014**.
11. Singh, U. *A History of Ancient and Early Medieval India: From the Stone Age to the 12th Century (PB);* Pearson Education: India, **2009**.
12. Parpola, A. *The Harappan Unicorn in Eurasian and South Asian Perspectives. Linguistics, Archaeology Hum. Past-Occasional Paper. 12;* Indus Project, Research Institute for Humanity and Nature, **2011**, 125−188.
13. Shuli, F.; Jarwar, A. H.; Wang, X.; Wang, L.; Ma, Q. Overview of the Cotton in Pakistan and its Future Prospects. *Pak. J. Agric. Res.* **2018**, *31*, 396.

14. Zafar, A.; Zaidi, S. Z. H.; Hayat, K.; Shahzad, M. A.; Kamran, R. Economic Analysis of Factors Affecting Cotton Production in Pakistan. *Res. J. Inno. Ideas Thoughts* **2016**, *4*, 12−18.
15. Dadgar, M. The Harvesting and Ginning of Cotton. In *Cotton Science and Processing Technology: Gene, Ginning, Garment and Green Recycling;* Wang, H., Memon, H., Eds.; Springer Singapore: Singapore, **2020**; pp 61−78.
16. Banuri, T. *Pakistan: Environmental Impact of Cotton Production and Trade;* International Institute of Sustainable Development, **1998**, 161.
17. Osakwe, E. *Cotton Fact Sheet: Pakistan* **2009**.
18. Rice, R. Nomadic Pastoralism and Agricultural Modernization. *Mid Am. Rev. Sociol.* **1981**, *6*, 71−92.
19. Simair, A. A.; Simair, S. P. Status and Recent Progress in Determining the Genetic Diversity and Phylogeny of Cotton Crops. In *Cotton Science and Processing Technology: Gene, Ginning, Garment and Green Recycling;* Wang, H., Memon, H., Eds.; Springer Singapore: Singapore, **2020**; pp 15−37.
20. Ashraf, J.; Zuo, D.; Wang, Q.; Malik, W.; Zhang, Y.; Abid, M. A.; Cheng, H.; Yang, Q.; Song, G. Recent Insights into Cotton Functional Genomics: Progress and Future Perspectives. *Plant. Biotechnol. J.* **2018**, *16*, 699−713.
21. Ali Hayat, G.; Hussain, M.; Qamar Khan, M.; Javed, Z. Textile Education in Pakistan. In *Textile and Fashion Education Internationalization: A Promising Discipline from South Asia;* Yan, X., Chen, L., Memon, H., Eds.; Springer Nature Singapore: Singapore, **2022**; pp 59−82.
22. Bakhsh, A.; Rao, A. Q.; Shahid, A. A.; Husnain, T.; Riazuddin, S. Insect Resistance and Risk Assessment Studies in Advance Lines of Bt Cotton Harboring Cry1Ac and Cry2A Genes. *Am.-Eurasian J. Agric. Environ. Sci.* **2009**, *6*, 1−11.
23. Sial, K. B.; Kalhoro, A. D.; Ahsan, M. Z.; Mojidano, M. S.; Soomro, A. W.; Hashmi, R. Q.; Keerio, A. Performance of Different Upland Cotton Varieties Under the Climatic Condition of Central Zone of Sindh. *Am.-Eurasian J. Agric. Environ. Sci.* **2014**, *14*, 1447−1449.
24. Wang, H.; Memon, H. Introduction. In *Cotton Science and Processing Technology: Gene, Ginning, Garment and Green Recycling;* Wang, H., Memon, H., Eds.; Springer Singapore: Singapore, **2020**; pp 1−13.
25. Bank, A. D. *Building Climate Resilience in the Agriculture Sector of Asia and the Pacific;* Asian Development Bank, **2009**.
26. Memon, H.; Ranathunga, G. M.; Karunaratne, V. M.; Wijayapala, S.; Niles, N. Sustainable Textiles in the Past "Wisdom of the Past: Inherited Weaving Techniques Are the Pillars of Sustainability in the Handloom Textile Sector of Sri Lanka." *Sustainability* **2022**, *14*, 9439.
27. Lo, J.; Wangchuk, P. C. Weaving Through Generations: A Study on the Transmission of Bhutanese Weaving Knowledge and Skills Over Three Generation. In *Textile and Fashion Education Internationalization: A Promising Discipline from South Asia;* Yan, X., Chen, L., Memon, H., Eds.; Springer Nature Singapore: Singapore, **2022**; pp 175−198.
28. Shafi, M. Sustainable Development of Micro Firms: Examining the Effects of Cooperation on Handicraft Firm's Performance through Innovation Capability. *Int. J. Emerg. Mark.* **2020**.
29. Shafi, M.; Sarker, M. N. I.; Junrong, L. Social Network of Small Creative Firms and its Effects on Innovation in Developing Countries. *SAGE Open.* **2019**, *9* 2158244019898248.

30. Yang, Y.; Shafi, M.; Song, X.; Yang, R. Preservation of Cultural Heritage Embodied in Traditional Crafts in the Developing Countries. A Case Study of Pakistani Handicraft Industry. *Sustainability* **2018**, *10*, 1336.
31. Sohail, S. *Labor force Survey 2017–18;* Gov. Pak., Ministry Statistics, Pakistan Bureau of Statistics: Islamabad, **2018**.
32. Forero-Montaña, J.; Zimmerman, J. K.; Santiago, L. E. Analysis of the Potential of Small-Scale Enterprises of Artisans and Sawyers as Instruments for Sustainable Forest Management in Puerto Rico. *J. Sustain. For.* **2018**, *37*, 257–269.
33. Rehmani, N. A.; Phulpoto, N. Ajrak as Symbol: The Fabric of Life and Cultural Affinity. *N. Horiz.* **2012**, *6*, 17.
34. Dutta, S.; Bansal, P. Textile Academics in India—An Overview. In *Textile and Fashion Education Internationalization: A Promising Discipline from South Asia;* Yan, X., Chen, L., Memon, H., Eds.; Springer Nature Singapore: Singapore, **2022**; pp 13–34.
35. Dhamija, J. *Encyclopedia of World Dress and Fashion: South Asia and Southeast Asia;* Oxford University Press, **2010**.
36. Dhamija, J. Berg. Encyclopediaof World Dress and Fashion:[English Edition]. Vol. 4. *South Asia and Southeast Asia;* Berg, **2010**.
37. Edwards, E. M. Ajrakh: From Caste Dress to Catwalk. *Text. Hist.* **2016,** *47*, 146–170.
38. Chen, L.; Qie, K.; Memon, H.; Yesuf, H. M. The Empirical Analysis of Green Innovation for Fashion Brands, Perceived Value and Green Purchase Intention—Mediating and Moderating Effects. *Sustainability* **2021,** *13*, 4238.
39. Memon, H.; Jin, X.; Tian, W.; Zhu, C. Sustainable Textile Marketing—Editorial. *Sustainability* **2022,** *14*, 11860.

# Digital textile printing innovations and the future

*Degu Melaku Kumelachew[1,2], Bewuket Teshome Wagaye[1,2] and Biruk Fentahun Adamu[1,2]*
[1]Textile Engineering Department, Ethiopian Institute of Textile and Fashion Technology, Bahir Dar University, Bahir Dar, Ethiopia, [2]Key Laboratory of Textile Science and Technology, Ministry of Education, College of Textiles, Donghua University, Shanghai, P.R. China

## 13.1 Digital textile printing advancement history

Printing of textiles has been a common trend with well-known analog printing techniques. Although conventional printing is a productive and well-known printing process, it has the following limitations that initiated digital printing innovation, that is, color smear, dye migration, burning, stains, pinholes, color accuracy, color consistency, inappropriate curing, and poor design registration. From the technology of roller printing to today's advanced printing system using a CAD system, the industry has progressed in stages to printer heads with deposits of fine nozzles ejecting fine droplets of individual-colored inks onto pretreated fabrics in digital printing.

While the Milliken Millitron digital carpet printer was one of the first transitions to digital printing form in the 1970s, the first inkjet textile printers on the market were released in the 1990s, with previous examples, including the Stork TrueColor Jet Printer (1991) and the Ichinose Image Proofer (1999). As the name implies, these early Ichinose machines were designed to produce samples and proofs of concepts for ultimate manufacturing on conventional rotary screen printers. The Milliken Millerton digital carpet printer was one of the first attempts at digital printing in the 1970s. The first commercial inkjet textile printers, however, appeared in the 1990s, with early examples, including the Stork TruColor Jet Printer (1991) and the Ichinose Image Proofer (1999).

In 1968, Lu University's CH Hertz filed the first continuous inkjet patent. Patents for thermal inkjet technology were granted to Canon and Hewlett-Packard in 1977. So far, most digital printing research and development has been focused on the official document and home printer industries. The print heads used in these markets have a short life duration, ranging from 500 hours to 1 year. As a result, most inkjet print heads on the market are insufficient for industrial or production printing. Fortunately, much progress has been made in digital textile machines in the previous 20 years. Several businesses are currently developing commercial printers that use inkjet technologies.

The early start of digital printing manufacturing started in Japan around the 1990s. Canon printers were the first to manufacture digital textile printers with the thermal printhead. Next, Seiren, a comprehensive textile manufacturing company, manufactured Viscotecs with piezo printhead technology. The first production digital printer, DReAM, was exhibited by Reggiani Machine of Italy at ITMA Birmingham in 2003. This printer became the benchmark for the digital textile printing industry printhead technology. The speed was so slow that it produced about 2.5 m/min.

One of the first breakthroughs was single-pass digital printing technology for textile printing, which uses a different mechanism than the traditional scanning type printing system, in which a carriage with multiple printheads moves from selvage to selvage of the textile substrate and prints a set of information (image data) in multiple passes. A single droplet of ink prints each piece of information (image data) from the printheads on the continuously moving textile substrates in the single-pass system. Multiple printheads are lined upwithout gaps on a long beam that spans the print bed width and are secured structurally. One set of colorants is assigned to each beam. Seven bays on the printing system are necessary to use seven colorants (e.g., C, M, Y, K, and three extra inks). MS Industry (formerly Dover/MS) of Italy debuted the La Rio printer, the first single-pass printing machine, at the ITMA in Barcelona, Spain, in 2011. The La Rio could produce up to 70 linear meters per minute in tests. This printing rate is acceptable.

The world's leading digital printing solution supplier with over 2000 customers, Atexco is committed to the flexible production of green printing and dyeing in the textile field. In 2009, Atexco was approved by the Ministry of Science and Technology of China to establish the National Digital Printing Engineering Technology Research Center, got the National Technology Invention Award in the years 2007 and 2017, respectively, and won the championships of textile digital printing machine evaluated by the Ministry of Industry and Information Technology in 2020. Atexco was successfully listed on the Shanghai Stock Exchange Star Market in 2021, becoming the first listed enterprise in the digital printing industry.

The invention of a hybrid printing technology is another advancement in digital textile printing. Some digital textile printing machines now include a digital printing system for primary colorations and an analog screen-printing mechanism for pretreatment and supplemental printing. Because digital printing alone does not allow for the printing of unique effects like metallic, pearl, flocking functional agents, Devoré, and Plissé, a hybrid method is advantageous. Synchronizing an analog screen-printing machine with a primary digital printing unit, however, allows for creating special effects in addition to digital printing. Additionally, print design features with discrete spot colors (e.g., fluorescent and out-of-gamut colors), splotch, and huge solid-color regions can be synchronized to analog screens.

Inkjet technology is the outcome of a collaboration between three major engineering fields. Chemical engineering is responsible for formulating new inks to achieve excellent printing on the substrate and the necessary surface functionality. Mechanical Engineering ensures that the substrate's printing process is appropriately coordinated to achieve the maximum possible printing quality. Electronic engineering is in charge of carrying out the digital development of the process.

### 13.1.1 Machine innovative advancements

As the above innovation history of digital textile printing indicated, the machine features have progressed through many advancements based on the competitive requirement. Automation, accuracy, product volume and flexibility, and production speed have changed through all those times. The main limitation of digital textile printing was the speed and initial establishment cost in every textile printing factory. At the same time, the application area is the other advancement in machine technology. Sublimation printing for synthetic printing with high speed is an excellent solution for synthetic printing industries. Its speed (180 $m^2$/h) and high-quality output. Mimaki's latest grand-format transfer printer gives high maximum productivity for the exhibition, retail, and home furnishing markets. One of the recent printers, Model X plus by Atexco, is shown in Fig. 13.1, can also print as fast as 250—465 m/h under different modes.

### 13.1.2 Ink formulation technology

To be jetted directly onto a range of media substrates while offering print quality that rivals offset printing, digital printing inks must have specific qualities. The necessary components must be synthesized to reach this level of quality and the benefits of digital printing, such as personalized, changeable data printing, print-on-demand, cost-effective small runs, and quick turnarounds. Inkjet inks are the most critical component in inkjet printing. The composition and chemistry of inks determine printing quality and jetting characteristics.[1] Digital printing has successfully altered textile printing in terms of print speed, print head technology, and color spectrum. Digital textile printing inks come in various forms, including dispersed,

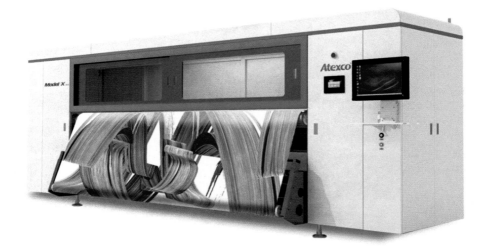

**Figure 13.1** Model X Plus by Atexco as a high-speed digital textile printing machine.

reactive, acid, and pigment. Textile reactive digital inks are a type of digital ink used on textiles. The most common type of ink is scattered ink.[2]

There has been a great application of artificial intelligence in all textile applications.[3] The increased interest in inkjet technology is primarily due to advancements in control algorithm research. Furthermore, the current development of novel materials with excellent adjustable surface properties and the depth of understanding of fluid dynamics that governs the production and ejection of drops makes this technology suitable for any industrial process. One of the disciplines of inkjet technology where increasing attention is being placed on current research is the formulation of new inks. On the substrates it is applied, inkjet technology creates new functional surfaces. This technology may produce prints with components that make the material conductive, magnetic, hydrophilic, superhydrophobic, and antibacterial, among other things. The main focus in ink formulation is to modify the current pad printing ink with good adhesion in the inkjet. Obtain injectable ink in Inkjet equipment with high adhesion, modifying pad printing ink to adjust to the density, viscosity, and surface tension necessary to be injectable. The ink formulation, the mechanical containment of inks on the machine, and the electrical accurate printing element are accommodated in a single system. Different machines would have different designs and arrangements. The schematic diagram of the secondary ink cartridge of a decorative material printer, that is, Deco Y, is shown in Fig. 13.2A, while Fig. 13.2B shows an actual image secondary ink cartridge on Deco Y.

The most common piezoelectric materials have been in piezoelectric inkjet printers, in which electric charges are supplied to the head, causing it to bend and resulting in ink drop production; however, current research has successfully produced piezoelectric inks with tiny piezoelectric crystals. This, combined with 3D printing technology, enables any coating or creation of materials with piezoelectric capabilities.[4]

The present invention is a novel high fixation ink composition for digital textile printing that includes at least one reactive dye compound with two reactive groups in an amount of 1%−50% by weight; an organic buffer in an amount of 0.05%−10% by weight; a humectant in an amount of 10%−50% by weight; and a solvent in the remaining amount. When the ink mentioned above is used in digital textile printing, the dye-fixing rate on materials is high.

With various dispersions, dispersants, resins, and surface modifiers, Lubrizol continues to enhance digital printing technologies for textile, film, paper, packaging, and ceramic substrates. Diamond Dispersions from Lubrizol is a water-based disperse dye and colored dispersions for inkjet ink milled to perfection using cutting-edge dispersing technology. Hyper dispersants are frequently used in UV-cured, eco-solvent, solvent-based, water-based inkjet ink and other digital print processes. Surface modifiers can be used successfully in inkjet inks to improve ink rub resistance and slip/blocking concerns. Lubrizol offers water-based inkjet ink products and formulation and application advice.

Nanoprinting inks for textile materials are developing and expected to boom in digital textile printing. Nanopigment printing is a type of inkjet printing that uses pigments tinted with pigments (powders) that have a diameter of fewer than

# Digital textile printing innovations and the future

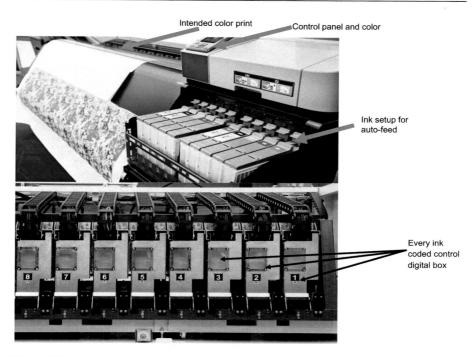

**Figure 13.2** Ink set upon advanced digital inkjet printers (A) schematic diagram of secondary ink cartridge on Deco Y (B) actual image secondary ink cartridge on Deco Y.

100 nm (typical solvent and UV (ultraviolet); curable inks are colored with powders with diameters of approx. 1000 nm or larger). The usage of nanopigments gives the ink several benefits. Nano inks consume heads and other printer ink system components more slowly than conventional inks. Even if the material was not meant for digital printing, such minuscule dye particles attach to the substrate naturally, employing nanoparticle attraction and other techniques. The advantages of the nano ink innovations are Printing on nearly any material; very high print quality, see Fig. 13.3; exceptionally high performance; flawless surface structure and high print flexibility; vibrant and high light reflectance, see Fig. 13.3; clear, saturated colors—excellent print lightfastness and ecology; operator comfort, environmental protection.

Waterborne dispersions of electrically conductive grades of carbon black pigment were turned into completed inks using a variety of polymeric binders. The electrical characteristics of the inks revealed that none of the dispersions showed considerable instability after letting downwith various binders. Inks created from dispersions of high surface pigments had lower endurance in washing and creasing, which could be due to the much lower binder content in these inks. The nonprinting ink grades of carbon black pigment can generate highly conductive inks using the conductive ink formulation process proposed in this study. Here are a few examples of innovative inks.

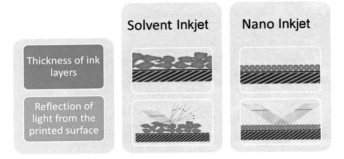

**Figure 13.3** Comparison of nano to conventional ink.

The NeoPigment Robusto Softener solution for Presto, Korn Digital's system for digital, pigment-based direct-to-fabric decoration, has been introduced. Kornit Digital is a global leader in digital textile printing technology Kornit's dedication to providing sustainable, on-demand (i.e., waste-free) decoration capabilities to all corners of the textile industry, is reflected in this technique, which removes a significant barrier between digital, pigment-based impressions and the fashion industry, a softer hand feel.

For interior decoration, Agfa introduces the Interiojet water-based inkjet printing system. Epson's Ultrachrome Dg Inks have received approval. Textile makers must demonstrate their eco-credentials as customers and brands grow more aware of the sources of the garments and fabrics they buy. Epson is thrilled to announce that the Global Organic Textile Standard has authorized their Ultrachrome DG inks and pretreatment solutions. Both are standard Epson's newly released SC-F3000 direct-to-garment printer, and the smaller SC-F2100 printer in the same linRicoh's new environmentally friendly collaboration with Farbenpunkt Inks for digital textile printing will save 90% of fresh water and 50% of energy.

### 13.1.3 Inkjet heads/print heads

The printing head and nozzles are the most critical features of an inkjet printer. For both the technologies, 10–100 p nozzle diameter with a resolution upto 720 dpi is generally used. In a drop-on-demand (DOD) system, drops are generated either by a thermal jet or a piezo crystal (Fig. 13.4). These printing heads are relatively simple and cheap; however, it has a short life and limited applications. Thermal jets are inexpensive and water-based ink is used, whereas Cotton-linen Silk Wool piezo crystals can use any dye. Disperse dye or pigment is more suitable for high-volume printing. On the other hand, continuous inkjet generates droplets continuously, but only the required amount.[5,6]

There are several issues that inkjet requires; Ink clogging, ink drying, and printer maintenance are all issues that need to be addressed. Printer heads are divided into two categories: fixed and disposable. A fixed head has the advantage of lowering printer maintenance costs. In this situation, the ink needs to be replaced, but the

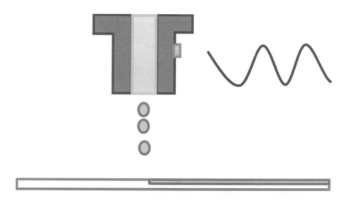

**Figure 13.4** Drop-on-demand print head.

printing head remains the same throughout the printer's lifetime. This printing head will, however, require well-defined cleaning and maintenance methods for clogging piezoelectric inks. Furthermore, pigmented inks are significantly more UV resistant than dye-based inks, taking far longer for noticeable fading and drying to occur, whereas, in the case of droplet transducer disposable heads, not only inks but also the printing head must be replaced.[5]

Textiprinters' most common frustration source was print head damage caused by their proximity to the substrate. The print heads used in textile printing are similar to those used in graphic design. Flat materials must be printed in this market, and the print heads must be placed close to the material. With thicker materials, print head damage occurs due to the proximity to the substrate, resulting in significant expenses for textile printers. SPGPrints has a solution in the form of Archer +, a unique printing system that helps printers improve quality by reducing drop-size range and increasing resolution (true 1200 dpi). The Archer + technology was developed in response to the rising demand for printers, particularly print heads ideal for digital printing.[6–8]

## 13.1.4 Limitless design (media and color management software)

Using software such as Inedit, Optitex, Gerber, Lectra, or Bronzwear, a designer can create virtual Toiles (sample garments) that only exist virtually and digitally on screen, removing the need for an actual sewed sample before approval. The Designer can adjust and fine-tune the clothes for fit and style, apply the printed artwork pattern using 3D simulations, and change the fabric or the scale of the printed design in real-time to see a virtual twin of the finished outfit on screen. The garment can be authorized remotely as a digital twin and then pushed into the manufacturing supply chain, saving time and resources.

Digital artwork has been a cornerstone in the design business for almost 30 years, but the industry increasingly wants seamless integration. The addition of embedded production data and the requirement for a new layer of image handling will ensure that the design integrity required f "print-on-dem" is achieved, minimizing waste in all elements of textile production and the supply chain. The final print is a replica of the design file created at the start of the design process—input equals output. Adobe Photoshop has become the software of choice for textile designers thanks to its extensive capabilities for converting and modifying pixel-based graphics.

There is no limit to the number of colors that one can use with digital, and one has more latitude with repeat pattern sizes. Realistic photo images can also be printed digitally on cotton and linen fabrics. With digital, the customer can provide any design that can be produced, unlike screen printing, where design limits such as repeats, color, and finesse might be imposed. One can use graphic design tools, such as Illustrator or Photoshop, to create designs or use images as commissioned prints on the fabrics. CAD, Inedit, C-design fashion, or other software could be applied to the designing tool.[2]

Textile designers have grown accustomed to the constraints imposed by traditional printing over time. These constraints are now completely integrated into the design process. A new generation of designers is, however, emerging, and they are becoming increasingly aware of the benefits that digital textile printing can provide. These benefits and how they can affect the way one creates textiles in the following sections: Designs that are one-of-a-kind and never repeated. One deals with screens in traditional textile printing, so one must always consider the possibility of a repeat in his design as a designer. This means that the screens' circumference determines the unique design's length. This means that the circumference of the screens determines the length of the unique design, and the design will be replicated for the rest of the cloth. The most typical displays have a 64 cm circumference, which means that every 64 cm of fabric has an identical design; however, with digital printing, one may build a one-of-a-kind design without a screen constraint. One can make his design as long as one likes. The file size one wants to produce is his only theoretical limitation. It implies that one can, for example, make a dress with a front and backside that do not duplicate (unique image from bottom to top).

For example, the latest AJet RIP can be used for typesetting and printing, as shown in Fig. 13.5. The different sections of the windows of AJet RIP are shown in Fig. 13.5. The section of file printing settings controls print length, width, or count settings.

Also, it can display the amount of ink needed to print a file in the RIP interface, as shown in Fig. 13.6. Thus, this color management software assists in printing the document used in the ink chromatic number and guides the amount of ink available.

There are other functions available in this software to meet customers' production needs and keep the software section simple for the reader; various functions and many small tools are not discussed in this section.

### 13.1.5 Surface imaging

Surface imaging, a concept created to revitalize the digital printing business, particularly the digital textile printing sector, could be one of these new options. Surface imaging implies democratizing digital printing systems by providing applications

Digital textile printing innovations and the future

**Figure 13.5** Different sections of AJet RIP color management software.

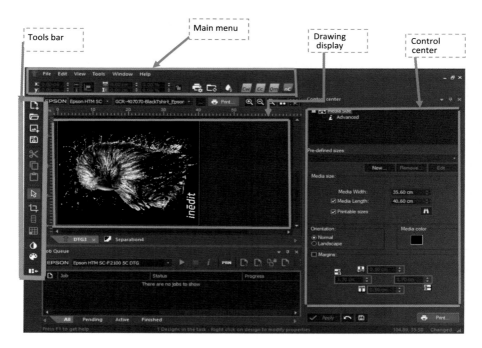

**Figure 13.6** Different sections of color print appeared at AJet RIP color management software.

for a broader range of users. It is characterized as a realm in which advanced digital printing technologies such as direct colorations, deposition, and subtraction printing are used to display images on a wide range of substrates, including porous, non-porous, rigid, and flexible surfaces. Any designs, including drawings, paintings, typographies, raster imageries, vector imageries, and photography from Adobe, can be digitally printed onto various substrates. Textile pattern designs, for example, can be digitally printed on a cloth for a garment or upholstery and drapery in the interior but also on wooded doors, glass windows, metal cabinets, ceiling panels, outdoor walls, flooring materials, or any other surface. While a roll-to-roll digital printing system can be used to print textiles, a similar printer with suitable ink sets can print on paper, polymeric films, and other flexible materials. On the other hand, current flatbed digital printers can easily support ridged substrates in average construction material sizes (4 ft by 8 ft and many inches thick) and print imagery on a wide range of hard surface materials.

As a result, what was once known as "textile design for surface imaging" has been renamed, allowing for a more extensive range of applications and functions. On the other hand, surface imaging stresses a comprehensive system that includes design, engineering, and business entities in the digital printing arena, rather than just design or engineering development. A lack of knowledge and communication among machine makers, software developers, printing operations, and application end-users is one of the current digital textile printing sector issues. The textile industry's design, engineering, and commercial entities have not been adequately integrated. In the industry, systems thinking must be understood and implemented. Surface imaging relies on systems thinking to integrate design, engineering, and business components. A typical product development cycle might include (1) ideation, (2) development, (3) production, (4) distribution, (5) marketing, and (6) sales, for example. A domain primarily contributes to the first stages of (1) ideation and (2) development. Second, in the middle of the (2) development, (3) production, and (4) distribution stages, (6) sales, and (5) marketing.

## 13.2 Printing process innovative advancements

The production process can be summarized as (1) design development by suitable design software; (2) fabric selection pretreatment; (3) printer setting; (4) printing; (5) curing or dye fixation, that is, posttreatment; and (6) final washing.[8]

### 13.2.1 Fabric preparation

It is critical to prepare the fabric before inserting it into the digital printing equipment. Pretreatment must be consistent across the fabric width and in a controlled environment, as this is the only method for producers to ensure that quality pretreatments are customized to meet the needs of the clients.

- Required textile surface treatments for ink fixation with acid and reactive inks;
- Ink image quality control fabric's motion influences registration between dots, resulting in banding or stitching lines;
- Design, color calibration, penetration, and capillary behavior reduction;
- Increased color build-up;

- Improved image/ resolution quality;
- Improved colorfastness and minor bleeding.

### 13.2.2 Product development creativity and inspiration

Textile sector customers are looking for a wide range of patterns and colors. On the other hand, traditional printing technologies cannot meet the new standards, forcing printers to look for innovative ways to meet client demands while reducing costs and waste. Inkjet printing technology, which was created for printing on paper, is increasingly being used in the fabric printing business to suit the demands of the emerging textile market.[9]

Designs are developed either through the digitization process or by scanning the pictures directly. The created file is then saved in printer-compatible format. All the changes in color combinations, the brightness of the images, etc., are done at this stage. Adobe Photoshop and CorelDraw Graphic Suite are used for design development. In the past, the designs used to be drawn first on Inspiration and mood board. A mood board is a collage consisting of images, words, and samples of objects in a composition. Inspiration boards have become a popular tool for attracting people, whereas a mood board is a collage consisting of images, text, and samples of objects in a composition. A mood board might be based on a specific topic, or it can be made up of any material chosen at random.[9] Now, these ideas are directly painted on software.

### 13.2.3 Volume and flexibility

The input data can be as big as one wants it to be, and it does not have to be repeated because the data is obtained from a computer file that is theoretically endless in size. The substrate is not touched at all. It can be printed on flat or curved surfaces, smooth or rough surfaces, and delicate or hard surfaces because the ink is deposited onto the substrate. In most cases, it is possible to create inks compatible with any specified surface.

Multicolor, Thousands of colors can be printed using the cyan-magenta-yellow-brick color gamut without a color kitchen. Extremely fast. Of course, printing rates are affected by resolution, the type of printing required, head technology, and other factors.

During printing, the ink moves, not a mechanical instrument. As a result, inkjet printers are built to be movable. Paper and other smooth surfaces have influenced the development of technologies. As they advanced, more options arose for them to be applied to a broader range of substrates. A dripping tap can illustrate the notion of digitizing the printing process. A drop of water falls onto the surface below with each drip. The drips of water can be made to wet a substrate in a controlled manner by controlling the location and movement of the faucet. The system closely resembles the digital printing process if more controls are added so that the drop can either fall or not fall for each position of the faucet. The drop sizes produced by digital printers are substantially smaller than those produced by a dripping tap. The system switches from DOD to DOD as the flow rate increases.

### 13.2.4 Speed of printing

Textile inkjet printers achieved speeds of 90 m/min at ITMA 2019, about 30 years after the technique was first demonstrated, putting it on par with rotary screen printing. Textile inkjet printing has a competitive advantage over rotary printing because it eliminates the need for screen preparation and enables the digital storage of print data until it is ready for production.

By eliminating the screen and jetting the colorant, inkjet printing allows for more flexibility in print repeat size, increases the number of colors without increasing cost, allows for finer line detail, and saves water and energy. Because there are no duplicate sizes, the print designer can produce a print that matches the text product's scale, form, and best position, allowing for new product variations. The fine line detail and increased number of colors on the textile substrate can provide a near-photographic image, and the elimination of repeat sizes allows the print designer to engineer a print within the scale, shape, and optimal position of the textile product, permitting new product varieties.

## 13.3 Benefits of innovation

### 13.3.1 The economics

Traditional textile printing is used for mass manufacturing and long run, to be feasible and lucrative. It is also used in a variety of manufacturing processes. On the other hand, short runs and fragmentary items can be accommodated by digital textile printing, as well as pretty eco-friendly runs that are much faster and less expensive. Moreover, the impact of legislation on business and customer requests for ecologically friendly products is rapidly becoming an issue. Until recently, the global digital textile printing market for garment, home décor, and industrial applications was growing at over 34% compound annual growth rate (CAGR).

### 13.3.2 The environment

As more consumers and merchants look for companies that provide environmentally responsible solutions, brand owners are typically stumped on where to start when defining their sustainability goals.[10,11] One might begin by looking at his product development as a brand owner. Examining his ingredient source, packaging, energy use, waste, and other concerns can assist one in aligning his activities with his goal. Contacting his present distribution and production partners, or finding new ones, to see what eco-friendly choices are available is a straightforward approach to start analyzing his options.

Flexible packaging is available in several eco-friendly choices. Brand owners can use recyclable, postconsumer recycled (PCR), or biodegradable packaging for their products. Brands can also collaborate with a packaging firm that uses digital printing, a more environmentally responsible way of printing. Some businesses are even able to offer both services. Here are four reasons digital printing is a green option for all types of businesses and why it can be a good fit for one.

### 13.3.2.1 Digital printing aids in emission reduction

It is vital to consider the carbon footprint and pollutants released during manufacturing regardless of the sort of packaging one chooses for his company. There has been emerging importance of carbon information disclosure by the firm as social responsibility.[12,13] The good news is that digital printers emit far less $CO_2$ than offset and rotogravure presses. At ePac, one uses carbon-neutral HP Indigo 20,000 presses to print pouches and roll stock digitally, which has a lower environmental impact than traditional printing. Emission reductions equal a happier planet.[14]

### 13.3.2.2 Digital printing reduces waste

The amount of waste produced by digital printing is significantly decreased. Unlike traditional printers, digital printers do not require printing plates for the creation, setup, and production of each stock keeping unit (SKU). Printing plates deteriorate with time, necessitating the creation of new plates whenever a packaging design is altered, resulting in more incredible waste. Digital printing saves plate waste by uploading files to the press and printing them just like a home printer. In the end, having no plates means using fewer materials and producing less waste.

### 13.3.2.3 Digital printing reduces inventory waste and obsolescent packaging

Compared to traditional printing techniques, brands can significantly reduce obsolescence (both waste and expenses) by printing on demand, printing numerous SKUs in one run, and placing small minimum orders. One will never have to throw out old packaging or lose money because a recipe, design, certification, or law has changed. Digital printing helps to keep unused plastic packaging out of landfills and more money in his pocket by reducing obsolescence.

### 13.3.2.4 Digital printing and eco-friendly flexible packaging materials

All components used in digital printing, from the inks to the films, are environmentally safe. HP digital presses are carbon-neutral; they employ polymer-based inks that do not emit hazardous pollutants into the atmosphere. These ecological inks use less energy and produce fewer greenhouse emissions than solvent-based or UV-curable inks frequently used in traditional packaging manufacturing. Since there has been attention to textile recycling,[15–17] one can also use digital printing to create environmentally friendly packaging, such as recyclable PE-PE films and films constructed from PCR resin. Both packing solutions keep plastic out of landfills, which benefits businesses and customers.

## 13.3.3 Optimizing the supply chain

Reduce complexity, improve agility, and increase adaptability by minimizing inventory, refer to Fig. 13.7. By streamlining replenishment cycles and order volumes, digital print is set to bring these benefits to packaging. They can reduce or eliminate their warehoused packaging stock and enjoy a more agile, responsive supply chain with lower minimum order numbers. Digital print jobs may be renewed many times throughout the product lifetime than traditional print jobs, allowing for more accurate order volumes based on actual sales and more relevant on-pack messaging, boosting even more sales, see Fig. 13.8.[14]

### 13.3.3.1 Integration into product lifecycle management

Most brands use a Product Life Cycle Management (PLM) system to plan for the next season. These technologies serve as aggregators for all the elements required to bring a new era of prosperity. From managing resources, enterprise resource planning, design components, collection, and ensembles to patterns and product photography, these collaborative platforms enable all functions and processes in creating seasonal products. A coordinated effort from brands, designers, textile

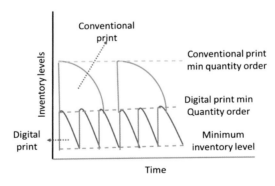

**Figure 13.7** Comparison of inventory levels of conventional versus digital textile printing.

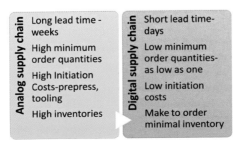

**Figure 13.8** Comparison of supply chain characteristics of conventional versus digital textile printing.

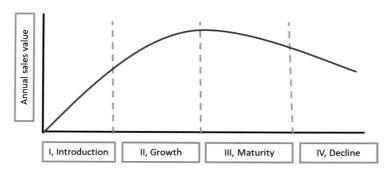

**Figure 13.9** Product lifecycle.

mills, and cut and sew operations to the logistics that move products to shelves or ship them out in packages—the product's life cycle (see Fig. 13.9) can be handled digitally from a data bank.

### 13.3.3.2 Just-in-time manufacturing

The just-in-time (JIT) or pull-based production system reduces inventory and increases flexibility. Backward schedules and operator control are used to achieve this. Nothing goes down the line until the operator is ready for it; that is, stock cannot be built up. While JIT production has technically existed since the 1960s, its applicability has recently increased. With JIT production, new businesses can bring their product lines to market in days or weeks rather than months. Larger companies may be able to respond more swiftly to the seasonal demands of the fashion industry. Textile companies can better serve their clientele by promptly having seasonal variants on the shelf.

## 13.4 The future of digital textile printing

As new and innovative technologies are developed, digital textile printing techniques change. Many textile entrepreneurs embrace these sustainable, efficient manufacturing procedures in every industry. The textile industry used a lot of energy and water before digital technology was adopted in textile printing. It had a detrimental influence on the environment since it used dyes and other toxic chemicals to manufacture textile materials, printing, inks, etc., and product finishing. The advent of digital printing technology in the textile industry, on the other hand, saved billions of liters of water worldwide. It provides a sustainable solution to the textile printing industry by pushing forward to replace unsustainable and wasteful manufacturing practices.[18]

As a result, the digital textile printing industry could be summarized as an industry developing and progressing rapidly, with new technologies making textile printing more straightforward, more reliable, and less water and energy intensive. The future of digital textile printing looks bright, thanks to the following developments.[18,19]

## 13.4.1 Technology adaption

The subsequent growth wave in the printing industry is driven by the advent of innovations and the expanding acceptance of digital technology solutions that enhance print speed, design, and efficiency. Digital printing technology is a comprehensive solution for textile designers and manufacturers. It consists of large-format digital printing technology, specialized software packages, and dyes and textiles explicitly created to meet the needs of the textile sector.[19]

## 13.4.2 Specialization

At ITMA 2015, hot topics included the development or refinement of printers for specialized markets such as grand-format printing, sublimation printing, printing/pretreatment systems, high-density fabric printers for carpet and fleece, product printers for sweaters, jeans, and leather goods, and high-quality printers for the luxury market.

## 13.4.3 Colors

Today, print technology and materials science are constantly improving, driving the usage of a wide range of inks for optimal results on various fabrics. With reactive dyes for natural fiber high-energy and low-energy sublimation inks, specialty inks for silk and nylon printing (i.e., acid), and the emergence of new generations of pigment inks that can print on most fabrics without the use of much water (as is the case with some textile inks), a whole new world of design freedom and color has opened up[8] (see Fig. 13.10).

Digital printing has progressed from being used solely for strike-off (i.e., proofing) in the early 1980s to today when some single-pass textile production printers can print at speeds of up to 200 ft per minute. The impact of print head advancements and cost savings on textile-specific transport systems, enabling production systems ranging from 1.8 to 3.2 m wide or more, has progressed the wide use of digital printing. Technology advancements in the textile sector have reached new heights, with up to 12 color channels and up to 64 print heads allowing for high throughput

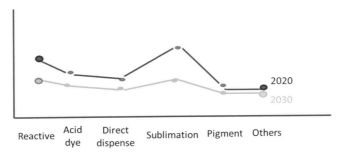

**Figure 13.10** Digital textile market share by ink type.

of thousands of square meters per hour. Many of these machines also feature "sticky be," which allows the fabric to be transported evenly through the machine.

## 13.5 Global market projections on digital textile printing

Different market studies have shown a different projection but a growth in market share soon. Research markets.com predicted USD 2.7 billion by 2026 (see Fig. 13.11). In August 2020, Grand View Research forecasted that the global textile printing market, valued at USD 146.5 billion in 2019, would grow to over USD 260 billion by 2025, with a CAGR of 8.9% during the projected period. There is high potential for textiles in south Asia[20]; changing consumer preferences in the Asia Pacific and the Middle East regarding printed textiles are expected to boost market growth.[21]

According to Grand View Research 2020, the rapid demand for digital printing technology in the apparel and advertising industries, combined with rising demand for sustainable printing, is expected to boost product demand over the forecast period. The increased need for speed of delivery, which is now possible thanks to the introduction of digital printers, is fueling market expansion. Market growth is projected to be aided by lower printing costs per unit and shorter product lifespans. Rapid technological advancements in the textile printing sector, combined with the advent of single-pass, big high-speed printers, have resulted in the upgradation of traditional textile printing machines, according to Grand View, complementing the expansion of the printed textile market.[21]

In terms of digital textile printing, Allied Market Research estimated in May 2020 that the global digital textile printing market, which was valued at $2.2 billion in 2019, would rise to $8.8 billion by 2027, quadrupling in size as it is expected to grow at a CAGR of 19.1% from 2020 to 2027. The global pandemic hiccupped the digital textile printing market, but overall growth was solid, with an 11.9% CAGR in value from 2016 to 2021 and a 12% CAGR in volume from 2016 to 2021. From 2021 to 2026, digital textile printed material will grow at a 13.9% CAGR, reaching 5.531 billion square meters. An in-depth look at the changing digital textile printing equipment market, focusing on essential growth categories.[21]

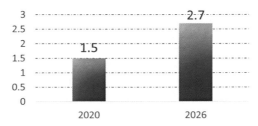

**Figure 13.11** Global Market in USD billions for digital textile printing is forecast to grow at a CAGR of 10.3%.

## 13.6 Conclusion

Digital textile printing is on its way to becoming a commercially viable textile printing technology with a bright future ahead. In terms of innovative design styles, mass customizations, JIT productions, agile manufacturing, and sustainable systems, digital textile printing outperforms analog printing technology. The sector is, however, beset by a flurry of technological conversions from analog to digital printing. Rather than incorporating digital textile printing technology into a new business model, current technology use has adapted its system and procedures into existing operations in the digital textile sector's time to make a paradigm shift. Surface imaging is one way to disrupt the current textile printing industry system and replace it with a more lucrative and holistic digital printing system in the future. With the nano ink development, the high-speed machine invention, the product development single-step process, and ease of design flexibility, the future printing market will be digital printing for textiles.

Digital textile printing is the way of the future, albeit no one knows when it will happen. The bottom line is that printers that fulfill the needs of both traditional textile printers and entrepreneurs are being developed. Digital printing saves time and money for traditional printers, allowing businesses to stay competitive in the ever-changing world of textile printing. For a business owner, digital printing provides all of the advantages listed above, as well as the freedom of unrestricted design and an actual vehicle for creative ideas.

## References

1. Memon, H.; Khoso, N. A.; Memon, S.; Wang, N. N.; Zhu, C. Y. Formulation of Eco-Friendly Inks for Ink-Jet Printing of Polyester and Cotton Blended Fabric. *Key Eng. Mater.* **2016**, *671*, 109–114. Available from: https://doi.org/10.4028/http://www.scientific.net/KEM.671.109.
2. Nabawy, A. M.; Meged, A. Effectiveness of Digital Design Thinking and the Evolution of Digital Textile Printing Technology. *J. Herit. Des.* **2021**, *1*, 1–4. Available from: https://doi.org/10.21608/jsos.2021.86160.1035.
3. Halepoto, H.; Gong, T.; Noor, S.; Memon, H. Bibliometric Analysis of Artificial Intelligence in Textiles. *Materials* **2022**, *15*, 2910.
4. Ugur Koseoglu, A. Innovations and Analysis of Textile Digital Printing Technology. *Int. J. Sci. Technol. Soc.* **2019**, *7*, 38–43. Available from: https://doi.org/10.11648/j.ijsts.20190702.12.
5. Kim, S. G.; Song, J. E.; Kim, H. R. Development of Fabrics by Digital Light Processing Three-Dimensional Printing Technology and using a Polyurethane Acrylate Photopolymer. *Text. Res. J.* **2019**, *90*, 847–856. Available from: https://doi.org/10.1177/0040517519881821.
6. Ali, M.; Lin, L.; Cartridge, D. High Electrical Conductivity Waterborne Inks for Textile Printing. *J. Coat. Technol. Res.* **2019**, *16*, 1337–1349. Available from: https://doi.org/10.1007/s11998-019-00214-5.
7. Karim, N.; Afroj, S.; Tan, S.; Novoselov, K. S.; Yeates, S. G. All Inkjet-Printed Graphene-Silver Composite Ink on Textiles for Highly Conductive Wearable Electronics

Applications. *Sci. Rep.* **2019**, *9*, 8035. Available from: https://doi.org/10.1038/s41598-019-44420-y.
8. Wang, M.; Parrillo-Chapman, L.; Rothenberg, L.; Liu, Y.; Liu, J. Digital Textile Ink-Jet Printing Innovation: Development and Evaluation of Digital Denim Technology. *J. Imaging Sci. Technol.* **2021**, *65*. Available from: https://doi.org/10.2352/J.ImagingSci.Technol.2021.65.4.040407 40407-40401-40407-40412.
9. Stempien, Z.; Khalid, M.; Kozicki, M.; Kozanecki, M.; Varela, H.; Filipczak, P.; Pawlak, R.; Korzeniewska, E.; Sąsiadek, E. In-Situ Deposition of Reduced Graphene Oxide Layers on Textile Surfaces by the Reactive Inkjet Printing Technique and their use in Supercapacitor Applications. *Synth. Met.* **2019**, *256*. Available from: https://doi.org/10.1016/j.synthmet.2019.116144.
10. Chen, L.; Qie, K.; Memon, H.; Yesuf, H. M. The Empirical Analysis of Green Innovation for Fashion Brands, Perceived Value and Green Purchase Intention—Mediating and Moderating Effects. *Sustainability* **2021**, *13*, 4238.
11. Memon, H.; Jin, X.; Tian, W.; Zhu, C. Sustainable Textile Marketing—Editorial. *Sustainability* **2022**, *14*, 11860.
12. Wu, D.; Memon, H. Public Pressure, Environmental Policy Uncertainty, and Enterprises' Environmental Information Disclosure. *Sustainability* **2022**, *14*, 6948.
13. Wu, D.; Zhu, S.; Memon, A. A.; Memon, H. Financial Attributes, Environmental Performance, and Environmental Disclosure in China. *Int. J. Environ. Res. Public. Health* **2020**, *17*, 8796.
14. Larsson, J. K. J. Digital Innovation for Sustainable Apparel Systems. *Res. J. Text. Appar.* **2018**, *22*, 370–389. Available from: https://doi.org/10.1108/rjta-02-2018-0016.
15. Memon, H.; Ayele, H. S.; Yesuf, H. M.; Sun, L. Investigation of the Physical Properties of Yarn Produced from Textile Waste by Optimizing Their Proportions. *Sustainability* **2022**, *14*, 9453.
16. Wang, H.; Memon, H.; Abro, R.; Shah, A. Sustainable Approach for Mélange Yarn Manufacturers by Recycling Dyed Fibre Waste. *Fibres Text. East. Eur.* **2020**, *28*, 18–22. Available from: https://doi.org/10.5604/01.3001.0013.9013.
17. Wagaye, B. T.; Adamu, B. F.; Jhatial, A. K. Recycled Cotton Fibers for Melange Yarn Manufacturing. In *Cotton Science and Processing Technology: Gene, Ginning, Garment and Green Recycling;* Wang, H., Memon, H., Eds.; Springer Singapore: Singapore, 2020; pp 529–546.
18. Elnashar, E. A. Applications of Mechatronics Opportunities in Textiles. *Int. Robot. Autom. J.* **2022**, *8*, 9–16. Available from: https://doi.org/10.15406/iratj.2022.08.00237.
19. Silva, M. C.; Petraconi, G.; Cecci, R. R. R.; Passos, A. A.; do Valle, W. F.; Braite, B.; Lourenco, S. R.; Gasi, F. Digital Sublimation Printing on Knitted Polyamide 6.6 Fabric Treated with Non-Thermal Plasma. *Polymers (Basel)* **2021**, *13*, 1969. Available from: https://doi.org/10.3390/polym13121969.
20. Yan, X.; Chen, L.; Memon, H. Introduction. In *Textile and Fashion Education Internationalization: A Promising Discipline from South Asia;* Yan, X., Chen, L., Memon, H., Eds.; Springer Nature Singapore: Singapore, 2022; pp 1–12.
21. Ujiie, H. *Digital Textile Printing: Status Report 2021. NIP & Digital Fabrication Conference*, 37. Thomas Jefferson University: Philadelphia, PA, USA, 2021; 47–52. Available from https://doi.org/10.2352/ISSN.2169-4451.2021.37.47.

# Index

*Note*: Page numbers followed by "*f*" and "*t*" refer to figures and tables, respectively.

**A**
Absorption coefficient, 198
Accuracy, 62, 243
Acetylation, 101*t*
Acid dye inks, 126−127
Acrylic monomers, 29
Acryl treatments, 101*t*
Adhesion, 195
Adobe Photoshop, 247−248, 251
AFM. *See* Atomic force microscopy (AFM)
Agfa, 65, 83, 246
Air-jet spinning, 211
Air vortex spinning, 211
AJet RIP, 69, 248, 249*f*
Ajrak, 229−230
    designed face masks, 234*f*
    design scarfs and stallers, 231*f*
    shape dupattas, 230*f*
    style female appraisers, 234*f*
    yardage, 232
Aleph, 65−66
Aluminum gallium arsenide (AlGaAs), 23
Analog or continuous-tone printing, 84
Analog printing techniques, 241
Analog screen-printing machine, 242
APP. *See* Atmospheric pressure plasma (APP)
Apparel, 213, 218
Apparel brands, 68
Application Specific Integrated Circuit (ASIC), 50
Aquario design, 67
    suites, 67
    textile design, 67
Aqueous inks, 28, 121
Archaeobotanical evidence, 226−227
Archer + technology, 247
Artificial intelligence, 21, 57−58, 88

Artisans, 232
Atexco, 242
Atexco digital, 69−70
Atexco printing software tools, 69
Atexco vision machine, 12, 17
ATmatcher, 69−70, 71*f*
Atmospheric plasma (AP), 148−151
Atmospheric pressure plasma (APP), 108, 139−140
Atmospheric-pressure plasma treatment, 193−194
Atomic force microscopy (AFM), 141−142, 146*f*, 153
ATR-FTIR analysis, 142−143, 144*f*
ATsoftproof, 69−70, 70*f*
Automated image analysis system, 200
Automated maintenances, 169−170
Automatic print quality analysis, 197, 202
Automation, 243

**B**
Back-shooter, 44−45
Balochistan provinces, 227
Bandwidth, 62
Baroque Print, 237
Bayer pattern, 78−79
Bend mode, 45−46, 45*f*
Benzoylation, 101*t*
β-glycosidases, 99
Binary deflection method, 47, 48*f*
Binary image, 77−78
Blending fibers, 113
Block-maker, 233
Block printing method, 1−2, 233
BMP, 79
Brands, 252−253
Bronzwear, 247
Bubble-free ink flow, 50

Bubble jet technology, 9–10, 164
Butyl-acrylate, 102

**C**
Canon, 31, 164
Canon Inc., 73
Carbon (C), 143–145
Carbon black pigment, 245
Carboxymethyl cellulose, 130
Carboxymethyl hydroxypropyl cellulose, 194
Caribbean, 220
Cationic agents, 127–128
C-design fashion, 248
Cellulase, 99
CH. *See* Chitosan (CH)
Chemical engineering, 242
Chitosan (CH), 110, 148–151
Chroma, 59
Chromix, 66–67
    ColorThink, 66
    ColorValvetPro, 67
Circular screen printing, 5
Clothing design, 11–12
Clothing fabric printing, 11
CMSs. *See* Color management systems (CMSs)
CMYK. *See* Cyan, magenta, yellow, and black (CMYK)
Coffee stain effect, 34–35
Cohesion, 195
Cold transfer printing machine, 8
Color, 57, 166–168
Colorants, 174–177, 194–196
    for digital printing, 120–122
    type, 174
Colorburst systems, 67
Color communication, 81–82
Colorfastness, 202
Color fiber-interaction technique, 190$t$
Color image segmentation, 199
Colorimeters, 63, 64$t$
Colorimetric values, 62
Color management systems (CMSs), 57, 64–70, 76
    Agfa, 65
    Aleph, 65–66
    Aquario design, 67
    Atexco digital, 69–70
    Chromix, 66–67

    classification of, 59–60
    Colorburst systems, 67
    color measurement, 60–63
        instrumental color measurement, 62
        measurement instruments, 62–63, 63$f$
    Datacolor, 69
    elements of, 59
    Ergosoft, 68
    MatchPrint II by Kodak polychrome, 69
    natural classification of, 60$f$
    textile print by adobe portable document format print engine, 67
    X-rite, 68–69
Color rendering index (CRI), 67
Colors, 256–257
Color selection, 12
ColorThink Pro, 66–67
ColorValvetPro, 66–67
Color wheel, 59
Commercial engraving, 4
Commission Internationale de l'Éclairage (CIE), 60–61
Complementary metal-oxide semiconductor (CMOS), 50–51
Compound annual growth rate (CAGR), 252
Computer-aided design (CAD), 57–58, 76
    applicable for editing textile print patterns., 76$t$
Conductivity, 132
Contact image sensor, 78
Contacting, 252
Continuous inkjet (CIJ) printheads, 163–164
Continuous inkjet (CIJ) technology, 25–26, 25$f$, 42$f$
Control algorithm research, 244
Conventional textile printing techniques, 208
Copper roller printing, 3
Copperplate printing, 2
CorelDraw Graphic Suite, 251
Cost of print head, 48–49
Cotton (*Gossypium hirsutum* L.), 227
Cotton Art transfer printing method, 8
Cotton-based fibers, 210–211
Cotton fabrics, 142–143, 188, 192–193
    surface element analysis of, 150$t$
Cotton farming, 226–227
Cotton fibers, 108
Cotton genus *(Gossypium)*, 226–227
Cotton-linen Silk Wool, 246

Cotton pollen, 226−227
Cotton seeds, 227
Cotton threads, 226−227
COVID-19 pandemic, 207−208, 218, 220−221, 233
Crafts, 231
CRI. *See* Color rendering index (CRI)
Croscolour CF, 96−97
Croscolour DRT, 96−97
Customization, 236
Cutting-edge 3D-printed clothing, 225
Cyan, magenta, yellow, and black (CMYK), 24, 28, 209

### D

Datacolor, 69
Data encoding, 79−81
Data interval, 62
Deco Y, 244, 245*f*
Devore techniques, 174
Dichlorotriazine (DCT), 125
Dielectric barrier discharge (DBD), 137
Dielectric Barrier Discharge Plasma treatment, 140
Dielectric properties, 132
Digital color management systems, 81−82
Digital cylinder printing, 24, 24*f*
Digital dyeing, 179−180
Digital finishing, 179
Digital image design
  color communication, 81−82
  digital color management systems, 81−82
  image capture, 77−79
  input and output devices color calibration, 83−84
  pattern data encoding compression and storage, 79−81
  pixel and image creation using inkjet printers, 84−86
  printing head performance monitoring, 87−88
  printing machine control, 87
  UV and latex curing methods, 86
Digital inkjet printing technology, 9−11, 93
Digital printing technology (DPT), 8−11, 21, 41, 73−75, 120
  advantages of, 33−34
  applications, 35
  barriers, 176*t*
  bottleneck of textile printing, 13−18
    attach importance to concept of technological innovation, 16−18
    bottleneck of traditional textile printing, 13−14
    cleaner production, 13
    production costs, 15−16
    products meet market demand, 14−15
    vast space for development in international application, 14
  classification of, 42−43
  cleaning mechanisms, 31−33
  clothing design, 11−12
  clothing fabric printing, 11
  colorants for, 120−122
  color selection, 12
  companies active in print head technology, 49−52
  continuous inkjet technology, 46−48
  disadvantages of, 34
  drop-on-demand inkjet technology, 44−46
  generation of droplets, 25−27
    continuous inkjet, 25−26
    drop-on-demand, 26−27
  important properties of inks, 130−132
    conductivity, 132
    dielectric properties, 132
    electrolytes, 131−132
    ink storage and stability, 132
    particle size, 131
    pH, 131−132
    surface tension, 130−131
    viscosity, 130
  methods, 22−24
    digital cylinder printing, 24
    fine art inkjet printing, 22−23
    notable digital laser exposure, 23
  pattern design, 12−13
  post-treatment for, 123
  preparation of substrate for, 122−123
  principle of, 41
  print head selection, 48−49
  printing heads, 30−31
    disposable head, 31
    fixed head, 30−31
  printing parameters and ink formulation, 27−30
    aqueous inks, 28
    dye sublimation inks, 29

Digital printing technology (DPT) (*Continued*)
- metal nanoparticle ink, 30
- solid ink, 30
- solvent inks, 28−29
- UV-curable inks, 29
- rapid development of, 226
- solidification mechanisms, 34−35
- of textiles, 132−133
- types of inks for, 123−130
  - acid dye inks, 126−127
  - disperse dye inks, 127−128
  - pigment inks, 128−130, 130*f*
  - reactive dye inks, 123−126

Digital textile pigment inks, 158

Digital textile printing, 68−69, 93, 103*f*, 119, 185
- advancement history, 241−250
  - ink formulation technology, 243−246
  - inkjet heads/print heads, 246−247
  - limitless design, 247−248
  - machine innovative advancements, 243
  - surface imaging, 248−250
- benefits of innovation, 252−255
  - economics, 252
  - environment, 252−253
  - supply chain, 254−255
- blended fabrics for, 113
- colors, 256−257
- enzymatic treatment for, 99−100
- global market projections on, 257
- other pretreatment processes for, 100−102
- plasma pretreatment for, 98−99
- pretreatment free digital printing, 113
- pretreatment of cotton fabrics, 98, 103−109
  - chitosan and acetic acid on color yield, 105−106
  - concentration of NaOH, 105
  - determination of color yield, 109
  - effect of cationization of cotton in in digital printing, 107
  - glycidyl trimethylammonium chloride on color yield of magenta ink, 105
  - hydroxypropyl methylcellulose pretreatment on, 107−108
  - impact of urea, 106
  - one-bath pretreatment for increased color yield of inkjet prints using reactive inks, 104−105
  - sodium bicarbonate on color yield, 106
- pretreatment of polyester fabrics, 110−111
  - on color saturation, 111
  - on color strength, 110
- pretreatment of silk fabric for, 111−112
  - amount of alkali, 112
  - amount of urea, 112
  - concentration of thickener, 112
  - effect of steaming time, 112
  - pH of pretreatment paste, 112
- pretreatment processes, 94−95
  - cationization for, 96−98
  - mercerization, 95−96
  - pigment digital printing, 102
- pretreatments of wool fabrics, 109−110
  - ammonium tartrate on color yield of, 110
  - sodium alginate on color yield of, 109−110
  - urea on the color yield of, 110
- printing process innovative advancements, 250−252
  - fabric preparation, 250−251
  - product development creativity and inspiration, 251
  - speed of printing, 252
  - volume and flexibility, 251
- quality, 186−196
  - colorants on, 194−196
  - fabric pretreatment on, 192−194
  - fabric type and, 191−192
  - fiber type and, 186−189
  - objective evaluation method of, 198−200
  - printing head on, 196
  - subjective evaluation method of, 197
  - yarn type and, 189−191
- quality assessment of, 185−186
- quality attributes, 186
- quality evaluation standards, 201−202
- quantitative image sharpness evaluation of, 200
- specialization, 256
- technology adaption, 256

Digital textile printing (DTP), 73, 207, 226
- COVID-19 impact on, 214−215
- fastgrowing vibrant sector in Pakistan, 233−236
- starting to make wise decisions, 236

technology, 235−236
market driving forces, 215−216
market overview, 212−214
   apparel and fashion, 213
   direct-to-garment printing, 213
   home furnishing, 212
   interior decoration, 212
   soft signage, 213−214
in printed textiles, 216−217
significant source of profit worldwide, 236−237
   Ezri collection, 2016, 237
   Forever 21, 2016, 237
   Urban outfitters, 2016, 237
and sustainability, 209−212
textile in Pakistan, 226−228
traditional clothing in Sindh, 228−233
Western markets, 217−220
   Caribbean, 220
   Europe, 218−219
   North America, 219−220
   South America, 220
Digital textile production, 119−120
Dip coating, 158
Direct inkjet printing, 93
Direct printing, 128
Direct-to-garment printing, 213
Disperse dye, 246
Disperse dye inks, 127−128
Disperse dyes, 195
Dots, 186
DPT. *See* Digital printing technology (DPT)
Dripping tap, 251
Drop-making technique, 163
Drop-on-demand (DOD), 25−27, 25$f$, 42$f$, 161, 163−164, 196, 246, 247$f$
   inkjet technology, 44−46
   piezoelectric DOD printing, 26−27
   thermal DOD printing, 26
Drop-test, 153
Drop volume, 27
Drum printing, 2−4, 23
Drying device, 7
Dutch Stork company, 5
Dye fixation, 51−52
Dyes, 166−168, 167$f$, 209
Dyes-based colorants, 119
Dye sublimation inks, 29
Dye-sublimation printing, 127

**E**

Eco-friendliness, 209
Eco-friendly plasma technology, 141
E-commerce, 215, 218
Economics, 252
Electrical conductivity, 30
Electrohydrodynamic (EHD) jet printing, 46$f$
Electroluminescence, 30
Electrolytes, 131−132
Electronic engineering, 242
Electronic engraving method, 4
Electronics, 171−173
Electrophotography, 236
Electrostatic digital printing, 23
Electrostatic field, 25
Electrostatic printing, 236
Embossed roller printing machine, 3−4
Encad, 164
Environment, 252−253
   digital printing aids in emission reduction, 253
   digital printing reduces inventory waste and obsolescent packaging, 253
   digital printing reduces waste, 253
   eco-friendly flexible packaging materials, 253
Enzymatic treatment, 99−100
Epihalohydrins, 96−97
Ergosoft, 68
Europe, 218−219
European market, 218
Eye-mind system, 57
Ezri collection, 2016, 237

**F**

Fabric
   cationization pretreatment of, 97$f$
   enzymatic pretreatment of, 100$f$
   plasma pretreatment of, 98$f$
   preparation, 250−251
   pretreatment, 192−194
   pretreatment and posttreatment of, 157−161
   treated/cationized, 98
   wicking property of, 191−192
Fabric type, 191−192
Fashion, 213
Fast fashion, 119−120
Fiber type, 186−189, 211$f$

Fine art inkjet printing, 22–23
Fixed-head printers, 30–31
Flatbed screen printing machine, 5
Flexibility, 243
Flexible packaging, 252–253
Forever 21, 2016, 237
Foveon X3 digital camera image, 78–79
Fuji, 83
Fully automatic flatbed screen printing, 5
Functional inks, 30

**G**

Gaiter-head printers, 30
Gamut mapping, 64
Geometry, 62
Gerber, 247
Global Organic Textile Standard, 246
Glutaraldehyde, 151
Glycerol, 130
Glycidyl trimethylammonium chloride (GTA), 104
Glycidyl trimethylammonium chloride (GTMAC), 111
Glycine (Gly), 110
Gray-level data, 77–78
Grayscale, 49
Grayscale printing, 84
Groove roller printing machine, 4
GTA. *See* Glycidyl trimethylammonium chloride (GTA)
GTMAC. *See* Glycidyl trimethylammonium chloride (GTMAC)

**H**

Halftoning, 84
Hand engraving method, 4
Handling motion systems, 170–171
Hand tool printing, 1
Hara invented inkjet technology, 9–10
Heat transfer printing machine, 8
Henniker plasma treatment device, 142*f*
Hertz method, 47
Hewlett-Packard Co., 73
Home furnishing, 212
Hot melt inks, 30
HPMC. *See* Hydroxypropyl methylcellulose (HPMC)
Hydroxypropyl methylcellulose (HPMC), 107

**I**

i1iO Automated Measurement Table, 68
i1Pro 3 Plus, 68
Ichinose Image Proofer (1999), 241
Ichinose machine, 157
Image analysis system
 digital print quality evaluation by, 198–199
 general principle of, 201*f*
Image capture, 77–79
Image compression, 79
ImageJ software, 199
Image-Pro, 200–201
Indian Subcontinent's desert regions, 231
Industrial Revolution, 2
Indus Valley Civilization, 226–227
Inedit, 247
Ink clogging, 246–247
Ink depository, 168
Ink droplets, 209
Inkjet, 236
Inkjet defect detection systems, 173*f*
Inkjet digital textile printing, 171–172
Inkjet inks, 121*f*
Inkjet printing, 14*f*, 21, 164, 209
 advantages and disadvantages of, 33–34
 color performance of, 108
 cotton substrate for, 108–109
 pixel and image creation using, 84–86
 polyamide fabric in, 107–108
 techniques, 43*f*
Inks, 119, 165
 important properties of, 130–132
  conductivity, 132
  dielectric properties, 132
  electrolytes, 131–132
  ink storage and stability, 132
  particle size, 131
  pH, 131–132
  surface tension, 130–131
  viscosity, 130
 storage and stability, 132
 supply system, 165–166, 166*f*
 types of, 123–130
  acid dye inks, 126–127
  disperse dye inks, 127–128
  pigment inks, 128–130, 130*f*
  reactive dye inks, 123–126
Input and output devices color calibration, 83–84

Index

Input data, 251
Integrated circuit (IC), 45
Interior decoration, 212
International Exhibition of Textile Machinery (ITMA), 185–186
ISO 13660 addendum, 201–202
ISO/IEC 13660, 201–202
Italian digital textile printing industry, 14

**J**
JBMIA SC28/WG4, 201–202
Just-in-time (JIT), 255

**K**
Khatris, 232
Khatris crafting block-prints, 231
Khyber Pakhtoon Khawah, 227
King Cotton, 227
Knitted structures, 191
Kodak, 31, 83
Kodak Enterprise Inkjet Systems, 50–51
Kodak's printing technology, 50–51
Kornit Digital, 246
Kubelka Munk equation, 198
Kyocera, 31

**L**
Laccase, 99–100
Latex curing methods, 86
Learning curve, 177–178
Lectra, 247
Lexmark International Inc., 73
Line quality compliance test, 201–202
Long-staple cotton, 227
Lord Kelvin, 9
Lubrizol, 244
Lyocell knitted fabrics, 151–152

**M**
MacDermid Colorspan, 164
Manual flatbed screen printing, 5
MatchPrint II by Kodak polychrome, 69
Match Textile, 69
MATLAB program, 199
MATLABVR, 200–201
Measurement instruments, 62–63, 63*f*
 spectrophotometer, 63
 tristimulus colorimeter, 62–63
Mechanical engineering, 242

Mechanical printing, 1
Medieval Era, 229–230
Mediterranean, 219
Medium Extended Air Defense (Mead), 50–51
Medium staple cotton, 227
Memjet, 51
MEMS. *See* Microelectromechanical systems (MEMS)
Metadata, 79–81
Metal nanoparticle ink, 30
Metal prints, 2
MetaVue VS3200, 68
Meteor eLab Pro user interface, 172*f*
Microelectromechanical systems (MEMS), 50, 162
Microencapsulated pigments, 195
Milliken Millitron digital carpet printer, 157, 241
Mingograph, 9
Ministry of Industry and Information Technology, 242
Model X plus by Atexco, 243, 243*f*
Modern flatbed scanners, 78
Monochlorotriazine (MCT), 124–125
Moroccan swimsuits, 237
Mottle, 131
Multinozzle printheads, 162–163

**N**
Nanopigment printing, 244–245
Nanoprinting inks, 244–245
National Digital Printing Engineering Technology Research Center, 242
NeoPigment Robusto Softener solution for Presto, 246
Nip Coating, 158
Nitrogen (N), 143–145
N,O-carboxymethyl chitosan (NOCS), 151
Normal medium-staple cotton, 227
North America, 219–220
Nozzle clogging, 168–169
Nozzle excitation, 45–46

**O**
Objective evaluation method
 digital textile print quality, 198–200
  automated image analysis system, 200
  color image segmentation, 199

Objective evaluation method (*Continued*)
    digital print quality evaluation by image analyzer system, 198–199
    general principles of image analyzing system, 199
    measuring digital textile print color values, 198
    quantitative image sharpness evaluation of digital textile print, 200
Objective evaluation method, 198–200
Online marketing, 218
Optitex, 247
Osiris's Isis printer, 163
Oxygen (O), 143–145

**P**

Pakistan cotton production, 228*f*
Pakistan's economy, 228–229
Pakistan's textile industry, 227
Particle-free silver ink, 141
Particle size, 131
Paste roller, 4
Pattern data encoding compression and storage, 79–81
Pattern design, 12–13
Permanganate treatment, 101*t*
Peroxidation, 101*t*
pH, 131–132
Photographic engraving method, 4
Photoionization, 138
Photolithography, 26
Photoshop 5, 81
PIAS software, 199
Piezoelectric continuous inkjet technique, 47
Piezoelectric DOD printing, 26–27
Piezoelectric effect, 9
Piezoelectric element, 45–46
Piezoelectric inkjet, 236
Piezoelectric on-demand inkjet technology, 9
Pigment-based direct-to-fabric decoration, 246
Pigment digital printing, 102
Pigment inks, 128–130, 130*f*, 195
Pigments, 15–16, 28, 166–168, 167*f*
Pixel, 73
Plasma activation, 137
Plasma coating, 137
Plasma deposition, 137
Plasma polymerization, 137

Plasma pretreatment, 98–99
Plasma surface modification, 153
Plasma technology, 137
Plasma treatment of varied fabrics to facilitate inkjet printing, 139–152
Plateau–Rayleigh instability, 25–26
Plate making, 4
Polyacrylamide (PAM), 111–112, 193–194
Polyacrylic acid (PAA), 111–112, 193–194
Polyamide (PA), 148–151
Polyester-based fibers, 210–211
Polyester fabric, 102, 127–128, 141, 145, 193
Polyethylene glycol (PEG), 130
Polyethylene terephthalate (PET), 99–100
Poly-ethylenimine, 151
Polyimide, 147–148
Polyvinyl alcohol, 127–128
Porous layer feed, 45–46
Portable Document Format (PDF), 67
Postconsumer recycled (PCR), 252–253
Post-treatment processes, 123
Precision, 62
Premium garments, 236
Prepared for Print (PFP), 158
Printed fabrics, 6, 197
Printheads, 161–164
    active mixing and switching, 180*f*
    continuous inkjet printheads, 163–164
    drop-on-demand, 163–164
    single-nozzle and multinozzle, 162–163
Printing
    digital printing technology, 8–11
        bottleneck of textile printing, 13–18
        clothing design, 11–12
        clothing fabric printing, 11
        color selection, 12
        pattern design, 12–13
    heads, 30–31, 196
        disposable head, 31
        fixed head, 30–31
        performance monitoring, 87–88
    history of, 1
    screen printing machine, 4–7
        flatbed screen printing machine, 5
        rotary screen printing machine, 5–7
    tools and equipment, 1–8
        block printing, 2
        drum and roller printing machine, 2–4

engraved printing, 2
transfer printing machine, 7–8, 7f
 cold transfer printing machine, 8
 heat transfer printing machine, 8
Printing machine control, 87
Printing unit, 6–7
Print-on-dem, 247–248
Product development creativity and inspiration, 251
Product Life Cycle Management (PLM) system, 254–255
Product volume, 243
Protruding fibers, 158
Push mode, 45–46, 45f

**Q**
Quality attributes, 186
Quality evaluation standards, 201–202
Quantitative evaluation method, 200–201

**R**
Radio frequency (RF), 138
Raster image processors (RIPs), 23, 68, 81
RAW format, 78–79
Rayleigh principle, 9
Reactive dye inks, 123–126
 chemical structure of, 125f
Reactive dyes, 192–193, 195
Reduced graphene oxide, 141–142
Reprographic printers, 85
Resin DWR, 97
Rheology, 130–131
Ring-spinning machinery, 211
Roller printing machine, 1–4
 embossed roller printing machine, 3–4
 groove roller printing machine, 4
Rolling engraving method, 4
Rose patterns, 2–3
Rotary screen printing machine, 5–7
Rotor spinning, 211

**S**
Scanning electron microscope (SEM), 139–140, 153
Scattering coefficient, 198
Screen printing machine, 4–7
 flatbed screen printing machine, 5
 rotary screen printing machine, 5–7
Seiko Epson Corp., 73

Semi automatic flatbed screen printing, 5
Semi-conductivity, 30
Semiconductor, 45
Shanghai Stock Exchange Star Market, 242
Shear mode, 45–46, 45f
Sidecar file, 79–81
Siemens' first drop-on-demand (DOD) printing, 9
Sieve, 4
Sindhi Ajrak, 232f
Sindhi culture, 229–230
Single nozzle printheads, 162–163
Single-pass printing, 171, 171f
Small batch production, 15–16
Small-sized printers, 73
Smart print, 66
Sodium alginate, 127–128, 130, 194
Sodium carboxymethyl cellulose, 194
Soft proofing software, 69–70
Soft signage, 213–214, 236
Software, 171–173
*Soleimani-organic*, 195
Solidification mechanisms, 34–35
Solid ink, 30
Solvent binders, 209
Solvent inks, 28–29
Source-destination-simulation model, 82
 destination profile, 82
 simulation profile, 82
 source profile, 82
 standard working space profile, 82
South America, 220
Spectracore, 67
Spectral range, 62
Spectrophotometer, 63, 64t, 198
 geometric arrangement of, 198f
Speed of printing, 252
Spray coating technique, 158–159
Squeeze mode, 45–46
Sticky rollers, 170–171
Stock keeping unit (SKU), 253
Stork's Amethyst printer, 163
Stork TruColor Jet Printer (1991), 157, 241
Subjective evaluation method, 197
Substrates, 170–171
Sudden Steam Printing technology, 9–10
Sulfur hexafluoride, 148
SuperFlex, 86

Supply chain, 254–255
  integration into product lifecycle management, 254–255
  just-in-time manufacturing, 255
Surface contact angle test, 147f, 153
Surface energy, 140
Surface modifiers, 244
Surface morphology, 142–143
Surface tension, 130–131
Surface Tension Driven Inkjet, 46
Surfactants, 130–131, 209
Sustainability, 209–212
Swiss markets, 217

**T**
Technological barriers to digital printing
  automated maintenances, 169–170
  awaiting chemistry, 179–180
    digital dyeing, 179–180
    digital finishing, 179
    technology advancements in deposition, 180
  clogging and monitoring of nozzles, 168–169
  concentrating on printing technology, 174–179
    digital textile printing-a threat, 178–179
    learning curve, 177–178
    opportunity, 178–179
    real-world example, 178
    risk, 178–179
  dyes, 166–168, 167f
  electronics, 171–173
  handling motion systems, 170–171
  ink depository, 168
  inks, 165
  ink supply system, 165–166, 166f
  pigments, 166–168, 167f
  pretreatment and posttreatment of fabric, 157–161
  printheads, 161–164
    active mixing and switching, 180f
    continuous inkjet printheads, 163–164
    drop-on-demand, 163–164
    single-nozzle and multinozzle, 162–163
  single-pass printing, 171, 171f
  software, 171–173
  substrates, 170–171
  textile printing, 168

Technology adaption, 256
Technology advancements in deposition, 180
Texprint, 14
Textile design for surface imaging, 250
Textile feeding mechanism, 170f
Textile industries, 160–161
Textile printing, 1, 168, 185, 192–193
Textiles, 21, 57
  digital inkjet printing of, 129–130
  enormous potential for, 21–22
Textiles printing inks, 190t
Textile surface procession, 211
Textiprinters, 247
Thermal DOD printing, 26
Thermal drop stimulation, 50–51
Thermal inkjet, 163
  printheads, 44f
Thermal transfer, 236
Thixotropy, 131
Thousand Flowers Diagram, 2
3-chloro-2-hydroxypropyltrimethyl ammonium chloride, 96–97
3D printing technology, 244
Traditional stainless steel scraper blade, 6–7
Traditional textile printing, 252
Traditional textile screen printing, 13
Transfer printing machine, 7–8, 7f, 93, 127–128
  cold transfer printing machine, 8
  heat transfer printing machine, 8
Triplet of colors, 59
Tristimulus colorimeter, 62–63
Tristimulus values, 61
T-shirt printing, 21–22
TWAIN, 78
2,3-epoxypropyltrimethyl ammonium chloride, 96–97

**U**
Urban outfitters, 2016, 237
User-friendly software, 62
UV, 86
UV-curable inks, 29

**V**
Very large-scale integration (VLSI), 50–51
Vinyl-acetate, 102
Vinyl sulfone (VS), 124–125
Viscosity, 130, 195

# Index

VLSI. *See* Very large-scale integration (VLSI)
VOCs. *See* Volatile organic compounds (VOCs)
Volatile organic compounds (VOCs), 28, 86, 121, 137
Voltage-based dynamic drop tuning technology, 87–88
Volume and flexibility, 251

## W

Water-based inks, 121–122
Waterborne dispersions, 245
Waterfall printing technology, 51
Western markets, 217–220
    Caribbean, 220
    Europe, 218–219
    North America, 219–220
    South America, 220
Wet postprocesses, 177
Wet processing, 211–212
White Gold, 227
Wool fabrics, 142–143
Wool fibers, 194
World Textile Information Network, 217

## X

Xaar, 50
Xaar AcuChp technology, 87–88
Xaar Drop Optimization Technology, 87–88
XaarGuard, 88
Xaar's business, 9–10
XaarSMART technology, 88
Xerographic printing technique, 23
XPS. *See* X-ray Photoelectron Spectroscopy (XPS)
X-ray Photoelectron Spectroscopy (XPS), 139–140, 147–148, 149*f*
X-rite, 68–69
    apparel brands, 68
    digital textile printing, 68–69

## Y

Yarn-spinning technologies, 211
Yarn type, 189–191
Yield value, 131

## Z

Zero-shear viscosity, 194
Zirconium titanate (PZT), 27

Printed in the United States
by Baker & Taylor Publisher Services